Developments in Economic Geology, 7

NUCLEAR METHODS IN MINERAL EXPLORATION AND PRODUCTION

Further titles in this series

1. *I.L. ELLIOTT and W.K. FLETCHER*
GEOCHEMICAL EXPLORATION 1974

2. *P.M.D. BRADSHAW*
CONCEPTUAL MODELS IN EXPLORATION GEOCHEMISTRY
The Canadian Cordillera and Canadian Shield

3. *G.J.S. GOVETT and M.H. GOVETT*
WORLD MINERAL SUPPLIES
Assessment and Perspective

4. *R.T. SHUEY*
SEMICONDUCTING ORE MINERALS

5. *J.S. SUMNER*
PRINCIPLES OF INDUCED POLARIZATION FOR GEOPHYSICAL EXPLORATION

6. *R.A. RICH, H.D. HOLLAND and U. PETERSEN*
HYDROTHERMAL URANIUM DEPOSITS

Developments in Economic Geology, 7

NUCLEAR METHODS IN MINERAL EXPLORATION AND PRODUCTION

Edited by

JEROME G. MORSE

Deputy Director, Colorado Energy Research Institute, Golden, Colorado, U.S.A.

Adjunct Associate Professor of Physics, Colorado School of Mines, Golden, Colorado, U.S.A.

ELSEVIER SCIENTIFIC PUBLISHING COMPANY
Amsterdam — Oxford — New York 1977

ELSEVIER SCIENTIFIC PUBLISHING COMPANY
335 Jan van Galenstraat
P.O. Box 211, Amsterdam, The Netherlands

Distributors for the United States and Canada:

ELSEVIER NORTH-HOLLAND INC.
52, Vanderbilt Avenue
New York, N.Y. 10017

Library of Congress Cataloging in Publication Data

Nuclear methods in minerals exploration and production.

(Developments in economic geology ; v. 7)
Bibliography: p.
Includes index.
1. Radioactive prospecting. 2. Oil well logging, Radiation. 3. Radioactive substances--Industrial applications. I. Morse, Jerome G.
II. Series.
TN269.N8 622'.15 77-23866
ISBN 0-444-41567-X

ISBN 0-444-41250-6 (series)
ISBN 0-444-41567-X (vol.7)

Printed in The Netherlands

To my wife Elizabeth

PREFACE

Nuclear methods now provide the basis for rapid identification of about two-thirds of the earth's naturally occurring chemical elements. The broad application of these techniques to exploration and production of minerals has been slow in developing. This unfortunate state of affairs may be due largely to a lack of awareness by the exploration geologist and production engineer of the burgeoning technologies in this segment of applied nucleonics.

The thrust of this book is to describe the current status of those applicable nuclear techniques which lend themselves to field use, in mobile laboratories or as portable devices. Some of the latter are designed as downhole probes for identifying and quantifying elemental compositions of rocks, discrete mineralogical deposits or delineating lithological features, all of which may lead to better definition of subsurface ore deposits. Others find use in control of mining operation, in beneficiation and recovery processes. Some of these techniques are now in field use, and the data presented are intended to provide a basis for assessing their advantages, as well as their disadvantages, in contrast with those of a more traditional nature.

The subject matter derives from interdisciplinary lecture series, presented by the editor to graduate students and faculty at the Colorado School of Mines. Audiences comprised the disciplines of geology, geochemistry, geophysics and physics. The considerable growth in literature related to this specialized application area of nuclear science virtually mandated inviting contributions from among those productively active in their current fields. It is intended that the book will find use among those professional engaged in the search for and production of metalliferous minerals, as well as bulk deposits of coal, oil, natural gas, phosphate, etc.

A number of people contributed to the preparation of this book and appreciation is expressed to all of them. Particular thanks are given to Drs. F.E. Senftle and A.B. Tanner (U.S. Geological Survey), Dr. J. Miskel, L. Germain, L.L. Schwartz, J. Hearst (Lawrence Livermore Laboratory), Dr. J. Burnett (Colorado School of Mines) and Rebecca Vories (Colorado Energy Research Institute) for reviewing portions of the manuscript.

J.G. MORSE

LIST OF CONTRIBUTORS

A.K. BAIRD	*Department of Geology, Pomona College, Claremont, California, U.S.A.*
B.C. CLARK	*Planetary Science Laboratory, Martin Marietta Aerospace, Denver, Colorado, U.S.A.*
D.W. HILCHIE	*Department of Petroleum Engineering, Colorado School of Mines, Colorado, U.S.A.*
K. KEIL	*Department of Geology and Institute of Meteoritics, University of New Mexico, Albuquerque, New Mexico, U.S.A.*
J.G. MORSE	*Colorado Energy Research Institute, Golden, Colorado, and Department of Physics, Colorado School of Mines, Golden, Colorado, U.S.A.*
S.S. NARGOLWALLA	*Nuclear Applications Research Laboratory, Scintrex Ltd., Concord, Ontario, Canada*
N.D. NGUYEN	*Nuclear Applications Research Laboratory, Scintrex Ltd., Concord, Ontario, Canada*
A.U. REHMAN *	*Nuclear Applications Research Laboratory, Scintrex Ltd., Concord, Ontario, Canada*
H.J. ROSE, Jr.	*U.S. Geological Survey, Reston, Virginia, U.S.A.*
M. SLAUGHTER	*Department of Chemistry and Geochemistry, Colorado School of Mines, Golden, Colorado, U.S.A.*
H.A. TEWES	*Lawrence Livermore Laboratory, University of California, Los Angeles, California, U.S.A.*
J. TOMAN	*Lawrence Livermore Laboratory, University of California, Los Angeles, California, U.S.A.*
P. TOULMIN, III	*U.S. Geological Survey, Reston, Virginia, U.S.A.*
H.A. WOLLENBERG	*Lawrence Berkeley Laboratory, University of California, Berkeley, California, U.S.A.*

* Now with Ontario Hydroelectric Commission, Nuclear Studies and Safety Department, Toronto, Ontario, Canada.

CONTENTS

Chapter 1

INTRODUCTION

JEROME G. MORSE

"Accuse not nature, she hath done her part: do thou thine. . ."
John Milton, Paradise Lost

Exploration for new mineral resources is but one of the alternatives available to man to satisfy his continuing and growing need for non-renewable raw materials. There are also other options which merit our serious concern now and into the forseeable future. They include measures to conserve and reduce our use of these materials, improved processing techniques for more effective recovery, reclamation and recycling of used materials, substitutions, etc., but it is to the subjects of exploration and production that this book is addressed. All options, however, require intensive study and in-parallel pursuit, and with significant shortages now anticipated worldwide, a reordering of many national priorities may well be inevitable.

We recognize that the viability of our civilization requires a steady supply of minerals as these substances virtually touch upon every facet of our lives. In fact, it has been said that without minerals and materials, we would have no gross national product. Further, the worldwide economic and population growth, as reflected in increasing affluence, urbanization and industrialization signals such forecasts for the period 1968—2000 of cumulative mineral values from \$ 2.0 to 2.9 trillion (in the U.S.) and from \$ 6.8 to 9.6 trillion (worldwide) in constant of 1968 dollars (U.S. Bureau of Mines, 1970). The magnitude of these numbers staggers the imagination. Although they are estimates only, refinements in these data must include inflation with escalating costs of energy for extraction and processing of these minerals. Energy costs will increase particularly in view of the need to recover resources of lower quality, and those costs necessary to return the disturbed environment to a regionally acceptable state.

The energy-minerals shortage is now a reality accepted within the borders of virtually all of the world's industrialized nations. Attention currently is being drawn nationally and internationally to a rapidly approaching non-energy-minerals shortage. For example, the U.S. Geological Survey (1975) stated that in 1974, the United States was more than 90% dependent on imports of such primary materials as manganese, chromium, titanium, niobium, strontium and sheet mica; 75—90% on aluminum, platinum, tin, tantalum, bismuth, fluorine, asbestos and mercury; 50—75% on zinc, gold, silver, tungsten, nickel, selenium and potassium. This dependency can be expected to increase with time unless other measures are taken, such as those noted earlier.

Solutions to these serious problems, as one would expect, are not simple. At a minimum, there is an urgent need to expand the continuing search for new mineral deposits using all technologies available to man to maximum advantage. It is the opinion of many that this is not now being done on a broad enough basis to put available bodies of knowledge to effective use. Reluctance to adopt change, particularly that inspired by technological advance, is often greatest among those industries least technology-oriented, and, unfortunately, this includes many of those involved with minerals. The National Academy of Sciences (1976) recommended recently, and rather strongly, the initiation of intensive research directed towards the lessening of U.S.' vulnerability to overseas supplies of vital materials. Included were techniques to be developed which run the gamut from exploration to recovery, along with the important consideration of using substitute materials where possible.

The techniques noted above will, of necessity, involve the evolutionary development of new instrumentation for measuring selected properties of minerals (or their elemental constituents), both existing and induced. We believe their performance in achieving effective use in geochemistry, geophysics and recovery will embrace the following criteria:

Exploration

(1) Speed of analysis — sufficiently rapid to enable a large number of analyses within a relatively short period of time.

(2) Amenability to field use — small, compact, light weight and stable under field conditions.

(3) Sensitivity — able to detect and accurately measure low concentrations of the elements being sought.

(4) Reproducibility of data.

(5) Reasonably low in cost.

(6) Simple to operate and maintain in the field.

Mining

(1) Compatibility with the mining environment, particularly with the newer rapid and more efficient mining technologies, in order to provide on-the-spot assessments of the constituents to be recovered.

(2) Versatility allowing them to be inserted as probes into sample holes, drilled in a predetermined grid pattern, to delineate a three dimensional model of the ore body, using computer-derived analyses, as appropriate. Optimally, analyses will be done entirely in the field obviating the need for sample retrieval and off-site analysis.

Processing

Capability of in-situ measurement in beneficiation and processing in such a way that rapid analyses may be coupled to computer control of reagent input to optimize the upgrading or recovery technique.

Since 1965, the International Atomic Energy Agency, recognizing the growing body of nuclear technology which could be applied to the areas of minerals exploration and recovery, hosted a number of conferences on this subject, the intent of which was to further and exchange relevant new information (IAEA, 1969, 1971). Growth in this area of nuclear applications may be attributed largely to the technologies derived from national and international programs in the space and nuclear fields.

Nuclear methods have already demonstrated many advantages over their non-nuclear counterparts, where specific applicability has been obtained. There are also some disadvantages to nuclear techniques and these will be identified. The following discussion briefly describes a variety of the techniques dealt with in this book.

Exploration for new mineral deposits, their evaluation, development and recovery are expensive and time consuming undertakings. Nuclear methods, where applicable (and their applicability is growing), have been shown to be significantly faster and often at least as accurate as conventional analytical methods. Both factors reflect chances for lower operating costs accrued in field work.

For example, Chapter 2 describes the use of down-hole spectroscopy in the characterization of uranium deposits. It reduces the need for sample recovery and subsequent laboratory analyses in the vital exploration phase. Other demonstrated approaches to uranium exploration are also reviewed.

Many chemical elements may be identified qualitatively and quantitatively down-hole, in core samples, on site in the field by using X-ray fluorescence (Chapter 3) or radioactivation (Chapter 5) methods. Further, radioactivation is now in commercial use for down-hole surveillance. All methods noted above are now in or available for field service, although growth in usage has been slow.

A specific X-ray fluorescence spectrometer is described in Chapter 4 as an example of a "space age" instrument to do geochemistry on a planetary surface (Mars). Embodying sophisticated advanced technologies, such developments often pave the way for earth-bound usage.

The oil industry for many years has been making considerable use of nuclear well logging and Chapter 6 reviews their current techniques.

Chapter 7 treats an interesting potentially useful method, currently latent in the U.S. but active in the U.S.S.R., for the recovery of shale-locked gas and oil through nuclear stimulation.

The exploration methods covered in the first chapters are not confined solely to this purpose, but are adaptable also to mineral recovery through beneficiation and processing. In many cases, they are found in current industrial use because rapid, accurate information at the site of the recovery operation will provide significant cost savings.

To assist those readers somewhat removed from nuclear terminology and technology, a refresher or background chapter on the fundamentals of atomic and nuclear physics is included as an appendix.

The disadvantages of nuclear methods are few in number. They include: (1) A single approach, such as thermal neutron activation will address a large and discrete number of chemical elements, but not the entire periodic chart. Prior knowledge of local geochemistry will enable proper instrument selection, and more than one may be required. (2) The use of radioactive materials in the field necessitates observing specific safety precautions plus adherence to the requirements of the local nuclear regulatory authorities. This may involve the use of cumbersome and heavy shields against radiation, monitoring of personnel to minimize radiation exposure, reporting exposure data, etc. (3) Skills necessary for the operation of instruments and interpretation of the results obtained may not be found normally in many manpower pools of the mineral industry. They are, however, readily obtained through the normal marketplace.

SUMMARY

It will be shown that in the relatively recent past nuclear techniques have made considerable progress in addressing the problems of mineral exploration and recovery, even surpassing those of a more conventional nature. Although the nuclear approach is evolving rapidly and many specific instruments are now both available commercially and in active field use, this approach may not provide the final or only answer. To attack the problem of worldwide mineral shortages, judicious application of all pertinent methods must continue to be developed and used to enable effective progress toward a solution.

REFERENCES

International Atomic Energy Agency, 1969. *Nuclear Techniques and Mineral Resources. IAEA Proceedings.* Vienna.

International Atomic Energy Agency, 1971. *Nuclear Techniques for Mineral Exploration and Exploitation. IAEA Proceedings.* Vienna.

National Academy of Sciences, 1976. *Mineral Resources and the Environment.* National Academy of Sciences—National Research Council, Washington, D.C.

U.S. Bureau of Mines, 1970. Mineral facts and problems. *U.S. Bur. Mines. Bull.*, 650.

U.S. Geological Survey, 1975. Mineral resource perspectives, 1975. *U.S. Geol. Surv., Prof. Paper*, 940.

Chapter 2

RADIOMETRIC METHODS

HAROLD A. WOLLENBERG

INTRODUCTION

This chapter will deal with methods to measure gamma radioactivity, primarily from the natural radioelements, ^{238}U, ^{232}Th, and ^{40}K. In seeking uranium resources it is necessary to characterize by its constituents the total gamma-radiation field observed, so that tell-tale ratios indicating, for example, above-normal concentrations of uranium, can be discerned. Similarly, radiometric methods can be used to detect the preponderance of other radioelements, such as potassium in alteration zones associated with base metal mineralization.

Superimposed on the natural radioelement field are gamma radiations from fission product fallout, the longest lived and, therefore, the most ubiquitous being ^{137}Cs. Autoradiographic techniques to determine the location and abundance of U and Th within the mineral matrix of rock samples will also be discussed, as will detection of ^{222}Rn, a gaseous decay product of the ^{238}U series, and often an indicator of deeply buried uranium concentrations.

To furnish the reader an idea of the breadth of variation in natural radioelement contents, average abundances of U, Th, and K in some representative rock types are listed in Table 2-I. Of the igneous rocks, radioelement contents are listed for the intrusives; values are comparable for their extrusive counterparts. The relative abundances of these radioelements are illustrated on a ternary diagram (Fig. 2-1), within which there are characteristic fields for broad categories of rock types (Wollenberg and Smith, 1968, 1969; Wollenberg et al., 1967). Most acidic and intermediate igneous rocks fall into sector *A*, as do most siliceous clastic sedimentary rocks. Ultrabasic to basic igneous rocks occupy a field which parallels the K-Th edge of the triangle, and ranges from sector *C* for the ultrabasics to sector *A* for the basic to intermediate rocks. The carbonate sedimentary rocks occupy a region parallel to the U-Th edge, ranging from sector *B* for the pure carbonates to sector *A* for carbonates more contaminated by siliceous material. Superimposed on these usual fields are rocks wherein one of the radioelements is concentrated: feldspar-rich pegmatites where K strongly predominates, monazite- and thorite-rich veins, monazite-rich sandstones, and some peralkaline igneous rocks where Th strongly predominates and, finally, the targets most sought after

TABLE 2-I

Mean values of radioelement contents of typical crustal rocks

Description	U (ppm)	Th (ppm)	K (%)	Reference
Igneous rocks				
Acidic	6	20	3.2	Wollenberg and Smith, 1968
Intermediate	3	11	2.1	
Basic	1.5	4	1.4	
Metamorphic rocks				
Granulite facies	0.6	2.5	2.5	Heier and Adams, 1965
High amphibolite facies	1.2	9.4	2.0	
Epidote amphibolite to low amphibolite facies	3.5	26.5	3.4	
Glaucophane schist and eclogite	0.5	1.4	0.5	Wollenberg et al., 1967
Sedimentary rocks				
Siliceous clastics	2.2	7.0	1.5	Wollenberg and Smith, 1969
Carbonates	0.7	1.0	0.2	

by radiometric methods, igneous and sedimentary rocks where U strongly predominates.

Also superimposed on these natural abundance patterns are the relative

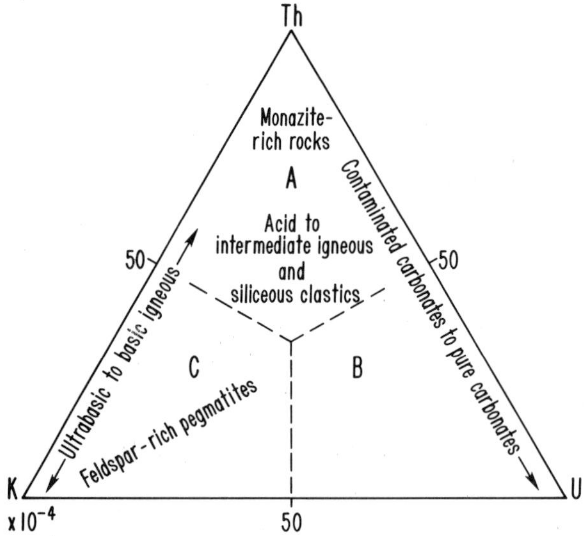

Fig. 2-1. Relative abundances of radioelements in different rock types. The field for uranium-rich rocks is near the U corner of sector *B*.

radioactivities of the natural radioelements, normalized on the basis of counts per minute (in the gamma-ray energy region above ~0.1 MeV) per gram of element: K : Th : U = 1 : 3.3×10^3 : 8.05×10^3. Therefore, in a "normal" siliceous rock where relative abundances of K : Th : U may be 1 : 5×10^{-4} : 1.5×10^{-4}, the contributions of U and Th to the samples' total gamma radioactivity may be respectively 1.2 and 1.6 that of K.

GAMMA-RAY SPECTROMETRY

The radiometric technique most useful in the detection and assessment of radioelement resources is gamma-ray spectrometry. It involves the detection of gamma rays of varying energy by scintillation or solid-state detectors, collection, sorting, and storage of energy spectra by single- or multi-channel pulse-height analyzers, and treatment of the spectral data, often by computers, to yield abundances of the individual radioelements. Gamma-ray spectrometers for field use are now available commercially. To illustrate the features of gamma-ray spectra we shall use spectra, generated in the laboratory, of standard radioelement mixtures. The subject of gamma-ray spectrometry, discussed necessarily briefly in this chapter, is covered at length by a recently published book by Adams and Gasparini (1970).

The gamma-ray spectrum

In its most general application, a gamma-ray spectrometer consists of a thallium-activated sodium-iodide [NaI(Tl)] scintillation detector, its accompanying photomultiplier, coupled to a single- or multi-channel pulse-height analyzer which stores the accumulated counts, and delimits discrete portions of the gamma-ray energy spectrum. The electronics of the spectrometer system are normally adjusted so that the spectrum encompasses gamma-ray energies from the region of a few tenths up to 3—4 MeV. This energy region contains the many characteristic gamma rays of the ^{238}U and ^{232}Th decay series, as well as the single gamma ray at 1.46 MeV of ^{40}K. Spectra from U, Th, and K standard materials are compared in Fig. 2-2. The 400-channel spectra here encompass the energy region 0.2—4 MeV. They were obtained from measurement of ~1000-g samples 6 inches (15.2 cm) in diameter and 1.5 inches (3.8 cm) thick, encased in plastic boxes, and centered on an 8-inch-diameter (20.3 cm) by 4-inch-thick (10.2 cm) NaI(Tl) detector.

In the U spectrum, prominent peaks from ^{214}Bi, at 0.61, 1.12, and 1.76 MeV are often used for analysis of U. Counts in channel intervals spanning the 2.62-MeV peak of ^{208}Tl, a decay product of ^{232}Th, are most often used for analysis of that element, while the prominent 1.46-MeV peak of ^{40}K is singularly diagnostic of K. In the cases of the U and Th decay series, it is generally assumed that the daughter products, upon which the analyses are

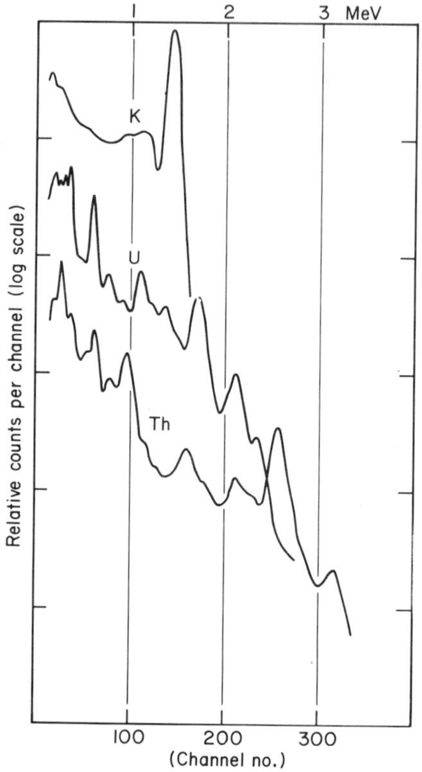

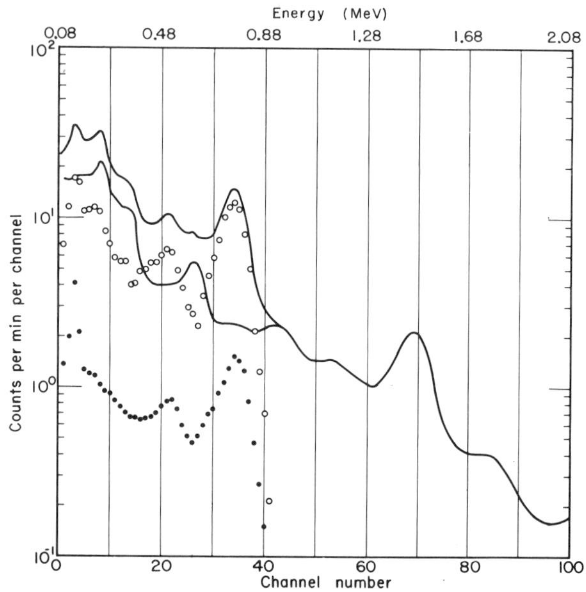

Fig. 2-2. Gamma-ray spectra for U, Th, and K standard materials.

Fig. 2-3. Observed, natural (solid lines), and fallout (open circles) gamma-ray spectra from a soil sample, background subtracted. Solid circles: a gamma-ray spectrum of dried weeds. (From Wollenberg et al., 1969.)

based, are in secular equilibrium with their parent radioelements. Equilibrium conditions are assumed to exist in samples of most rocks and soils. However, in many instances this is not the case; disequilibrium is often present in the U decay series, especially in uranium ore minerals. Disequilibrium is difficult to discern with a NaI(Tl) system, because one cannot resolve relatively low-energy peaks of the first five members of the U decay series, whose ratios are diagnostic of equilibrium conditions. Much better resolution of low-energy gamma-ray peaks is afforded by a system based on a lithium-drifted germanium [Ge(Li)] detector, permitting determination of the status of secular equilibrium between parent radioelements and their decay products.

In many cases, gamma-ray spectra of rocks or soils contain contributions from fission products, deposited in fallout from atmospheric nuclear bomb tests. The longest-lived (half-life ~28 years) and therefore the most ubiquitous gamma-emitting fission product is ^{137}Cs, with an energy of 0.66 MeV. Also present, depending on proximity of the sample location to the site of atmospheric nuclear testing, in time and space, may be the gamma activity from ^{95}Nb-^{95}Zr; an energy of 0.76 MeV. This peak will initially partially mask that of ^{137}Cs, but will decay relatively quickly (the effective half-life of ^{95}Nb-^{95}Zr is approximately 50 days) leaving the 0.66 MeV peak of ^{137}Cs as the sole fallout contributor. The effect of fallout on the 100-channel gamma-ray spectrum of a sample of soil from California, collected in 1963 (Wollenberg et al., 1969) is shown in Fig. 2-3. The set of open circles is the fallout component, the upper solid curve the total spectrum, the lower solid line the "natural" spectrum without the fallout component. The set of solid circles near the bottom of the diagram is the fallout deposited on a clump of dried weeds; the only activity in the weeds is from fission products. At present, the effect of fallout is to only add a few percent to the total gamma-ray counting rate measured in the field.

Data reduction

The most straightforward and most commonly used method to quantitatively determine radioelement contents from gamma-ray spectra, is to compare directly the sample spectra with those of standards, containing known amounts of the individual radioelements. In the laboratory, where hundreds of channels of pulse-height analyzer data are available and computers are at hand, sample spectra can be compared channel by channel with standard spectra, using least-squares fitting techniques, to yield highly accurate assays of U, Th, and K. In the field, and also in the laboratory, where hundreds of channels are usually not available in pulse-height analyzers, relatively broad intervals spanning diagnostic peaks in standard and sample spectra are used [methods and applications of field gamma-ray spectrometry are discussed later, and are covered in detail in articles by Løvborg et al. (1968, 1971)]. Most commonly, the energy regions examined are those centered at 2.62 MeV, the ^{208}Tl peak of the Th decay series; the 1.76-MeV peak of ^{214}Bi in the U decay series; and the region encompassing the 1.46-MeV peak of ^{40}K. A portable gamma spectrometer, the Scintrex GAM-1, is shown in Fig. 2-4. It utilizes a single-channel analyzer with rate-meter display. The detector shown has a 2-inch-diameter by 2-inch-thick NaI(Tl) crystal; detectors of varying sizes as well as one for borehole logging are available.

In laboratory analyses, pulse-height analyzer data must first be corrected for background: the nearly featureless gamma-spectral continuum upon which the gamma-ray energy peaks of the samples are superimposed. The peak areas are integrated by summing their counts. After background counts are subtracted from each interval, the net counting rates are treated as follows. Three simultaneous equations are solved for Th, U, and K; in this case the counting rates proportional to the abundances of the three radioele-

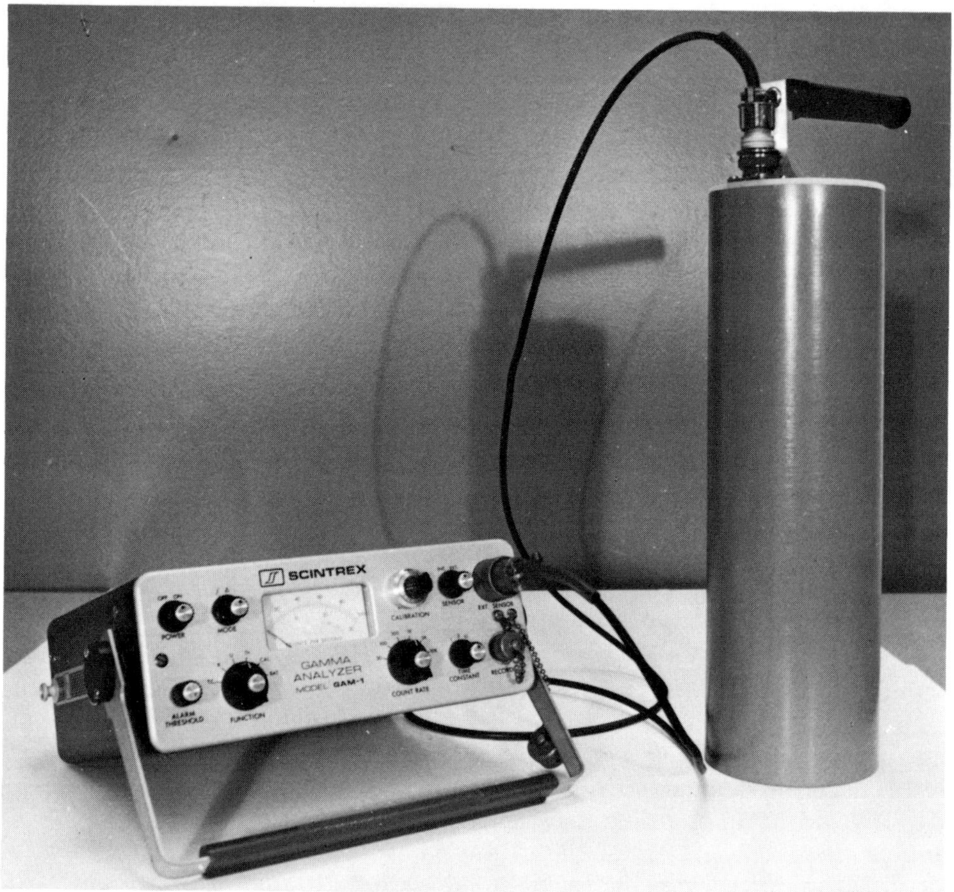

Fig. 2-4. The Scintrex GAM-1 portable gamma-ray spectrometer system.

ments:

$$N_1 = C_1\mathrm{Th}$$

$$N_2 = C_2\mathrm{Th} + C_3\mathrm{U}$$

$$N_3 = C_4\mathrm{Th} + C_5\mathrm{U} + C_6\mathrm{K}$$

where C_i are constants determined by calibrations with standards, and N_1, N_2, and N_3 are net counting rates (after background is subtracted) in the energy intervals spanning the 2.62-MeV ^{208}Tl, 1.76-MeV ^{214}Bi, and 1.47-MeV ^{40}K peaks, respectively. The simultaneous equations are required because the net counting rate from the gamma-activity of the sample (or from the effective sample volume in the case of field measurements) within each of the energy intervals, contains a component from gamma rays of higher energy.

Thus, for example, the prominent ^{40}K peak contains counts from scattering of the U series gamma rays of 1.76, 2.2, and 2.4 MeV, as well as counts from scattered 2.62-MeV gamma rays of the Th series. In field measurements it is usually assumed that there is no contribution from radioelements other than those of the Th decay series in the N_1 interval, centered at 2.62 MeV. Actually, there is a small contribution to this interval from high-energy "tails" of the 2.2- and 2.4-MeV peaks of the U series; this contribution is normally taken into account in high-precision analyses in the laboratory.

A simple computer program for calculation of U, Th, and K contents from three-interval data is normally employed. Inputs are weight of sample, length of run, and the interval counts from the 2.62-, 1.76-, and 1.46-MeV peaks. The counts are converted to count rates, corrected for background, three simultaneous equations solved incorporating net count rates and calibration constants, and sample weight applied to yield radioelement contents. Errors are calculated, based on counting statistics of the interval totals. Therefore, to achieve statistically significant results (errors due to counting statistics are less than 5% for each radioelement) in a minimum amount of time, one must optimize length of run and the number of counts accumulated in the intervals for each sample (or, in the case of field measurements, for each location). The sequence of calculations is listed in Table 2-II.

Laboratory standards, upon which the calculations are based, are easily made by blending equilibrium U- and Th-rich materials (for example, precisely analyzed diluted pitchblende ores and monazite furnished by the New Brunswick Laboratory of the U.S. Energy Research and Development Administration) with a matrix of low radioelement content, such as finely ground ultramafic rock. Potassium standard material may be a chemically pure salt such as KCl. Standards to provide calibration constants for field measurements may consist of large concrete slabs within which one radioelement predominates (Løvborg et al., 1971). If such standards are not readily available, field calibration constants can be derived, based on data

TABLE 2-II

Sequence in calculation of radioelement abundances

1. Inputs: weight, time, counts in three-channel intervals encompassing 2.62-, 1.76-, and 1.46-MeV peaks
2. Counts converted to counting rates, and background subtracted in the three intervals
3. Grams of U, Th, and K calculated, using three simultaneous equations incorporating constants based on standards
4. Concentrations calculated: Th, U, in ppm, K in %. Element ratios calculated
5. Errors in concentrations calculated, based on counting statistics
6. Results: Th, error; U, error; K, error; Th/U

from radioelement analyses of samples collected at many field measurement sites.

Detector systems

The NaI(Tl) and Ge(Li) detector systems will be briefly described and compared. As mentioned under "The gamma-ray spectrum" the detectors most used for gamma spectrometry of geologic materials are NaI(Tl) scintillation crystals. Gamma rays (photons) interacting with the detector material produce light flashes (scintillations) which are in turn detected by a photomultiplier tube optically coupled to the crystal. The amplitudes of the voltage-pulse outputs of the photomultiplier are directly proportional to the energy expended in the detector by the interacting photons. A pulse-height analyzer is then used to detect, classify, and store the pulse heights and, on command, to present the stored information in the form of a gamma-ray spectrum. The most widely used NaI(Tl) crystals are cylindrical, and are available in several sizes, ranging from a few centimeters in diameter by a few centimeters thick up to tens of centimeters in diameter and thickness. Also widely used are well-shaped crystals, permitting insertion of the sample so that it is nearly surrounded by the detector material. This configuration furnishes higher counting efficiency than a cylindrical detector, for a given detector volume. The configuration of the sample should be optimized with the size and shape of the available detector, to achieve the best efficiency most economically. Therefore, for example, a large flat cylindrical sample best suits a large cylindrical detector, while an equidimensional or elongate cylindrical sample best suits a well-type detector. Sodium-iodide detectors for portable field gamma spectrometers are limited by size and weight of the apparatus, and generally do not exceed 3 inches (7.6 cm) in diameter by 3 inches thick.

With appropriate sample-detector configurations, NaI(Tl) systems may have absolute peak-detection efficiencies of 20—40%, for gamma-rays of the order of 1 MeV. As shown in Fig. 2-5, peak-detection efficiency decreases as gamma-ray energy increases above $\sim$0.2 MeV.

A laboratory system, employing an unusual sample-detector geometry, was developed for continuous or step-scanning for U, Th, and K contents in diamond drill cores (Løvborg et al., 1972). Cores are placed on a movable tray which passes between two 2-inch by 3-inch (5.1 cm by 7.6 cm) NaI(Tl) crystals. Output from the detectors' photomultipliers is summed, and sent to a multi-channel pulse-height analyzer, where radioelement contents are determined. Resulting computer plots are compared with lithologic and downhole gamma logs in Fig. 2-9.

Over the past decade, the advent and increasing use of high-resolution lithium-drifted germanium or silicon [Ge(Li), or Si(Li)] semiconductor detectors has had profound effects on gamma-ray spectrometry. The greatest

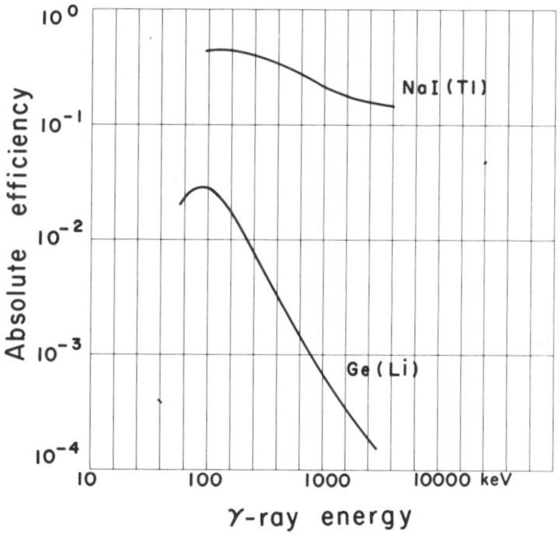

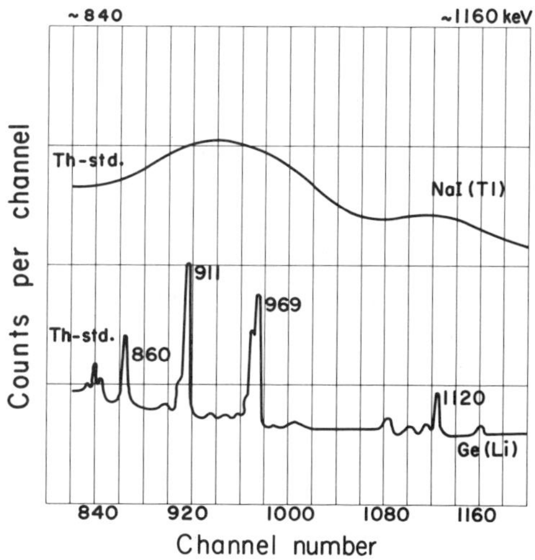

Fig. 2-5. Comparison of absolute peak detection efficiencies for a 10-cm^3 Ge(Li) and an 8-inch-diameter by 4-inch-thick NaI(Tl) gamma-ray spectrometer detector. (From Smith and Wollenberg, 1972.)

Fig. 2-6. Comparison of spectral shapes from NaI(Tl) and Ge(Li) gamma-ray spectrometers, showing portion of a gamma-ray spectrum on an expanded energy scale. (From Smith and Wollenberg, 1972.)

advantage of these detectors is that resolutions of gamma-ray peaks are at least an order of magnitude higher than those of NaI(Tl), disclosing fine detail in the structure of gamma-ray spectra. A comparison of NaI and Ge(Li) resolutions is illustrated in Fig. 2-6. A disadvantage is the relatively low absolute peak detection efficiency of Ge(Li) with respect to NaI(Tl). Efficiencies of an 8-inch (20.3 cm) diameter NaI and a 10-cm^3 Ge(Li) are compared in Fig. 2-5. A larger Ge(Li) would have somewhat greater efficiency than the 10-cm^3 detector, especially for the higher gamma energies, but NaI(Tl) efficiencies would still be over an order of magnitude greater than those of Ge(Li). An excellent review of Ge(Li) gamma-ray spectrometry was written by Heath (1969); the article emphasizes a system combining a small computer for data collection and analysis. A discussion of high-resolution gamma spectrometry of Th and U decay series has been recently published (Smith and Wollenberg, 1972).

Instead of using scintillations in the detector such as NaI, the principle of operation of semiconductor detectors is to collect the electric charge released in the solid medium by the absorption of gamma rays. The drifting of lithium into silicon or germanium crystals, under an electric field at temperatures of 100—400°C, provides a sensitive volume to permit their use in gamma spectrometry. Ge(Li) detectors must operate at liquid nitrogen temperatures to minimize the leakage current from thermal excitation of electrons. Because of its significantly higher density and atomic mass number, Ge has a higher absorption coefficient than Si; thus, Ge is better suited to gamma spectrometry than Si.

A Ge(Li) system is comprised of a detector, ranging in size from a few to over 100 cm^3, cooled in a vacuum by its accompanying cryostat, a field-effect transistor preamplifier (usually cooled in the same enclosure as the detector), and an accompanying pulse-height analyzer. Because of their large bulk, Ge(Li) systems do not lend themselves as readily to mobile field operation as do the more compact NaI(Tl) systems. However, car- and truck-mounted Ge(Li) systems are used especially to measure, by virtue of their high resolution, gamma rays of relatively low energy in the region 100—300 keV (Anspaugh et al., 1972).

Field applications

Surface measurements

Probably the most widely used and readily available systems for assessment of U and Th resources in the field are portable single- or multi-channel gamma spectrometers, coupled to NaI(Tl) detectors. Such a system was described briefly in a previous section, and illustrated in Fig. 2-4. To achieve statistically significant U and Th analyses from on-the-outcrop measurements, one must consider the source-detector geometry; in other words, the effective sample measured by the detector. But first one must decide if one is to use a detector collimated by an effective shield against extraneous gamma rays, such as lead (which may inhibit mobility in the field), or if field conditions warrant use of the detector uncollimated (Adams and Fryer, 1964, discussed a collimated detector for field surveys). Detectors may be used uncollimated at locations where terrain is relatively featureless over several square meters where the rock surface is well exposed, grain size fairly regular, and radioelements expected to be distributed rather homogeneously. At locations where one or more of these factors is significantly not met, one should consider collimating the detector. For example, the counting rate in an uncollimated detector placed at the base of a sharp step-like break in terrain, may be half again as great (for material of given radioelement content) as the counting rate measured by the detector placed on a flat surface. On very coarse-grained or pegmatitic surfaces, counting rates, especially in the 1.46-MeV interval, may vary by factors of two from position to position, because of the presence of large feldspar phenocrysts containing abundant potassium.

Serious thought has been given to the concept of effective sample, exemplified by the work of Løvborg et al. (1971). In that paper the effective sample was defined as the volume of rock in which the variance (the square of the standard deviation) of a particular radioelement (whose concentration is considered a random variable with constant mean value; Matheron, 1963), is equal to the estimation variance of a gamma-ray spectrometric determination of that radioelement. Fig. 2-7 shows a cross-section of the effective sample volumes for a collimated and uncollimated isotropic detector whose center is ~5 cm above the surface. The volume of material comprising the effective sample contributes approximately 60% of the measured gamma rays; it is roughly 1/10 the volume which contributes 90%. Table 2-III summarizes the mass of the effective sample (M_0), its surface radius (R_0), and maximum vertical extension (Z) for energies of 1.76 and 2.62 MeV, for collimated and uncollimated detectors whose centers are 5 cm above the surface. This comparison indicates that the effective mass sample is significantly larger for the uncollimated case, but that the vertical extensions are nearly the same for collimated and uncollimated geometries.

Field surveys incorporating on-site gamma-ray spectrometry for the natural radioelements have been conducted on exposed rock surfaces at many locations. Noteworthy studies were conducted by Richardson (1964) and colleagues at Rice University on the Conway Granite, New Hampshire; by Doig (1968) in the Canadian Shield; by Moxham et al. (1965) and Davis and Guilbert (1973) in hydrothermally altered zones associated primarily with porphyry copper deposits; and surveys in south Greenland, described by Løvborg et al. (1968, 1971). Fig. 2-8 shows a field measurement for U and Th in this latter study, on peralkaline rocks of south Greenland. A Pb collimator, surrounding the sides of the 3-inch-diameter by 3-inch-thick NaI(Tl) detector, was used for these measurements, primarily because of the coarse granularity of the rock, expected dishomogeneity of radioelements, and abrupt changes of relief on the rock surface. The detector is connected

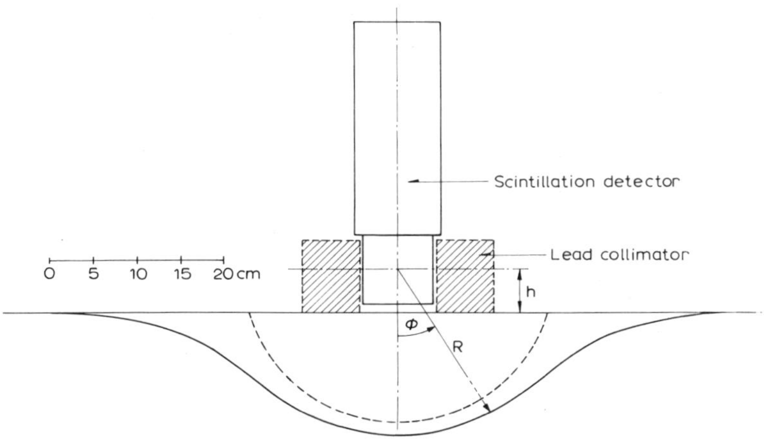

Fig. 2-7. Vertical cross-section showing calculated effective samples obtained by a collimated (dashed curve) and uncollimated (solid curve) detector placed on a flat rock surface. Parameters: E_γ = 2.62 MeV, μ = 0.111 cm^{-1}. (From Løvborg et al., 1971.)

TABLE 2-III

Mass, surface radius, and depth of effective sample measured by collimated and uncollimated detectors in the configuration shown in Fig. 2-7

	E_γ (MeV)	M_0 (kg)	R_0 (cm)	Z (cm)
Uncollimated NaI detector	1.76	37	40	12
	2.62	49	42	14
Collimated NaI detector	1.76	14	16	11
	2.62	21	17	13

to a single-channel pulse-height analyzer, with scaler readout. Such a system, measuring on closely spaced intervals, can obtain scores of analyses per day, furnishing data for contour maps of radioelement contents. In the case of

Fig. 2-8. A typical field setup on an exposed bedrock surface. (From Løvborg et al., 1971.)

the south Greenland measurements, U and Th contents were at least several hundred parts per million, permitting statistically significant analyses from counting times of 50—100 seconds. Lower contents would, of course, require proportionately longer counting times, using a comparable field spectrometer system. Løvborg (1973) calculated counting times required to obtain a standard deviation of 15% in a field U measurement on rock in which the Th/U ratio ~3. Using a 3 × 3 NaI crystal, 10 minutes is required for U contents of ~1 ppm, and 1 minute if U ~ 30 ppm.

Subsurface measurements

Downhole radiometric logging, sometimes incorporating gamma-ray spectrometry of drill holes, is quicker and significantly less expensive than coring and subsequent laboratory radioelement analysis. In its simplest form, radiometric logging utilizes a geiger tube or NaI(Tl) detector system on a cable attached to a winch and depth counter, to furnish gross-gamma counting rates, measured by a count rate-meter at the surface. The excellent concordance between in-hole gamma count rates and U and Th is illustrated in Fig. 2-9 (Løvborg et al., 1972). There is a great time advantage to downhole logging; the gamma log shown in Fig. 2-9 took a few hours to measure, while the accompanying U and Th core scans required eleven days.

It is preferred that the hole be uncased for logging, and essentially free of drilling mud or other contaminants, though the relatively high-energy gamma rays of the U and Th series are detectable through normal steel casing thicknesses. Where significant anomalies are furnished solely by U, measurements of total gamma activity are adequate to define, if not absolutely, at least the relative abundances of the radioelement. However, as in the case of field gamma spectrometry, one must assume that secular equilibrium exists between parent U and its daughters. With proper calibrations, based either on analyses of cores from drill holes, or on measurements in holes in full-scale standards (Dodd and Eschliman, 1972), abundances of equivalent U or uranium oxide can be calculated from gross-gamma count rates. The relationship between gamma activity and grade of uranium mineralization has been determined empirically (Dodd and Eschliman, 1972): $\bar{G}_\gamma T = KA$, where $\bar{G}_\gamma$ is the average grade in percent U_3O_8, T is the thickness of mineralization, K is a constant determined by instrument characteristics, and A is the area under the gamma-ray log curve.

Gamma spectrometry of downhole radiometry is more definitive, and more useful in terranes where U may not be the predominant radioelement. Single- or multi-channel pulse-height analyzers, used for field gamma spectrometry in surface measurements, are also useful for downhole spectrometry (for example, the Scintrex system, shown in Fig. 2-4, and the Risø analyzer, shown in Fig. 2-8). Recently, Sprecher and Rybach (1974) described a well-packaged borehole gamma-ray spectral logging system for holes of diameter greater than 2.5 cm. Calibration for this system was furnished by measure-

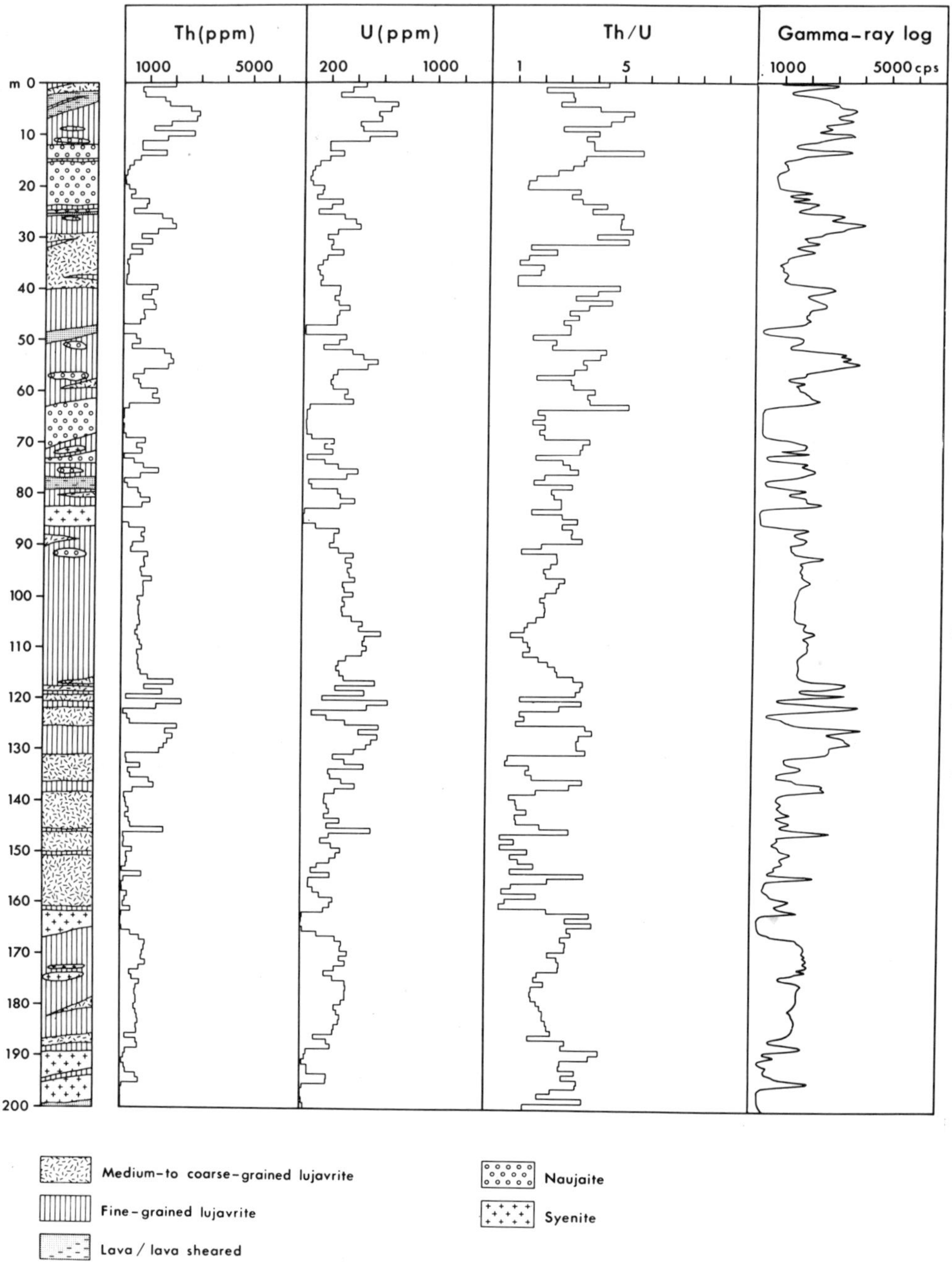

Fig. 2-9. Geologic, radioelement, and logging diagram of a borehole in the Ilimaussaq intrusion. Radioelement diagrams are based on continuous scans of 1-m-long core sections. (From Løvborg et al., 1972.)

ments in standards of known U content, and confirmed by U contents of analyzed drill cores.

Gamma spectrometry of borehole radioelements is complicated by the accentuation of the Compton continuum in the energy region below 1 MeV. This results from gamma rays having been scattered within the large mass of material effectively sampled, before they reach the detector. Thus, the ratio of counts in the low-energy region to those in the region greater than 1 MeV, is enhanced in comparison with a spectrum obtained in a laboratory counting configuration. Experience has shown (Dodd and Eschliman, 1972) that the enhanced Compton continuum strongly masks any full-energy peak in the region below 1 MeV. However, this phenomenon may be turned to advantage, because the ratio of the low-energy to high-energy portions of the spectrum may be diagnostic of equilibrium conditions in the U decay series.

In zones of ore concentration (0.2% U_3O_8 or greater) the relatively high equivalent atomic number (Z) of the ore zone material, compared to the Z of common rocks (in the range 10—20), has an attenuating effect on gamma rays, especially those with energies less than 0.4 MeV. The error in determining downhole U concentrations caused by this so-called Z-effect is, therefore, greatest in gamma probes most sensitive to low-energy gamma rays. Dodd and Eschliman (1972) indicate that such errors increase with grade of the ore, from approximately 10% in ores of 1% U_3O_8 to over 20% in 2.5% ores.

As in other facets of gamma-ray spectrometry, downhole spectral logging requires that the system be stabilized to prevent drift in channels or spectral windows assigned to specific gamma energies. A gain-stabilization system for borehole spectral logging has been described recently by Key and Eschliman (1974). A ^{54}Mn source provides a gamma peak of 0.835 MeV, which is used as the "benchmark" for stabilization. This energy was chosen because it is below the energy region usually examined for the natural radioelements (>1.3 MeV).

Dodd and Eschliman (1972) point out the value of combining information from gamma-ray logs with logs of other parameters. Especially useful are comparisons of gamma-ray, resistance, and spontaneous potential logs, which accurately define lithologic units in sedimentary sequences. There is also increasing utilization of neutron logs, which respond to changes in hydrogen content, reflecting the abundance of water and, indirectly, porosity. Magnetic susceptibility logs may disclose the inverse relationship between this parameter and some occurrences of uranium. Because uranium-bearing fluids may alter magnetic minerals, localized negative magnetic anomalies may be associated with uranium ore zones.

Airborne measurements

Considerable exploration for uranium has employed, and is presently employing, airborne gamma-ray surveys as the primary surveying method. In the early days of uranium exploration, small, single-engine aircraft, carrying a geiger counter or primitive scintillation detectors "flew the rims" of the canyons of the southwestern United States, occasionally detecting outcrops of ore (and not infrequently failing to clear the rims). Today, much more sophisticated airborne gamma systems, incorporating volumes of several thousand cubic inches of NaI(Tl), survey in multi-engine aircraft on evenly spaced flight lines, 150—200 m above ground surface (Darnley, 1972, 1973). For more detailed surveys at lower altitudes and slower speeds, helicopter-mounted systems of smaller NaI volume are often employed (Adams, 1968).

As with large-area surveys on the surface, the most up-to-date airborne methods incorporate spectral analyses, measuring the regions encompassing the ^{40}K (1.46 MeV), ^{214}Bi (1.76 MeV), and ^{208}Tl (2.62 MeV) peaks. Four "windows" centered on each of these peaks, and one spanning nearly the entire spectrum are often employed. Data concurrently obtained includes altitude (by radar altimeter), position (by doppler radar, inertial navigation, or continuous imaging of the ground by video or photography) and, often, accompanying geophysical sensors, such as a magnetometer. Gamma spectral and ancillary data are stored on multi-track magnetic tape for processing by on-board computer, or by ground systems, subsequent to the flight.

Airborne gamma surveys lend themselves to several stages of an exploration program. Large-area regional surveys are best carried out by air, especially over inaccessible terrain. Reconnaissance surveys may utilize the total gamma spectrum, on comparatively widely spaced flight lines, with follow-up surveys over areas of interest on closer spaced lines, incorporating spectral interval data. Even more detailed surveys may be done by helicopter, permitting a geologist to land and conduct ground radiometry and sampling.

Parameters considered most important by Darnley (1973) are the ratios U/Th and U/K. These ratios can be calculated in "real time" by an on-board computer, or from the three-window interval data at the conclusion of the flight. An example of results from such a system is shown in Fig. 2-10 (from Darnley, 1972). Calibration of gamma spectrometers by flights over terrain of known radioelement content, or by positioning the aircraft on concrete pads of given content, permits actual concentrations of U, Th, and K to be calculated from airborne spectral count rates (Grasty and Darnley, 1971). If such absolute assays are required, it is often necessary to employ ground-truth teams, sampling rocks and soils for moisture content and typical radioelement values, concurrently with flight operations.

Operational parameters specified by the Geological Survey of Canada (Darnley, 1973) are considered necessary to achieve statistically significant counting rates to differentiate between U, Th, and K over terrain of "normal" radioelement content: For a mean flying height of 400 ft (123 m) and an air speed of 60 mph (97.5 km/hour), the volume of NaI(Tl) crystals must total approximately 1500 cu. inches (24581 cm^3); for an air speed of 120 mph (198 km/hour) total crystal volume should be approximately 3000 cu. inches (49161 cm^3). Detectors are stabilized for single-channel analyzer

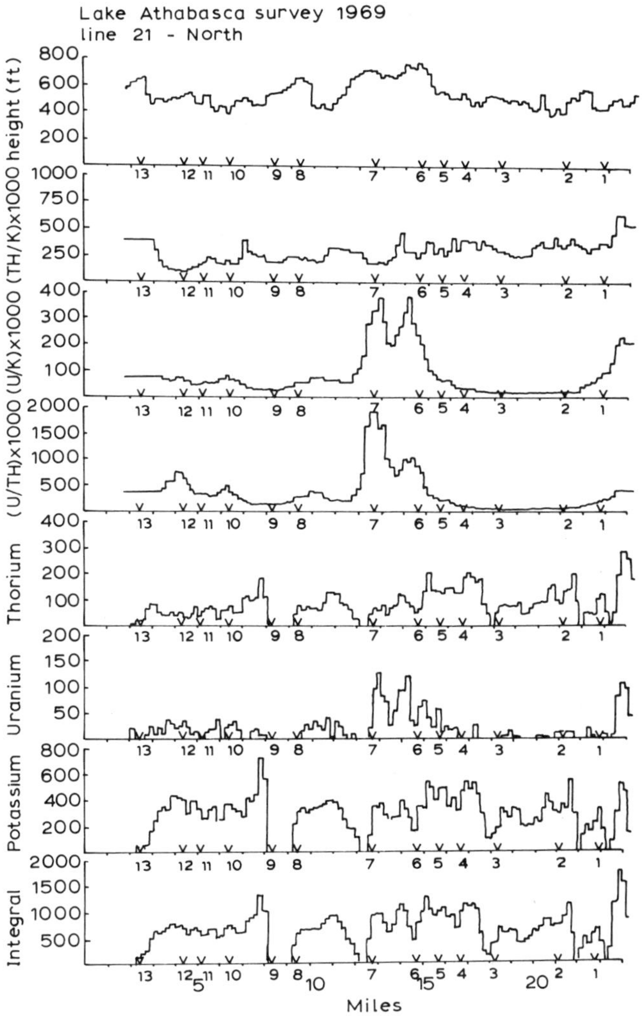

Fig. 2-10. Airborne gamma spectral plots from a survey over the Canadian Shield. The uranium anomaly is essentially hidden in the integral plot. (From Darnley, 1972.)

operation, and individual crystals should be no smaller than 5 inches (12.7 cm) in diameter by 4 inches (10.2 cm) thick. A lower-flying, slower helicopter-borne system may employ a considerably smaller detector volume. In comparing the effectiveness of different systems, Darnley (1973) employs as the figure of merit the square root of the normalized counting rate of each system. Then detector volumes, ground speed, and altitude are directly related to the figure of merit.

Before data reductions are complete, corrections are made for the cosmic-ray contribution to the gamma spectrum (based primarily on flights over water in the survey region), and the abundance of atmospheric ^{222}Rn. Several methods have been attempted to correct for Rn: one employs an upward-looking set of NaI detectors, shielded from the ground contribution by a layer of lead, whose summed count rate is subtracted from that of the downward-looking detectors (C. Jupiter, EG & G Inc., private communication). Another Rn correction method involves periodically collecting an air sample and counting it on board the aircraft for Rn daughters (Burson et al., 1972). A third method, presently being tested (P. Dodd, private communication) incorporates the collection of Rn daughters on a charged wire continuously circulating outside the aircraft, and counting its gamma radioactivity as the wire is wound around a take-up spool.

AUTORADIOGRAPHY OF URANIUM AND THORIUM

For many decades alpha-track and, more recently, fission-track autoradiographic techniques have been utilized to determine the location of radioactive minerals in thin or polished sections of rock.

Alpha autoradiography

Alpha particles radiating from U- and Th-bearing zones near the surface of a thin section are registered on an overlying photographic emulsion. When the emulsion is developed, sites of radioactivity are located and their alpha activity quantitatively determined. Bowie (1954) described geologic applications of alpha radiography while, more recently, Ragland (1964) presented an excellent review article on the technique and its applications. Though it is effective in registering radioactive mineral grains, a drawback of the alpha radiographic method is the long time required to obtain results. Minerals with relatively high radioactivity (several hundred to several thousand parts per million U or Th) require exposure times of the order of ten days to yield statistically significant numbers of countable alpha tracks for quantitative determination of radioelements.

Furthermore, in a population of alpha tracks it is very difficult to differentiate between those from the Th decay series and those from the U decay series. Also, it is possible that secular equilibrium does not exist between ^{238}U and its daughters which are the principal alpha-particle emitters. Therefore, unless the ratios of Th to U within the radioactive minerals are known, either through independent measurements, or are successfully assumed, alpha-particle autoradiography does not furnish independent determinations of the radioelements.

The fission-track method

This method, developed in the mid 1960's by Fleischer et al. (1965), permits determination of the location and abundances of U and, under favorable conditions, Th, in thin or polished sections after short exposures (generally a few minutes) to reactor neutrons. Over the past several years, geologists have applied the method to study U contents, ranging from tens to thousands of parts per million, in thin and polished sections containing accessory and rock-forming minerals — see, for example, Kleeman and Lovering (1967), Bowie et al. (1972), and Wollenberg (1972). U contents as low as a few tens of parts per billion have been measured in finely powdered preparations of whole rock ultramafic samples (Fisher, 1970).

The basic procedure to obtain fission tracks from U and/or Th is quite simple. Solid-state track detectors, usually mica or plastic, are mounted on a series of samples and standards, which are then exposed to reactor neutrons. On completion of the exposure, the detectors are removed and etched in the appropriate reagents. The resulting tracks are then viewed and counted with the aid of a microscope. (Fig. 2-11 is an example of a fission-track field.) In choosing the neutron exposure appropriate for the expected concentration of uranium, one employs the equation which governs density of fission tracks in a detector placed directly on the flat surface of material containing U, and irradiated with neutrons:

$$\rho = \phi\sigma E n \qquad [1]$$

where ρ is the track density in tracks cm^{-2}, ϕ is the integrated neutron flux in cm^{-2}, σ is the neutron-induced fission cross-section (580 barns for thermal neutron fission of U; 1 barn = 10^{-24} cm^2), E is the efficiency of detection, and n is the number of ^{235}U nuclei per square centimeter of thin section surface within the effective range of fission fragments. In the usual case of a flat detector-sample interface, E is independent of materials, and is 0.5; half of the fission events will register in the detector. For fragments from the fission of U and Th the effective range in most silicate and phosphate minerals is ~3 mg cm^{-2}. Thus:

$$n = (3Ca/A) \times 10^{-3} \qquad [2]$$

where C is the concentration of radioelement in grams per gram, a is Avogadro's number, and A is the atomic mass number of the radioelement.

In choosing a fission-track detector, one must weigh the advantages and disadvantages of different materials. The two detectors most commonly used in geologic applications are high-quality muscovite mica and polycarbonate plastic (Lexan or Makrofol). These are compared briefly in Table 2-IV, and their tracks illustrated in Fig. 2-12. Tracks in mica are easier to recognize than those in plastic. However, Lexan's principal advantage is that it has essentially no background U, whereas the highest quality mica has a background of tracks from the spontaneous and induced fission of its inherent ^{238}U content. Therefore, if a low content of U is expected in the sample, it may be preferred to use a Lexan detector, or expose a mica detector "blank" to obtain the effective background value.

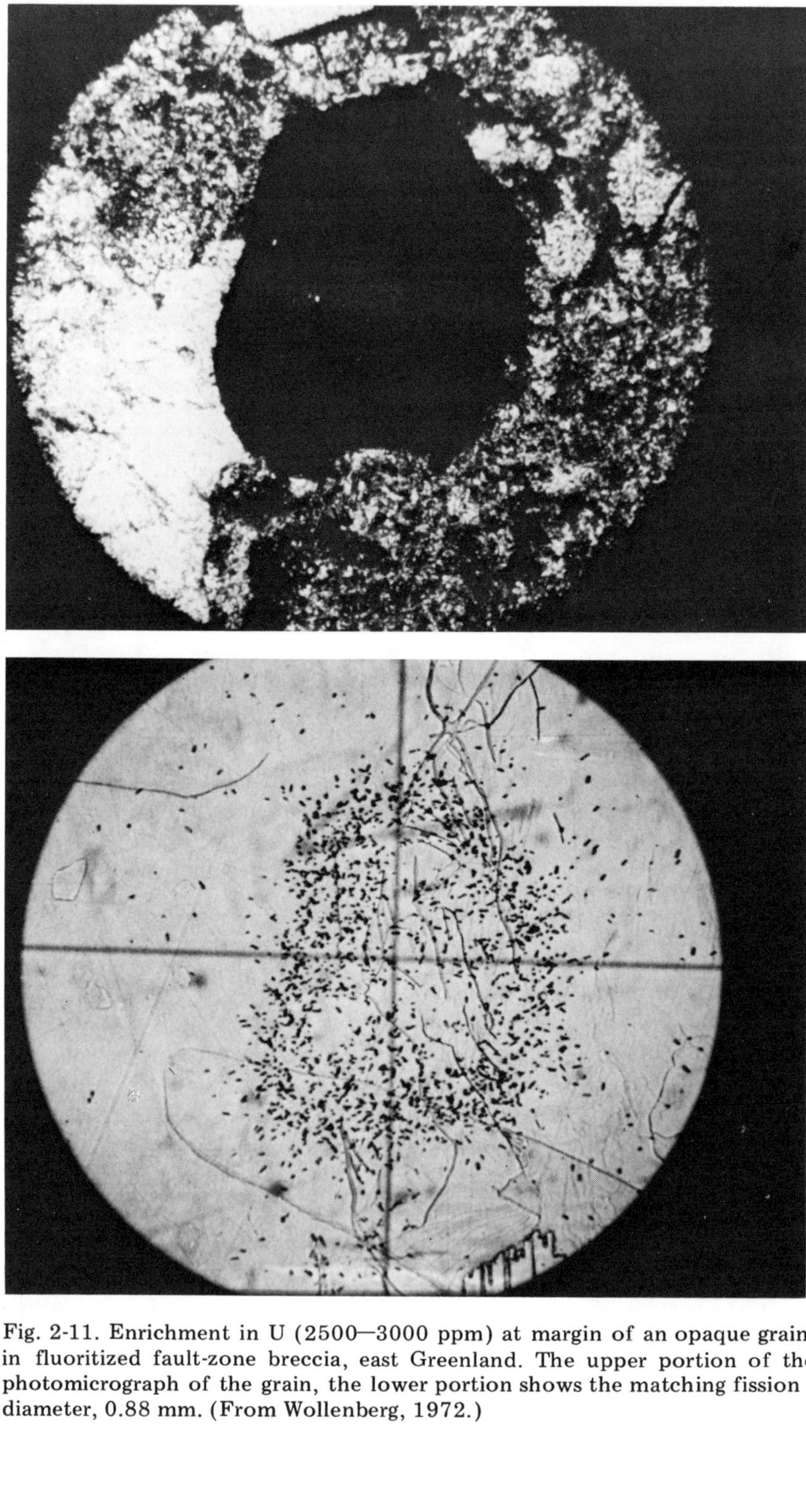

Fig. 2-11. Enrichment in U (2500—3000 ppm) at margin of an opaque grain (hematite?) in fluoritized fault-zone breccia, east Greenland. The upper portion of the figure is a photomicrograph of the grain, the lower portion shows the matching fission tracks. Field diameter, 0.88 mm. (From Wollenberg, 1972.)

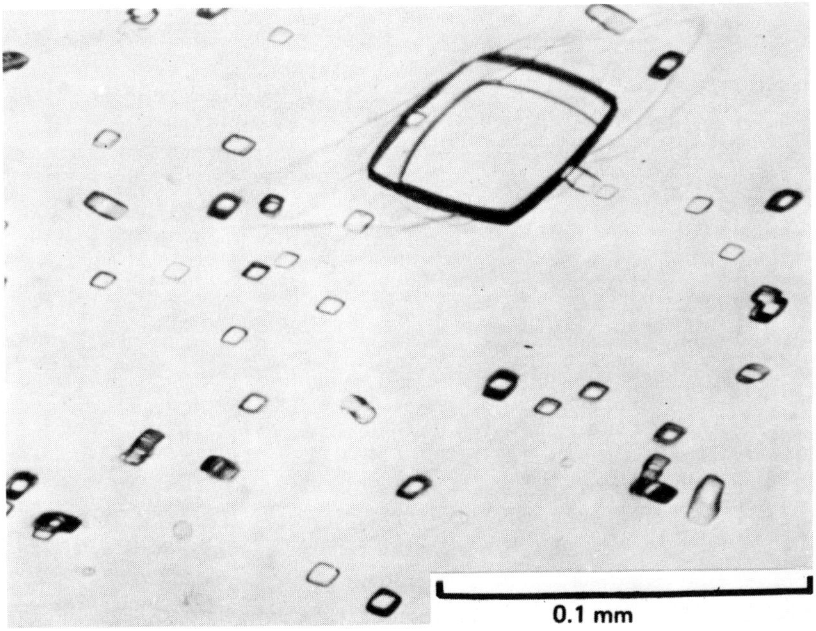

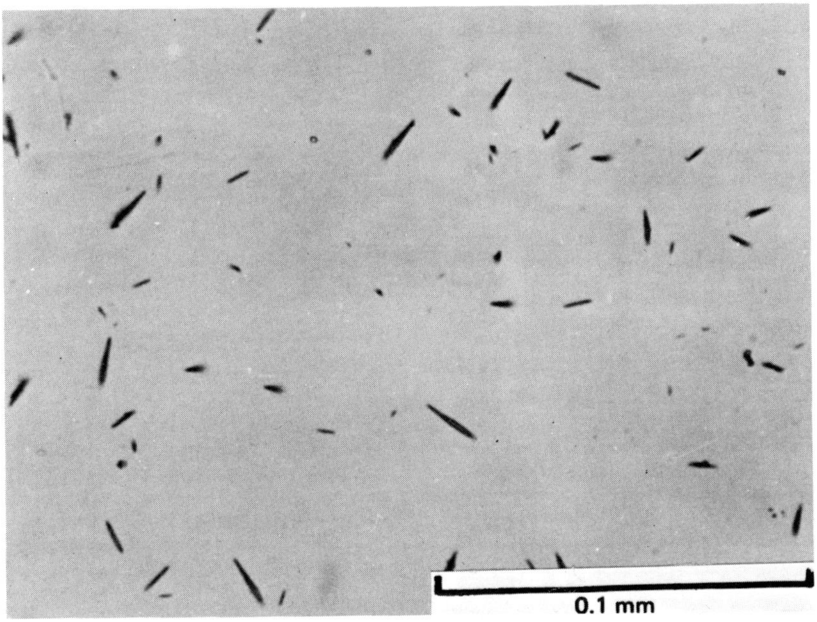

Fig. 2-12. Fission-track etch pits in muscovite (top) and Lexan polycarbonate (bottom). Track density, 2.25×10^5 tracks cm^{-2}. Mica was etched in 48% HF for ~1 hour, the Lexan in 6*N* KOH at 60°C for 15 minutes. (From Wollenberg, 1972.)

TABLE 2-IV

Comparison of detector materials

Natural muscovite	Polycarbonate plastic
Very durable	Surface scratches easily
Tracks easy to recognize, readily distinguished from scratches	Tracks perpendicular to surface appear as small dots, easily confused with scratches
Etching time, $\frac{1}{2}$—4 hours	Etching time, 10—15 minutes
Track size regulated by varying etching time without danger of appreciable surface loss	Surface removed readily as etching progresses; not possible to enlarge tracks after 10—15 minutes
U content plus geologic age yield background tracks from spontaneous and induced fission. This governs detection limit (1—5 ppm U)	Essentially no background tracks, detection limit governed by neutron exposure

Exposure, processing, and etching conditions for different U concentrations are summarized in Table 2-V. The choice of irradiation conditions depends upon the element to be determined and its concentration. If U alone is desired, thermal neutrons are employed for the essentially exclusive fission of ^{235}U. The samples are exposed in the thermal column of the reactor (a position outside the core, surrounded by graphite to exclude fast neutrons), where the ratio of thermal to fast neutrons is highest. Conversely, a fast-neutron irradiation is used to detect Th. The sample-detector package is placed as close as possible to the core of the reactor, and is encased in at least a 3-mm thickness of cadmium to permit as pure a flux of fast neutrons as possible. In the flux estimate for a Th irradiation, one must include the large contribution to the observed track density from ^{238}U. The estimated

TABLE 2-V

Exposure, processing and observation conditions for various evenly distributed uranium concentrations

U content (ppm)	Neutron exposure (n cm^{-2})	Track density (tracks cm^{-2})	Etching time (hours)	Magnification	Tracks/ field
50	2×10^3	1.2×10^4	3—4	× 36	~500
500	1×10^3	6×10^4	2—3	× 80	~450
5000	1×10^3	6×10^5	1—2	× 270	~400

* Etching of high-quality muscovite mica in 40—48% HF at room temperature.

integrated neutron flux is then:

$$\phi = \rho / E(\sigma_{Th} n_{Th} + \sigma_U n_U) \qquad [3]$$

where ρ is the desired track density, E the efficiency of detection, σ_{Th} and σ_U the fast neutron fission cross sections for ^{232}Th (0.3 barn) and ^{238}U (~1.1 barn) respectively, and n_{Th} and n_U are the estimated concentrations of Th and ^{238}U nuclei.

If the Th/U ratio of a mineral is great enough, the use of fast neutrons permits a determination of Th content. Since practically all Th-bearing minerals also contain some U, it is necessary first to determine U by a

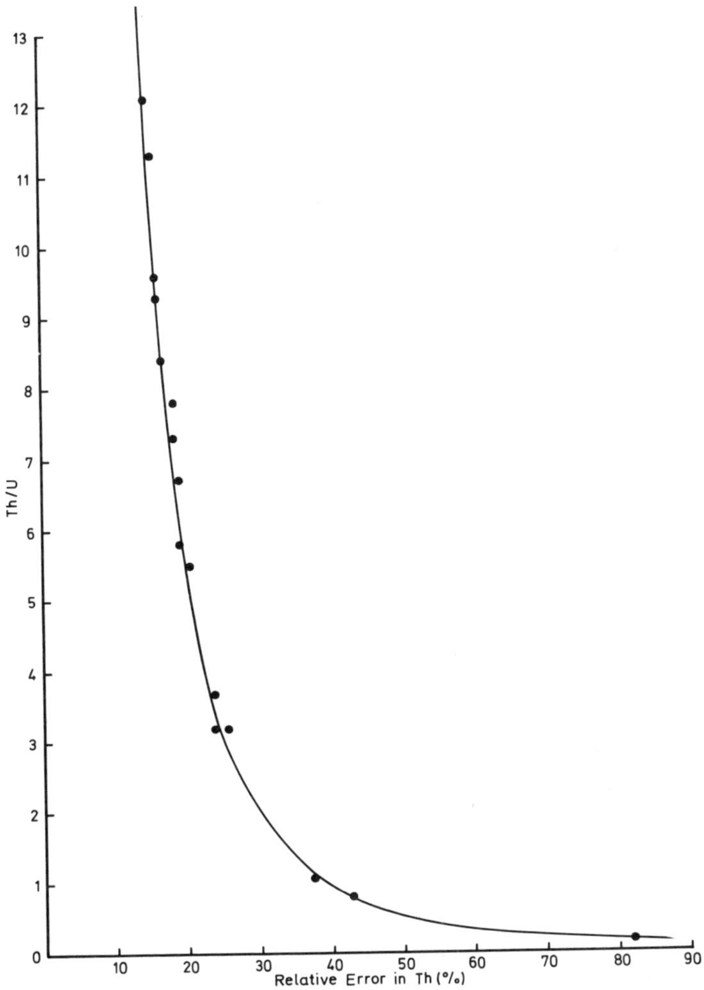

Fig. 2-13. Th/U ratio versus relative error in Th for typical exposure and counting conditions. (From Wollenberg, 1972.)

thermal neutron exposure. Then the sample, with a new detector and U and Th standards is exposed to a fast neutron flux. The resulting track density is from the fission of ^{238}U and ^{232}Th, but knowing the U content, and having simultaneously exposed a U standard, the U track contribution is determined. The contribution from Th is then the difference between the total track density and that from U.

The main difficulty of this method of Th determination is that the relative errors due to counting statistics are rather large; the cross-section ratio for fast neutrons favors U over Th by nearly a factor of 3. Fig. 2-13 relates Th/U ratios and relative errors for typical exposure and counting conditions of accessory-mineral samples. Relative errors are ~25% if Th/U is greater than ~3; for ratios <3 the errors increase rapidly and are unacceptably high (~40%) for Th/U < 1.

RADON DETECTION

^{222}Rn, a gaseous decay product of the ^{238}U series, has been successfully used as a diagnostic parameter for locating U beneath a thickness of cover which would otherwise conceal the ore from detection by surface or air-borne gamma-ray surveys.

Physical parameters

Because it is an inert gas, Rn diffuses from its source sites, mineral grains containing abundances of parent ^{238}U, without chemical interaction with constituents of rock or soil. The Rn daughter isotope of the ^{232}Th series, ^{220}Rn (termed thoron), has a much shorter half-life (55 seconds) than ^{222}Rn (3.8 days). Therefore, the U series daughter travels significantly farther in the soil before decaying, and is consequently present at or near the earth's surface in greater abundance than thoron.

A parameter termed the emanation coefficient, the ratio of Rn emanated by a mineral to the Rn actually produced in the mineral, varies widely for given minerals and their physical character (Barretto et al., 1972). Rn losses of 1—20% are not uncommon for rock materials crushed to identical grain size fractions, and values of 10—50% were observed for soils. Accessory minerals with high U contents have relatively low emanation coefficients; Rn loss depends upon whether the mineral grains are fresh or badly damaged by alpha particles, weathering, or other factors. By virtue of closing pore spaces through which Rn may move, soil moisture tends to inhibit diffusion and exhalation. The movement of Rn can, however, be enhanced by groundwater flow (Tanner, 1964).

Israel (1958) defined the exhalation rate of a radioactive gas from the ground: $E = \alpha(D/\lambda)^{1/2}$, where α is the rate of generation of gas atoms into soil pore spaces in Ci $cm^{-3}\,s^{-1}$, D is the bulk diffusion coefficient through the porous medium in $cm^2\,s^{-1}$, and λ is the radioactive decay constant in s^{-1}. Megumi and Mamuro (1974) list exhalation rates of Rn and thoron for various size fractions of soil originating from weathered granite. Calculated exhalation rates for Rn were approximately seven times greater for minus 200-mesh sizes than for the 5—10-mesh fraction of a granite-derived soil.

^{222}Rn decays through its daughters to ^{210}Pb, with the emission of 3 alpha particles of 5.48, 6.0, and 7.69 MeV, each distinguishable by alpha spectroscopy. Within the decay series, ^{214}Pb and ^{214}Bi emit gamma rays detectable in the energy regions encompassed by normal gamma-ray spectrometry.

Sampling and analysis

Most instruments for Rn measurement, to date, have been developed for assessment of the element's environmental effects. A valuable compendium of information on these measurement systems has been prepared by Budnitz (1973). In both the environmental and prospecting applications, Rn assessment utilizes primarily, alpha-particle counting. A relatively new method, described subsequently, employs solid-state alpha-track detectors.

Alpha counting

Field instrumentation for on-site measurement of Rn in soil gas incorporates extraction of the gas and immediate counting of its alpha activity. Such a lightweight, portable instrument, an alpha probe, is described in a review article by Miller and Ostle (1973). A sharp-pointed bar is driven 60 cm into the ground then withdrawn. The suction of withdrawing the tube forces soil air into the hole, which is then cased. Insertion of the alpha probe the hole forces air into the measuring chamber. Alphas are counted by interaction with silver-activated zinc sulfide, a phosphor-coated detector, viewed by a photomultiplier tube. The total alpha radioactivity of Rn and its daughters is measured, and results are normally expressed as counting rates.

Other systems utilize pumping of soil gas from uncased holes, through an alpha counter. A portable pumping system, the Scintrex ETR-1, which measures alpha activity in an ionization chamber, is shown in Fig. 2-14. Some pumped systems employ closed cycles, where the exhaust from the detector chamber is circulated back into the capped hole, others exhaust the gas from the chamber to the atmosphere. The latter draw gas from a volume considerably larger than the hole. Pumping systems may introduce appreciable dust and moisture, thus they require filtration which may remove some Rn daughters which adhere to dust particles.

Because Rn daughters adhere to dust particles, detector chamber contamination levels build up with successive measurements, despite flushing with atmospheric air. This contamination requires that ZnS detectors be

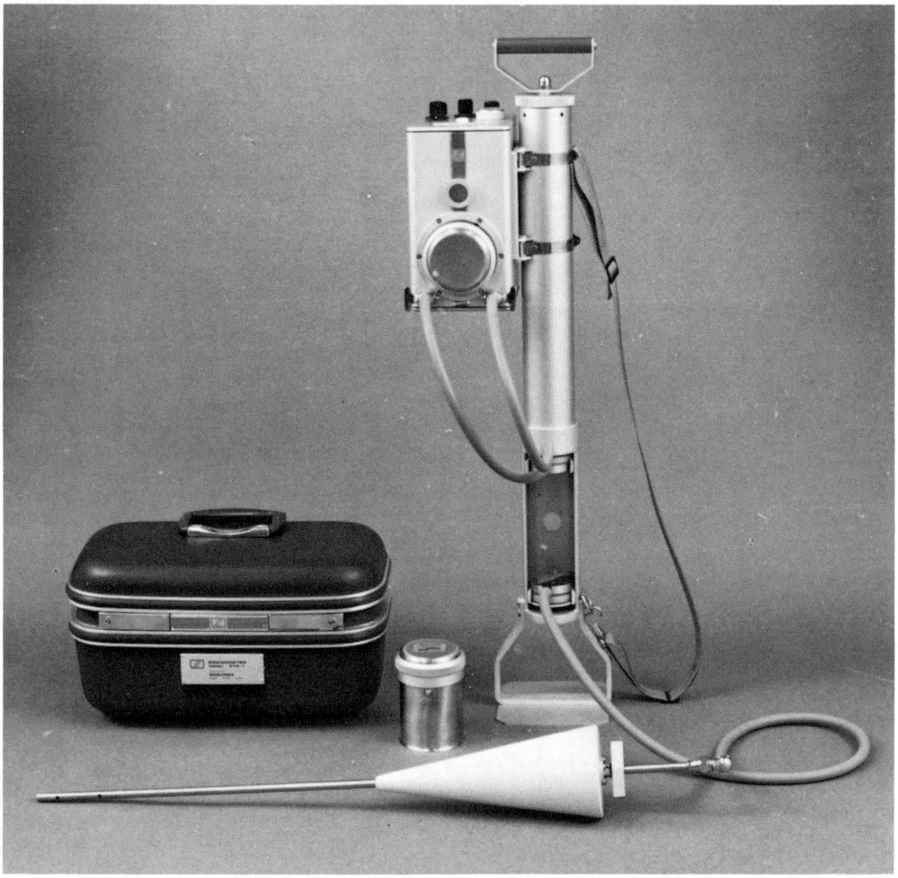

Fig. 2-14. The Scintrex ETR-1 Emanometer for analysis of soil-gas ^{222}Rn.

changed frequently. Filtering reduces, but does not completely prevent, accumulation of daughters.

Field soil-gas measurement systems range in size from the afore-mentioned probe, to systems employing vehicle-mounted augering and pumping equipment. Results of measurements are generally expressed as alpha count rates, rather than absolute Rn concentrations. In areas where ^{232}Th is abundant, ^{220}Rn (thoron) may be an important constituent of soil gas. Its effect can be minimized by retaining the air sample in the detector chamber for several half-lives (~10 min) before counting, allowing thoron to decay.

Rn contents of stream and lake waters may be indicators of nearby U mineralization (Dyck, 1972). Rn is extracted from the water sample by bubbling air or an inert gas through the sample, and collecting the gas-Rn mixture in a ZnS-coated cell for alpha counting. If a low-background gamma-counting system is close at hand, water samples can be analyzed directly for

^{222}Rn using the predominant ^{214}Bi gamma rays (Wollenberg, 1974). Results of water analyses are normally expressed as picocuries per liter (10^{-12} Ci l^{-1}).

Alpha-track detectors

The registration of alpha particles in plastics has been successfully applied to the search for U in recent years. The basic concept — called Track Etch

Fig. 2-15. Plastic cup with alpha-track detector attached. (From Gingrich, 1975.)

because it etches the alpha tracks in the plastics — was discovered by scientists at the General Electric Research and Development Laboratory (Fleischer et al., 1972). It was subsequently modified and patented by scientists of the Terradex Corporation to detect alpha particles from the decay of ^{222}Rn Alter et al., 1972).

Inverted plastic cups with a specially treated dielectric alpha-track detector attached inside (Fig. 2-15) are placed, each in an approximately 0.5-m-deep hole, in arrays of varying extent. The walls of the cup are thick enough to exclude alpha particles from Rn in the air outside the cup and other alpha emitters in the soil surrounding the cup; the only contribution is from the Rn immediately inside the cup. After exposure of several weeks, the cups are retrieved, the detectors removed and etched revealing track densities proportional to the emanating Rn flux. A major advantage of the Track Etch method is this long exposure which averages the large variations (sometimes as large as 100-fold in 24 h) that occur in Rn levels. Fig. 2-16 shows a Rn contour map based on this technique (Gingrich, 1975) where subsequent drilling led to the discovery of a sizeable orebody which would have otherwise gone undetected by surface gamma-ray surveys.

The principal advantage of the track detector method over the direct measurement on-site of ^{222}Rn, is that the track method integrates over a period of time, and therefore smooths out effects of fluctuations in emanation due primarily to atmospheric phenomena.

Quite recently, a new method of field Rn detection has been developed by

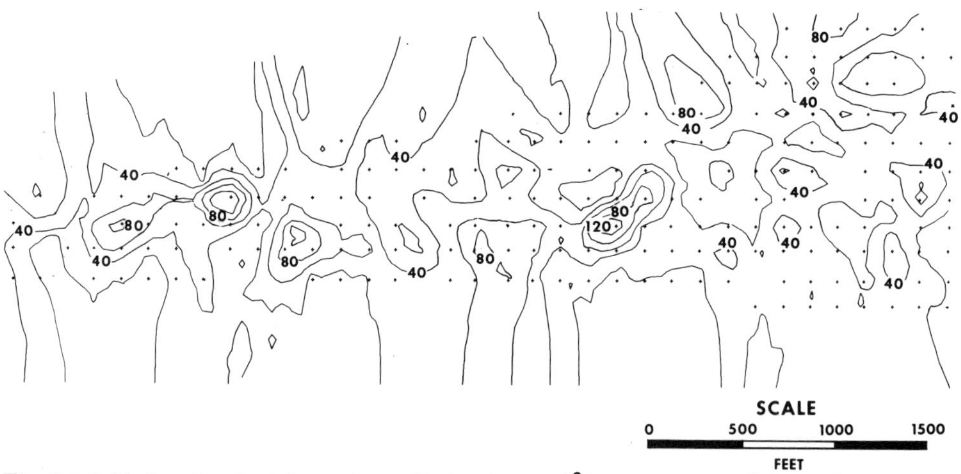

Fig. 2-16. Radon track-etch contours (in tracks mm^{-2}), over a potential uranium resource area. (From Gingrich, 1975.)

Exxon Eastern Inc., and Alpha-Nuclear Company. A plastic cup with a silicon-semiconductor detector is placed in a hole, similar to the track-etch detectors. Alpha particles from the decay of ^{222}Rn in the cup are registered on the semiconductor, whose pulses are stored by an electronic unit, also in the hole. The integrated counts are then read out periodically by a separate, portable unit. Though initial costs greatly exceed those of the track-etch cups, the electronic Rn detection units can be reset and used indefinitely, making them cost-effective over a long period.

CONCLUSIONS

Radiometric methods are applicable to all aspects of the search for uranium resources, the definition of reserves, and their development. A three-phase program incorporating radiometric methods is generally employed: exploration for target areas; evaluation of targets, leading to definition of reserves; followed by development and production. Table 2-VI lists the radiometric methods which fit into these categories. It is emphasized that data from radiometric methods alone, are not a sufficient base upon which to decide the course of a resource development program. Of great impor-

TABLE 2-VI

Radiometric methods applicable to phases of uranium resource development

1. *Exploration for target areas:*

   Airborne total gamma and gamma-spectral surveys

   Groundborne field gamma spectrometry on broad-scale surveys, and laboratory gamma spectrometry of selected samples

   Soil and water ^{222}Rn surveys over large areas

2. *Evaluation of targets:*

   Detailed field total gamma and gamma spectral surveys over target areas

   Laboratory gamma spectrometry of specimens from intensely samples areas

   Rn alpha-track surveys of target areas

   Downhole total gamma and gamma-spectral measurements of targets delineated by surface surveys

   Autoradiography of selected thin or polished sections

3. *Development and production of resource areas:*

   Downhole total gamma and gamma-spectral measurements for grade control

   Field total gamma and gamma-spectral measurements of working faces, benches, and stockpiles, and laboratory gamma spectrometry of samples from these areas

tance are economic considerations and a strong knowledge of the geologic setting, coupled with a thorough understanding of non-radioactive geochemical and geophysical parameters. These, then, combined with information from radiometric methods, furnish a firm basis for evaluation of uranium resource potential.

REFERENCES

Adams, J.A.S., 1968. Total and spectrometric gamma-ray surveys from helicopters and vehicles. In: *Symposium on Use of Nuclear Techniques for Prospecting and Developing Mineral Resources, Buenos Aires.* IAEA, Vienna.

Adams, J.A.S. and Fryer, G., 1964. Portable gamma-ray spectrometer for field determination of thorium, uranium, and potassium. In: J.A.S. Adams and W. Lowder (Editors), *The Natural Radiation Environment.* University of Chicago Press, Chicago, Ill.

Adams, J.A.S. and Gasparini, P., 1970. *Gamma-Ray Spectrometry of Rocks.* Elsevier, Amsterdam, 295 pp.

Alter, W. et al., 1972. Radon detection using track-registration materials. *U.S. Patent No. 3,665,194* (issued May 23, 1972).

Anspaugh, L.R., Phelps, P.L., Gudiksen, P.H., Lindeken, C.L. and Huckabay, G.W., 1972. The in-situ measurement of radionuclides in the environment with a Ge(Li) detector. In: J.A.S. Adams, W. Lowder and T. Gessell (Editors), *The Natural Radiation Environment II. U.S. ERDA Rep. Conf.*, 720805-P2.

Barretto, P.M.C., Clark, R.B. and Adams, J.A.S., 1972. Physical characteristics of radon-222 emanation from rocks, soils and minerals: its relation to temperature and alpha dose. In: J.A.S. Adams, W. Lowder and T. Gessell (Editors), *The Natural Radiation Environment II. U.S. ERDA Rep. Conf.*, 720805-P2.

Bowie, S.H.U., 1954. Nuclear emulsion techniques. In: H. Faul (Editor), *Nuclear Geology.* Wiley, New York, N.Y., pp. 148—174.

Bowie, S.H.U., Simpson, P.R. and Rice, C.M., 1972. Application of fission-track and neutron activation methods to geochemical explorations. In: M. Jones (Editor), *Geochemical Exploration, 1972.* Institution of Mining and Metallurgy, London.

Budnitz, R., 1973. Radon-222 and its daughters. In: *Instrumentation for Environmental Monitoring. Lawrence Berkeley Lab. Rep.*, LBL-1.

Burson, Z.G., Boyns, P.K. and Fritzsche, A.E., 1972. Technical procedures for characterizing the terrestrial gamma radiation environment by aerial surveys. In: J.A.S. Adams, W.M. Lowder and T.F. Gessell (Editors), *Natural Radiation Environment II. U.S. ERDA Rep. Conf.*, 720805-P2.

Darnley, A.G., 1972. Airborne gamma-ray survey techniques. In: S.H.U. Bowie et al. (Editors), *Uranium Prospecting Handbook.* Institution of Mining and Metallurgy, London, pp. 174—211.

Darnley, A.G., 1973. Airborne gamma-ray survey techniques — present and future. In: *Uranium Exploration Methods.* IAEA, Vienna, pp. 67—108.

Davis, J.D. and Guilbert, J.M., 1973. Distribution of the radioelements potassium, uranium, and thorium in selected porphyry copper deposits. *Econ. Geol.*, 68: 145—160.

Dodd, P.H. and Eschliman, D.H., 1972. Borehole techniques for uranium exploration and evaluation. In: S.H.U. Bowie et al. (Editors), *Uranium Prospecting Handbook.* Institution of Mining and Metallurgy, London, pp. 244—276.

Doig, R., 1968. The natural gamma-ray flux: in-situ analysis. *Econ. Geol.*, 33: 311—328.

Dyck, W., 1972. Radon methods of prospecting in Canada. In: S.H.U. Bowie et al.

(Editors), *Uranium Prospecting Handbook*. Institution of Mining and Metallurgy, London.

Fisher, D.E., 1970. Homogenized fission track determination of uranium in whole-rock geologic samples. *Analyt. Chem.*, 42: 414—416.

Fleischer, R.L., Price, P.B. and Walker, R.M., 1965. Tracks of charged particles in solids. *Science*, 149: 383—393.

Fleischer, R.L., Alter, H.W., Furman, S.C., Price, P.B. and Walker, R.M., 1972. Particle track etching. *Science*, 178: 255—263.

Gingrich, J., 1975. Results from a new uranium exploration method. *Trans. Soc. Min. Eng., Am. Inst. Min. Eng.*, 258: 61—64.

Grasty, R.L. and Darnley, A.G., 1971. Calibration of gamma-ray spectrometers for ground and airborne use. *Geol. Surv. Can. Paper*, 71-17.

Heath, R.L., 1969. Gamma-ray spectrometry and automated data analysis systems for activation analysis. In: J.R. DeVoe (Editor), *International Conference on Modern Trends in Activation Analysis, 2. Natl. Bur. Stand. (U.S.), Spec. Publ.*, 312.

Heier, K.S. and Adams, J.A.S., 1965. Concentration of radioactive elements in deep crustal material. *Geochim. Cosmochim. Acta*, 29: 53—61.

Israel, H., 1958. Die Naturliche Radioaktivität in Boden, Wasser, und Luft. *Beitr. Phys. Atmos.*, 30: 177—188.

Key, B.N. and Eschliman, D.H., 1974. Preliminary evaluation of a spectral logging method for the borehole. Unpublished report, Lucius Pitkin Inc. for ERDA Resource Division.

Kleeman, J.D. and Lovering, J.F., 1967. Uranium distribution in rocks by fission-track registration in Lexan plastic. *Science*, 156: 512—513.

Løvborg, L., 1973. Future development in the use of gamma-ray spectrometry for uranium prospecting on the ground. In: *Uranium Exploration Methods*. IAEA, Vienna.

Løvborg, L., Kunzendorf, H. and Rose-Hansen, J., 1968. Use of field gamma spectrometry in the exploration of uranium and thorium deposits in South Greenland. In: *Nuclear Techniques and Mineral Resources*. IAEA, Vienna, pp. 197—211.

Løvborg, L., Wollenberg, H., Sorensen, P. and Rose-Hansen, J., 1971. Field determination of uranium and thorium by gamma-ray spectrometry, exemplified by measurements in the Ilimaussaq alkaline intrusive, South Greenland. *Econ. Geol.*, 66: 368.

Løvborg, L., Wollenberg, H.A., Rose-Hansen, J. and Leth Nielsen, B., 1972. Drill-core scanning for radioelements by gamma-ray spectrometry. *Geophysics*, 37: 675—693.

Matheron, G., 1963. Principles of geostatistics. *Econ. Geol.*, 58: 1246—1266.

Megumi, K. and Mamuro, T., 1974. Emanation and exhalation of radon and thoron gases from soil particles. *J. Geophys. Res.*, 79: 3357—3360.

Miller, J.M. and Ostle, D., 1973. Radon measurement in uranium prospecting. In: *Uranium Exploration Methods*. IAEA, Vienna.

Moxham, R.M., Foote, R.S. and Bunker, C.M., 1965. Gamma-ray spectrometer studies of hydrothermally-altered rocks. *Econ. Geol.*, 60: 653—670.

Ragland, P., 1964. Autoradiographic investigations of naturally occurring materials. In: J.A.S. Adams and W. Lowder (Editors), *The Natural Radiation Environment*. University of Chicago Press, Chicago, Ill.

Richardson, K.A., 1964. Thorium, uranium and potassium in the Conway Granite, New Hampshire, U.S.A. In: J.A.S. Adams and W. Lowder (Editors), *The Natural Radiation Environment*. University of Chicago Press, Chicago, Ill.

Smith, A.R. and Wollenberg, H.A., 1972. High-resolution gamma-ray spectrometry for laboratory analysis of the uranium and thorium decay series. In: J.A.S. Adams, W. Lowder and T. Gessell (Editors), *The Natural Radiation Environment II. U.S. ERDA Rep. Conf.*, 720805-P2.

Sprecher, C. and Rybach, L., 1974. Design and field test of a scintillation probe for gamma-logging of small siameter boreholes. *Pageoph*, 112: 564—570.

Tanner, A., 1964. Radon migration in the ground: a review. In: J.A.S. Adams and W. Lowder (Editors), *The Natural Radiation Environment.* University of Chicago Press, Chicago, Ill.

Wollenberg, H., 1972. Fission-track radiography of uranium and thorium in radioactive minerals. In: M. Jones (Editor), *Geochemical Exploration 1972.* Institution of Mining and Metallurgy, London.

Wollenberg, H., 1974. Radioactivity of Nevada hot spring systems. *Geophys. Res. Lett.*, 1: 359—362.

Wollenberg, H.A. and Smith, A.R., 1968. Radiogeologic studies in the central part of the Sierra Nevada batholith. *J. Geophys. Res.*, 73: 1481—1495.

Wollenberg, H.A. and Smith, A.R., 1969. Radioactivity and radiogenic heat production in prebatholithic rocks of the central Sierra Nevada. *J. Geophys. Res.*, 75: 431—438.

Wollenberg, H.A., Smith, A.R. and Bailey, E.H., 1967. Radioactivity of Upper Mesozoic graywackes in the northern Coast Ranges, California. *J. Geophys. Res.*, 72: 4139—4150.

Wollenberg, H.A., Patterson, H.W., Smith, A.R. and Stephens, L.D., 1969. Natural and fallout radioactivity in the San Francisco Bay area. *Health Physics*, 17: 313—321.

Chapter 3

X-RAY ANALYSIS IN MINERAL EXPLORATION

M. SLAUGHTER

INTRODUCTION

Mineral exploration is the art of finding and defining economically desirable natural resources. When executed logically, mineral exploration encompasses a narrowing sequence of deductive and inductive predictions of presence, quality and quantity of ore. The entire exploration process from regional geological analysis to detailed chemical analysis rests on an analytical base. Mineral exploration is a complex art requiring many types of analytical tools. Two of the tools to determine chemical composition and mineral content of rocks and ores are X-ray fluorescence analysis and X-ray diffraction analysis. Both methods of X-ray analysis add vital information to a pool of knowledge necessary for mineral finding and exploitation.

X-ray fluorescence analysis determines chemical composition. An X-ray source excites X-ray emission in a specimen. X-rays emanating from the specimen indicate its elemental composition. X-ray fluorescence analysis is particularly suited to mineral exploration because it determines many elements in rocks and ores simultaneously, rapidly and accurately as well as determining elements that are difficult or expensive to measure by other means.

X-ray diffraction analysis uses X-rays to determine mineralogical content of a specimen. An X-ray source impinges X-rays on a specimen which diffracts them much as light diffracts through a fine wire mesh, the angles of diffraction depending on the minerals in the specimen.

X-ray fluorescence and X-ray diffraction analysis, although powerful tools for mineral exploration, serve best in a particular niche of the exploration sequence. Out of their niche, both methods give but supporting help to the exploration process and may not be useful at all. How are these methods used in mineral exploration? We shall try to show the exploration niche of both methods of analysis by developing a simplified example of exploration for uranium. Application of the X-ray methods is similar in exploration for all types of metal and non-metal ores.

The geologist begins looking for a uranium deposit perhaps by choosing an area in sedimentary rocks flanked by a granitic terrain. He knows that uranium frequently concentrates in sedimentary rocks from a source in granites. Should the literature lead him to a promising area, he may initiate field inspection for potential ore sites. The geologist might choose to look for both suitable source and host rocks for his initial investigation. He might visually analyze potential source-rock lithology for a likelihood of containing uranium. He might sample run-off waters from the potential source rock or sample the source rock itself for minute quantities of uranium or related daughter products using gamma-ray spectrometry. Normally the geologist would not use surface radiometry because of low concentration of radioactive elements.

During this preliminary phase of source-rock inspection chemical and mineralogical analyses by X-ray methods have limited use. The geologist specifies analytical methods to determine chemical concentration in parts per billion or parts per trillion, out of the useful range of X-ray fluorescence analysis. Selected spot samples, subjected to chemical analysis by X-ray fluorescence and qualitative mineralogical analysis by X-ray diffraction, might help to establish a chemical and mineralogical framework of the potential source rocks.

Concurrently with source investigation, the geologist might investigate surface rock exposures to find a suitable lithology for uranium transport and concentration. Again the geologist analyzes the rocks visually. Should he find a sedimentary sequence where a not-too-pervious sandstone is over- and underlain by impervious shales, having the granite as a sediment source, he may proceed to analyze for uranium indications down-dip from the sediment source. He may analyze the surface using radiometric surveys, radon analysis, or trace element analysis to detect leakage to the surface of daughter products or "pointer" elements from a subsurface uranium deposit. He may even analyze plants for radiometric elements or trace elements such as molybdenum which may be associated with a uranium deposit. Should these indications, coupled with analyses of sediment nearer the source, show relative depletion of uranium and daughter products, a new phase of exploration begins — exploration drilling.

The beginning of the drilling phase of exploration marks a transition in the usefulness of X-ray analytical methods. Initial investigation of potential uranium accumulation sites use trace element analysis, radiometric measurements of various sorts and visual analysis. Neither X-ray diffraction nor fluorescence contributes except perhaps for occasional analyses to determine some indirectly useful rock parameter.

From the beginning of the drilling phase of exploration or its equivalent for other minerals to the completion of drilling prior to mining, X-ray fluorescence and X-ray diffraction analysis can contribute systematically to the finding and definition of orebodies. For example, during the drilling phase of exploration of a given site, one company submitted 100,000 samples for X-ray analysis to help find and define its ore. X-ray fluorescence defined metal pointer elements, rock-forming elements and uranium.

The exploration team begins drilling to look for ore. Rotary drilling holes spaced relatively widely allow the uranium geologist a small window to the subsurface. Through this window he determines stratigraphy, structure, lithology, texture, porosity, permeability, alteration products of migrating groundwater solutions, uranium indicator elements, and uranium itself. He might find in one hole that a potential ore-bearing bed had insufficient permeability to carry ore-bearing solutions. If sufficiently permeable he might find zeolites by X-ray diffraction analysis to tell him that leaching solutions carrying uranium had not reached his drill hole. Conversely, mineralogical analysis by X-ray diffraction could show an abundance of kaolinite and no zeolites indicating acid groundwater leaching up-dip from a potential deposit. Closer to a uranium deposit, chemical analysis provides a framework leading to ore. X-ray fluorescence analysis may determine anomalously high copper or vanadium amounts and may detect and quantify uranium itself.

Systematic X-ray diffraction and X-ray fluorescence analyses of samples from each hole provide a mineralogical and chemical framework to, at first, supplement and then to supplant visual analysis and downhole logging methods. Both analytical X-ray methods give a basis for quantitative mapping as did radiometric surveys during the previous exploration phase. Subtle indicators of a uranium deposit or the direction to a deposit can be mapped by X-ray analysis even with contaminated cuttings from rotary drilling.

The geologist having found a uranium orebody now enters the final phase of exploration. With a combination of core and rotary drilling he defines the boundaries, quantity and quality of ore. During earlier phases of exploration he used analyses of various types to find ore. During this phase he uses analyses to establish the ore inventory and the lithologic and structural framework preliminary to mining. X-ray fluorescence and X-ray diffraction become very powerful analytical tools during this final phase. The combination of methods can quantitatively document the most important elements and lithologies inexpensively and accurately, providing a good three-dimensional model of the orebody. The deposit may now be mined.

From the inception of exploration, X-ray fluorescence and X-ray diffraction analysis have progressively more importance in the exploration process. At first neither method has much application. Later both methods serve occasionally to help correlate chemical or mineral composition with visual or other methods of analysis. Finally in the later stages of exploration both X-ray fluorescence and X-ray diffraction analysis can contribute significantly to mineral exploration and orebody definition. Either or both methods can be the most powerful laboratory analysis methods for the final phase of exploration drilling.

The purpose of this chapter is to develop the theory, instrumentation and application of X-ray analysis to mineral exploration. As with most analytical methods, complications and limitations require sophistication and competence for the analyst and a reasonable degree of knowledge for the exploration team to know when and how to use each technique. We therefore present the theory of X-ray fluorescence and X-ray diffraction in sufficient detail to serve as a primer for the user to select and judge the kind and quality of results he needs. X-ray instrumentation not only serves various purposes in exploration analysis, but different instruments give different kinds, qualities and rates of analyses. To aquaint the potential X-ray user of these differences, we present short sections on X-ray instrumentation. Finally we introduce the reader to selected applications of X-ray analysis to raw materials exploration. Where appropriate, we mention strengths, weaknesses and pitfalls encountered when analyzing rock and ore materials.

For more complete treatments of X-ray fluorescence analysis method refer to an excellent short text by Jenkins (1974) or to a comprehensive reference text by Bertin (1975). X-ray diffraction analysis is treated comprehensively by Klug and Alexander (1974).

X-RAY FLUORESCENCE ANALYSIS

X-ray fluorescence analysis determines the chemical composition of solids and the partial chemical composition of liquids. The basic principle of X-ray fluorescence analysis is simple. An X-ray source generates a range of wavelengths of X-ray photons much as an incandescent bulb generates white light, a range of photons wavelengths in the visible region of the electromagnetic spectrum. The X-ray photons impinge on a smooth specimen surface where they cause ejection of electrons from atoms in the specimen. To maintain electrical neutrality other electrons re-occupy the vacant sites in the specimen atoms, emitting new X-ray photons characteristic of each atom. A photon detection system measures the numbers and wavelengths of photons emanating from the specimen. The wavelengths and numbers of photons detected translate into atomic kind and amount.

In the following sections we develop the X-ray fluorescence analysis process beginning with X-ray generation and ending with methods of translating raw X-ray photon counts into chemical analyses of rock and mineral materials.

Generation of X-rays

The continuous spectrum

A very simple X-ray-generating tube produces primary X-rays used to excite secondary X-rays in an analyte sample. Fig. 3-1 shows the construction of the X-ray tubes. A low-voltage (7—14 V), high-current circuit heats a tungsten filament to cause emission of thermal electrons. A large potential, usually between 30 and 75 kV, accelerates the thermal electrons emitted from the filament cathode toward a target anode. A metal cup below the

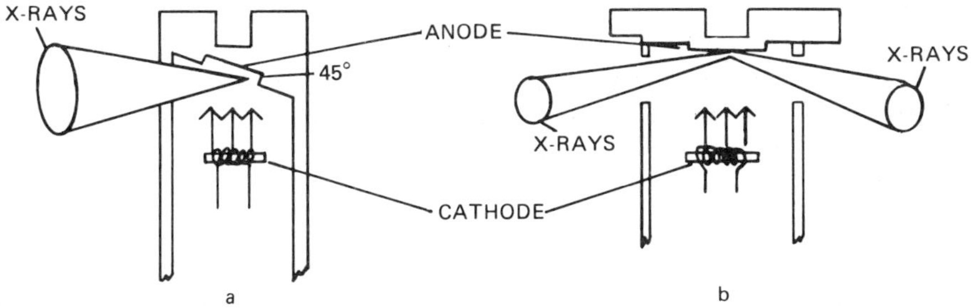

Fig. 3-1. X-ray fluorescence tube (a) and diffraction tube (b). Arrows indicate direction of accelerated electrons. Cones indicate beams passing through windows.

filament traps thermal electrons moving in the direction opposite the target. Electrons accelerated toward the metal target material strike the target with an energy $E = Ve$ where V is the voltage and e is the charge on the electron (1.602×10^{-19} C).

The electrons, upon entering the metal target material, accelerate as they pass through the force fields of the atomic nuclei and the diffuse electron clouds about the nuclei. The accelerating (slowing) electrons in the field of nuclei and electron clouds lose all or part of their energy with emission of one or more photons of energy $h\nu$. Ve establishes the upper limit of energy loss or emitted photon energy, that is:

$$E = Ve = h\nu_{max}$$

where h is Planck's constant (6.6256×10^{-34} J s) and ν is the frequency of the emitted energy. If the electron accelerates to zero velocity instantaneously:

$$\nu_{max} = c/\lambda_{min}$$

where c is the speed of light (2.998×10^8 m s^{-1}) and $\lambda_{min} = hc/Ve = 12400$ Å$/V$. Any electron striking a target thus produces a minimum wavelength of photon radiation when it loses all its energy at once.

Most electrons do not lose all their energy at once but in increments producing a distribution of photon energies. This distribution, a function of accelerating voltage, is pictured in Fig. 3-2a. The spectral distribution resembles black-body radiation and a derivation similar to Planck's derivation of black-body radiation explains the distribution. The wavelength of maximum intensity of photons emanating from the target is about 1.5—2.0 times λ_{min}. We term this spectral distribution the "white" or continuous spectrum.

The continuous spectrum radiates from the water-cooled target material in all directions but passes from the tube through a berylium window to the analyte specimen. The continuous spectrum striking the sample excites electrons in the atoms to generate a characteristic spectrum.

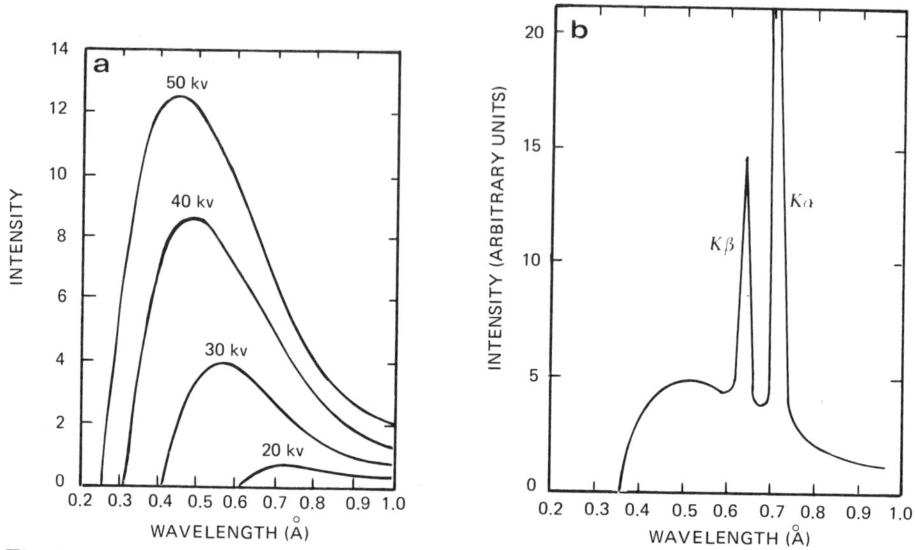

Fig. 3-2. Intensity distribution versus wavelength of the X-ray spectrum. (a) Continuous distribution of tungsten radiation at several voltages. (b) A portion of continuous and characteristic spectrum of molybdenum.

The characteristic spectrum

The electrons in atoms of analyte material occupy specific and more or less well-defined energy levels. We designate these energy levels by a set of numbers called quantum numbers. Each electron energy becomes a function of the designating quantum numbers.

Specifying the pertinent quantum numbers we have:

$n =$ principal quantum number; specifies the total electronic energy exclusive of orbital momentum and electron spin; values are 1, 2, ... in a hydrogen-like atom.

$l =$ specifies orbital angular momentum; values 0, 1, 2, ... n.

$m_s =$ specifies the spin angular momentum of the electron; values are $\pm\frac{1}{2}$ corresponding to spin in one direction or the other.

We may combine the orbital and spin terms to obtain a new quantum number which contains both orbital angular momentum and spin in the same term (total angular momentum), thus $j = l + m_s = l \pm \frac{1}{2}$. A function of the new set of numbers n, l, j also specifies the total electron energy. Fig. 3-3 shows energies of electrons in a first-row transition element.

If we impinge a continuous spectrum on atoms containing the energy levels of Fig. 3-3, photons may eject (excite) electrons from energy levels of those atoms. A photon from the continuous spectrum whose energy is equal to or greater than the negative of the energy of an atomic electron can eject an atom's electron, losing its own energy in the process. When an X-ray

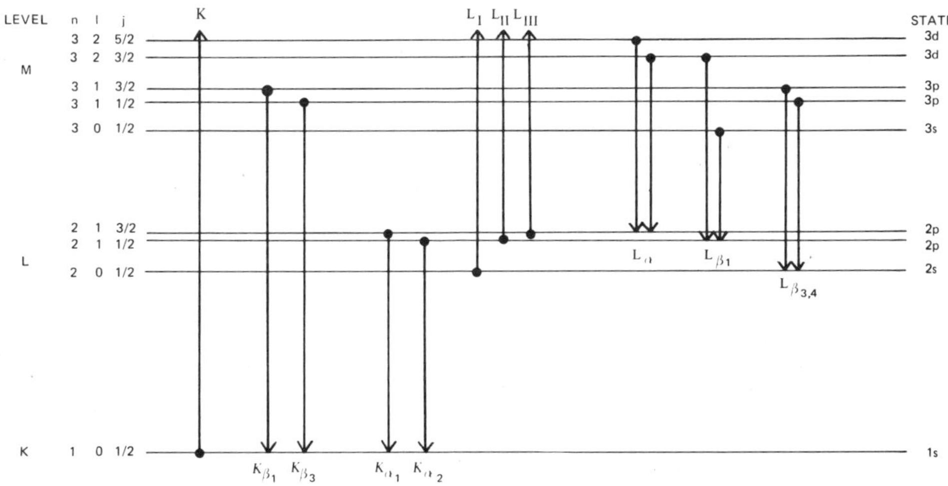

Fig. 3-3. Electronic excitations and transitions yielding a characteristic spectrum of a hypothetical first-row transition element. Arrows pointed downward represent transition energies. Arrows pointed upward represent excitation energies.

photon excites an electron from an atom, an electron from a higher energy level in the same atom enters the vacated level generating a new photon. The newly emitted photon will have energy $E = h\nu_2 - h\nu_1$, that is, the difference between the energies of the two levels.

Empirical selection rules govern allowed transitions from higher energy levels to lower levels and simplify the emitted spectrum:

$$\Delta n = 0; \quad \Delta l = \pm 1; \quad \Delta j = \pm 1 \text{ or } 0.$$

The arrows in Fig. 3-3 show allowed electron transitions within that atom. The length of the arrows are roughly proportional to the amount of energy of the transition, hence energy of the emitted photons. Photons emitted as a result of these transitions comprise the characteristic spectrum. Fig. 3-2b shows part of the characteristic spectrum from the target material of an X-ray tube superposed on its continuous spectrum where accelerated electrons eject atomic electrons.

Each element has a differently charged nucleus and each has electrons occupying different energy levels, consequently the characteristic spectrum of each element differs from the spectrum of all other elements. Table 3-I shows examples of energies required to eject electrons from different atoms with energies of photons emitted from the different atomic transitions. Complete lists of energies are given by Bertin (1975).

The X-ray photon flux from an X-ray tube provides the energy for X-ray fluorescence analysis or X-ray diffraction. If we wish to analyze for the elements in a specimen we use the spectrum without modification. If we wish

TABLE 3-I

X-ray excitation energies, mean emission energies and fluorescent yields for common rock- and ore-forming elements

Element	Excitation energy (keV)				Emission energy (keV)				Fluorescent yield	
	K	L_I	L_{II}	L_{III}	$\overline{K}_\beta$	$\overline{K}_\alpha$	$\overline{L}_\beta$ **	$\overline{L}_\alpha$ ***	ω_K	ω_L
C	0.283	—	—	—	—	0.279 *	—	—	0.001	—
Na	1.080	0.055	0.034	0.034	—	1.041	—	—	0.013	—
Mg	1.303	0.063	0.050	0.049	—	1.255	—	—	0.019	—
Al	1.559	0.087	0.073	0.072	1.55	1.487	—	—	0.026	—
Si	1.838	0.118	0.099	0.098	1.838	1.739	—	—	0.036	—
P	2.142	0.153	0.129	0.126	2.142	2.014	—	—	0.047	—
S	2.470	0.193	0.164	0.163	2.468	2.307	—	—	0.061	—
K	3.607	0.341	0.297	0.294	3.589	3.312	—	—	0.118	—
Ca	4.038	0.399	0.352	0.349	4.012	3.690	0.344	0.341	0.142	0.001
Ti	4.964	0.530	0.460	0.454	4.931	4.508	0.458	0.452	0.197	0.001
V	5.463	0.604	0.519	0.512	5.427	4.949	0.519	0.510	0.227	0.002
Cr	5.988	0.679	0.583	0.574	5.947	5.411	0.581	0.571	0.258	0.002
Mn	6.537	0.762	0.650	0.639	6.494	5.895	0.647	0.636	0.291	0.003
Fe	7.111	0.849	0.721	0.708	7.059	6.400	0.717	0.704	0.324	0.003
Co	7.709	0.929	0.794	0.779	7.649	6.925	0.790	0.775	0.358	0.004
Ni	8.331	1.015	0.871	0.853	8.265	7.472	0.866	0.849	0.392	0.005
Cu	8.980	1.100	0.953	0.933	8.907	8.041	0.948	0.928	0.425	0.006
Zn	9.660	1.200	1.045	1.022	9.572	8.631	1.032	1.009	0.458	0.007
Mo	20.002	2.884	2.627	2.523	19.633	17.443	2.395	2.164	0.749	0.039
U	115.991 *	21.753	20.943	17.163	111.786 *	97.143 *	17.218	13.000	0.960	0.478

* Values not attainable or measurable with standard equipment.

** Emission from M_{IV} to L_I levels for all but U.

*** Emission from M_V, M_{IV} to L_{III} levels for all but U.

to identify minerals by X-ray diffraction we monochromatize the X-ray spectrum eliminating almost all radiation except the characteristic K_α radiation from the target.

Secondary X-rays — fluorescence

X-ray photons of a given wavelength passing into a material eject electrons from the atoms of that material because photons have momentum:

$$mc = h/\lambda = h\nu/c$$

Like electrons, X-ray photons to eject an electron from an atom must have energy equal to or greater than the electron's energy. If a photon ejects an electron, an electronic transition occurs as in the target of the X-ray tube. If there is a distribution of photon energies, electronic ejection followed by electronic transitions in the specimen material produces all lines of the characteristic spectrum of the atom of the specimen material. Photons have no electrical charge and cannot interact with the electrical field of specimen atoms, therefore, cannot generate a continuous spectrum. The spectrum so generated is called the fluorescent spectrum and is the basis for X-ray spectrometric analysis.

Considering only fluorescence, the number of X-ray photons of a specific electronic energy transition emitted from a specimen element would be proportional to the weight fraction of the element in the sample. We should have only to compare the intensity of a characteristic spectral line of an element from the specimen, $I_{i(\mathrm{u})}$, with the same radiation from a known amount of the same element in a standard, $I_{i(\mathrm{s})}$, to accurately estimate the amount of the element in the unknown specimen:

$$w_{i(\mathrm{u})} = w_{i(\mathrm{s})} \frac{I_{i(\mathrm{u})}}{I_{i(\mathrm{s})}} \qquad (1)$$

The weight fraction of element i in the unknown and standard respectively are $w_{i(\mathrm{u})}$ and $w_{i(\mathrm{s})}$.

Absorption and secondary fluorescence

Although equation (1) is sometimes a good approximation for estimating amounts of elements in a sample, usually it is not. Some X-rays emitted from an element in a specimen have enough energy to eject electrons from another atom of the same specimen. These X-rays can be absorbed. Absorption decreases the intensity of X-rays from the emitting element. On the other hand, the element whose electron is ejected by the photon emitted from another element also emits a photon resulting in "secondary fluorescence". Secondary fluorescence enhances the intensity of X-ray lines caused by photons from the X-ray tube.

The X-ray absorption spectrum

Lambert in 1792 found that the fraction of radiation absorbed, dI/I, passing through a thin layer of matter is proportional to the thickness of material:

$$dI/I = -\mu_l \, dx$$

For finite thicknesses:

$$\int_{I_0}^{I} \frac{dI}{I} = -\mu_l \int_0^x dx$$

or:

$$I/I_0 = e^{-\mu_l x} \qquad (2)$$

The linear absorption coefficient, μ_l, depends on the state of matter. We may substitute a new coefficient μ^* that is independent of the solid, liquid or gaseous state:

$$\mu^* = \mu_l/\rho \qquad (3)$$

where ρ is the density of the material. Equation (2) becomes:

$$I/I_0 = e^{-\mu^* \rho x} \qquad (4)$$

Absorption of a photon of a given wavelength depends in equation (4) on μ^* for that wavelength, the density of the absorbing element and the thickness of the absorber.

The quantity, μ^*, called the mass absorption coefficient, varies with the wavelength of radiation. Fig. 3-4 shows a plot of μ^* against wavelength for platinum. Absorption is at a local maximum when the photon striking the material has just enough energy to eject an electron from a particular shell of the absorbing element. Absorption is at a local minimum when the photon has slightly less energy than enough to eject an electron from a particular shell. The probability that a photon will eject an electron decreases the more the photon energy exceeds the ejection energy. We know the mass absorption coefficients as a function of wavelength fairly accurately for all but the very light elements. Therefore, if we know each of the other parameters in equation (4), we can accurately predict the amount of absorption in a material for any given wavelength of impinging radiation.

Secondary fluorescence

Elements in a specimen absorbing either the primary beam or radiation from other atoms in the specimen must re-emit radiation. When radiation from one element of a specimen causes absorption and re-emission from another element of a specimen, the second element is said to be secondarily fluorescing. K_α radiation from copper in a specimen is absorbed by iron and the iron atoms re-emit iron-K_α radiation. The copper-K_α intensities are

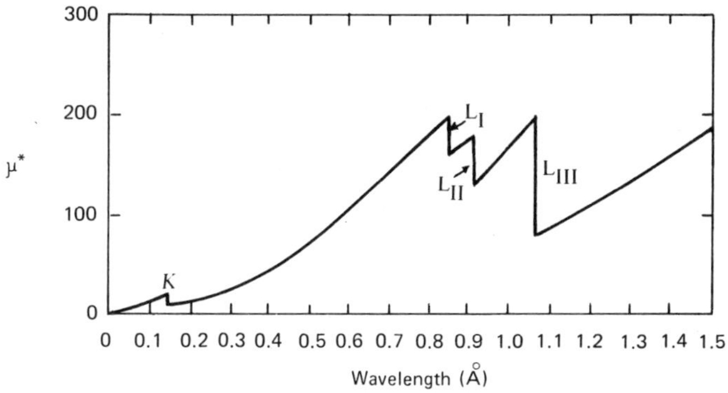

Fig. 3-4. The mass absorption coefficient, μ^*, versus wavelength for platinum showing K and L absorption edges.

diminished by iron absorption while the iron-K_α intensities are enhanced by secondary fluorescence. Both effects are illustrated in Fig. 3-5 in comparison with the expectation of equation (1).

A fairly simple order exists in the periodic table to predict which element will absorb and fluoresce K-radiation. In the first and second rows of the

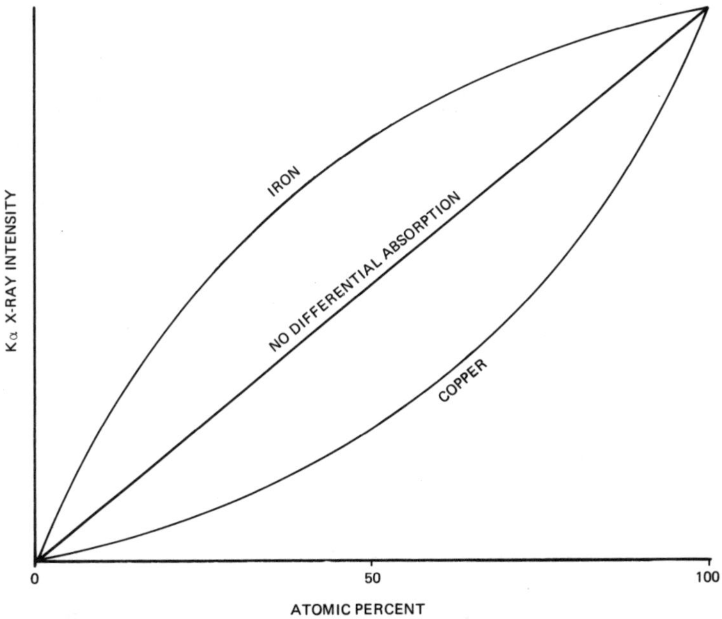

Fig. 3-5. Relative X-ray intensities versus atomic percent for a copper-iron mixture showing relative effect of absorption of copper radiation by iron and absorption and secondary fluorescence of iron radiation by copper.

periodic table an element strongly absorbs K-radiation from an element one higher in atomic number. Aluminum strongly absorbs silicon K-radiation and silicon strongly absorbs sulfur K-radiation. In turn, the aluminum fluoresces K-radiation from absorption of silicon and sulfur K-radiation. In the third row of the periodic table of elements, an element absorbs very strongly the K-radiation from an element two higher in atomic number. Two or three elements successively lower in atomic numbers than the very strongly absorbing element also absorb strongly. Cobalt very strongly absorbs copper K-radiation with iron, chromium and manganese successively less strongly absorbing the copper K-radiation. In the fourth-row elements the relation is similar except the most strongly absorbing element is three down in atomic number from the fluorescing element. The absorbing element emits K-radiation as secondary fluorescence. The amount of emitted secondary radiation is less than the absorbed radiation because some of the absorbed radiation ejects L, M, etc., electrons.

Lighter elements absorb and re-emit less strongly than heavier elements. However, the distribution of primary X-rays from an X-ray tube yields a relatively small number of photons to excite light elements so any absorption of the radiation from light elements profoundly limits analytical sensitivity of the light elements from aluminum down. Absorption of light-element radiation effectively limits detectable radiation to that emitted from very near the surface of a specimen.

Qualitative fluorescence analysis

Obviously, for good exploration analysis results, the exploration analyst must account for the several interrelated physical processes that produce his analytical determinations: fluorescence, absorption and secondary fluorescence. It is the quantitative accounting (or lack of it) of these processes that determines the utility and quality of fluorescence analysis in mineral exploration. Because the physical processes condition results so completely, we proceed by explaining qualitative analysis followed by a more precise development of quantitative analysis using the physical geometry of the qualitative analysis. Quantitative analysis is the primary use of X-ray fluorescence in mineral exploration.

Qualitative X-ray fluorescence analysis for major and minor elements in rocks and ores is rapid and simple. Two types of instruments furnish the exploration analyst with his analytical tools.

Fig. 3-6a schematically illustrates the scanning crystal X-ray spectrometer. In this instrument the X-rays emanating from the X-ray tube strike the sample causing X-ray fluorescence in the specimen. Part of the radiation from the specimen passes through a collimator to produce a parallel X-ray beam. The collimated beam of X-rays strikes an "analyzing" crystal surface where it diffracts into a second collimator, thence into a detector. Diffrac-

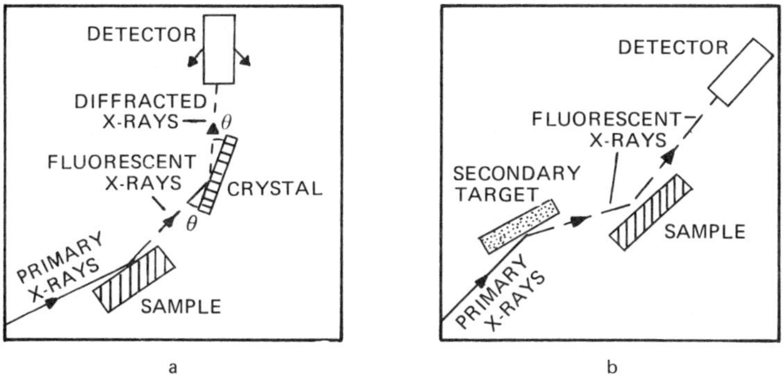

Fig. 3-6. Two types of X-ray spectrometer geometry. (a) Scanning or "crystal" spectrometer. (b) Energy dispersive spectrometer.

tion from the analyzing crystal follows the equation:

$$n\lambda = 2d_{\text{cryst}} \sin \theta \tag{5}$$

(See section "X-ray diffraction analysis" for derivation of this equation.) If the analyzing crystal is rotated about an axis at one-half the speed of rotation of the detector assembly about its axis, we may measure diffraction maxima corresponding to values of λ or $n\lambda$, where n is a positive integer. We choose d_{cryst}, the spacing of a particular plane in a tabular crystal to allow measurement of a range of wavelengths. Each instrument has a selection of collimators, analyzing crystals and detectors to cover all wavelength ranges. Target material of the X-ray tubes can also be selected for most efficient operation over a particular element or wavelength range. For example, chromium or rhodium targets are efficient for general rock analysis (lighter elements) whereas tungsten or platinum targets are efficient for lead, copper, or zinc ores. Spectra from a scanning spectrometer analysis resemble the spectrum shown in Fig. 3-7. Values of 2θ for each line can be correlated through equation (5) with the wavelength of the radiation. Tables of radiation wavelengths (White and Johnson, 1970) for each element identify elemental kind as in Fig. 3-7a. The scanning fluorescence spectrometer, the most widely used type of X-ray spectrometer, gives the best spectral sensitivity, resolution, and flexibility of all X-ray spectrometers for exploration analysis.

Fig. 3-6b illustrates schematically a second type of X-ray spectrometer, usually called an energy dispersive X-ray spectrometer. The geometry of this spectrometer is simpler than the scanning spectrometer. Typically an X-ray tube impinges a beam of X-rays on a specimen. The specimen emits fluorescent X-rays which pass through a collimator to eliminate stray radiation. The collimated X-rays then pass into a solid-state detector (usually lithium-drifted silicon) maintained at liquid nitrogen temperatures. For each photon the detector gives a pulse whose voltage is proportional to the energy of the

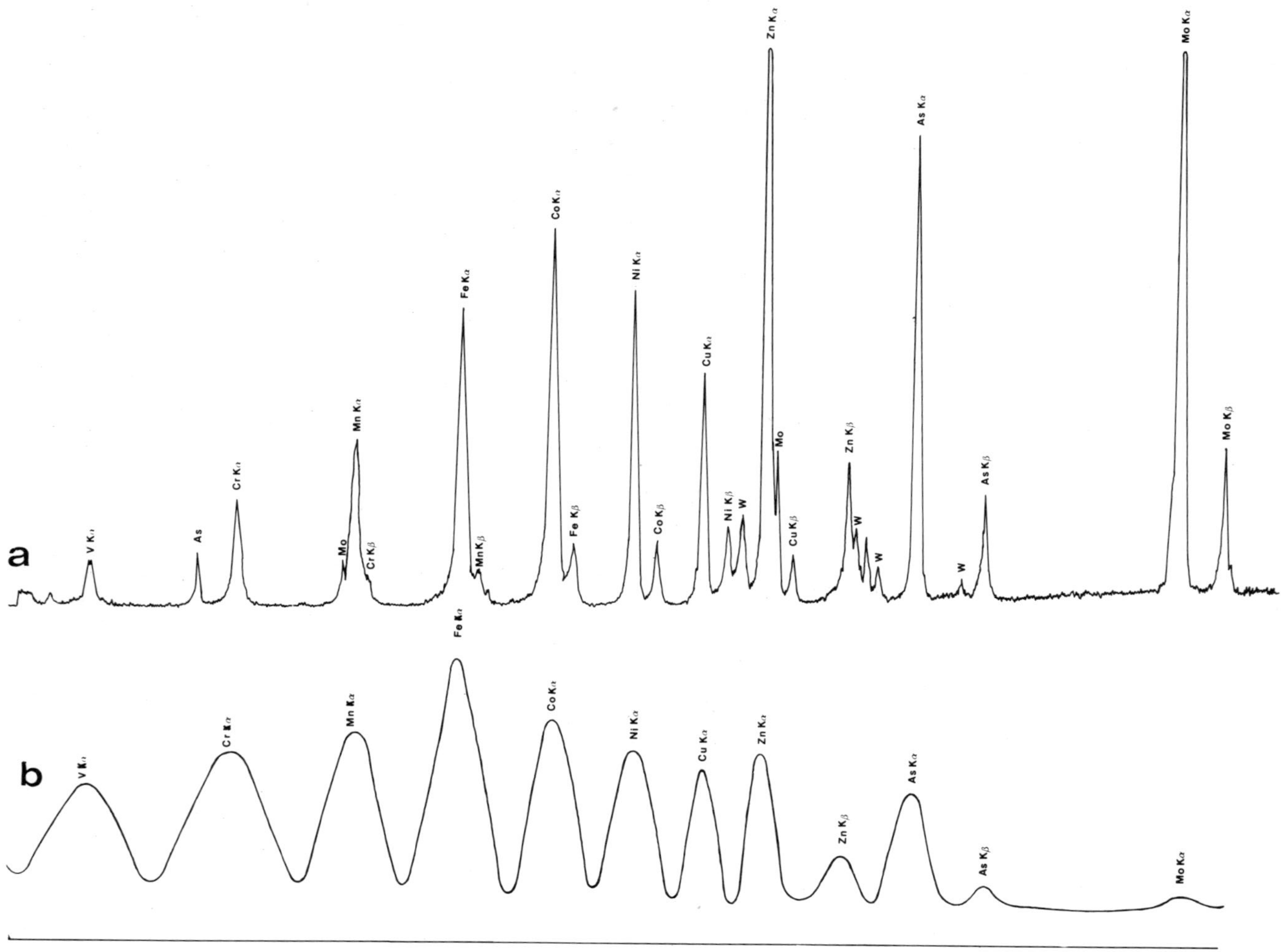

Fig. 3-7. X-ray spectra of elements each with 0.8% concentration. (a) Spectrum from scanning spectrometer, lithium fluoride analyzing crystal and scintillation detector, 40 kV. (b) Spectrum from energy dispersive spectrometer redrawn to scale of (a), 30 kV. Each scanned the same length of time.

photon. An analogue to digital converter converts the pulse to digital form and the digital signal is accumulated in a multi-channel analyzer according to voltage. When pulses have accumulated for a few seconds or minutes, the entire elemental energy spectrum can be displayed, printed or otherwise shown to the analyst (Fig. 3-7b).

The energy dispersive spectrometer has several advantages for exploration analysis. It is fast — up to ten times the analysis speed of the scanning spectrometer. It is less expensive to purchase than scanning spectrometers. It is mechanically simpler than scanning spectrometers. Lines of higher value of n than one in equation (5) do not clutter the spectrum. To balance the advantages, energy dispersive spectrometers have disadvantages too. Energy dispersive X-ray spectrometry can resolve energies to an average of about 150 eV while crystal X-ray spectrometry can resolve to about 3 eV. The poorer resolution of the energy dispersive spectrometer frequently requires elaborate mathematical unfolding of the spectrum to remove line overlaps for elements in the common ore minerals, making qualitative and quantitative analysis less precise. Although the entire spectrum is being collected simultaneously, the detector allows a maximum linear photon count rate of about 10 times less than detectors used in scanning crystal spectrometers, again necessitating elaborate correction for detector "dead-time" at the cost of analytical sensitivity and precision. Finally, a supply of liquid nitrogen must be available to continuously cool the detector.

There are other noteworthy aspects of both types of spectrometers. Quantitative analytical sensitivity under normal operating conditions ranges at best from about 1 ppm for heavier elements in a light-element matrix to 0.05—0.1 wt. % for light elements in a light-element matrix. As the matrix elements become heavier, sensitivity decreases. First-row transition elements in most rocks usually can be detected down to about 10 ppm and light elements such as F and Na can be detected to only about 0.5—1%. Special instrumentation configurations and scanning procedures are required to attain the best sensitivity. Equation (1) permits qualitative estimates of the amount of each element in a specimen, but the accuracy of the estimates can range from an error of a few percent of the amount of the element in a specimen to two or three orders of magnitude of the amount of the element. It is hazardous to put much reliance on quantitative estimates of elemental amounts in exploration samples unless the analyst has experience with both X-ray theory and geological materials. One special qualitative analysis method is useful on automatic scanning spectrometers. Automatic spectrometers can look for specific elements and at the same time give good intensity measurements for those elements.

Qualitative analysis, although useful in mineral exploration, usually serves to establish data for quantitative analysis. Quantitative analysis provides the data upon which the geologist can make exploration and exploitation decisions.

Quantitative fluorescence analysis

Quantitative X-ray fluorescence analysis has proved a very powerful, precise and accurate analytical tool in the minerals production and processing industries. However, accurate quantitative analyses for geological exploration impose more stringent conditions than typical in mineral production and processing. Exploration specimens vary widely and frequently unpredictably in composition from sample to sample. The more difficultly measurable low-atomic-number elements constitute major components. Specimen character, such as particle size and shape, introduces special problems, usually requiring specific and more elaborate procedures for maintaining accuracy. The problems not withstanding, automatic X-ray fluorescence analysis using rapid semi-automatic sample preparation, automatic data collection and computer correction programs requiring few analytical standards, makes possible low-cost accurate exploration analysis. The discussion following emphasizes the more stringent restrictions of X-ray analysis of geological samples and the use of computer automation techniques and features to overcome exploration analysis problems.

Line intensities with absorption and fluorescence corrections

Mass absorption coefficients signify more than just how an absorbing element diminishes the intensity of the radiation from the X-ray tube or from another emitting element. If we separate the curve of Fig. 3-4 into the mass absorption coefficients for each electronic shell K, L, M, etc., each of the separated curves μ^*_{K}, $\mu^*_{L_I}$, etc., would be proportional to the probability that a photon of a given wavelength of radiation will eject an electron from a particular shell. Should a photon eject an electron from a level and a transition occur, the amount of radiation leaving the atom from a particular transition is ωg. The term ω, the fluorescent yield (Table 3-I), defines the fraction of photons leaving an atom after the series of transitions to a given level such as the K-level. We define g as the relative intensity of a line in its series. Neglecting absorption and secondary fluorescence in the sample, for the ith element kind:

$$dI_{i,\lambda_J} = kI_{t,\lambda_p}\,\mu^*_{i,J}\,\omega_J g\,\rho_i dt \qquad (6)$$

where I_{t,λ_p} is the intensity of the primary beam from the tube at a depth t in the sample (Fig. 3-8).

Substituting Q for the constants in equation (6):

$$Q_i = k\mu^*_{i,J}\omega_J g$$

With the geometry of Fig. 3-8:

$$I_{t,\lambda_p} = I_{0,\lambda_p} \exp -(\mu^*_{\lambda_p}\,\rho t \text{ cosec } \phi)$$

where I_{0,λ_p} is the intensity of a given wavelength of the primary beam

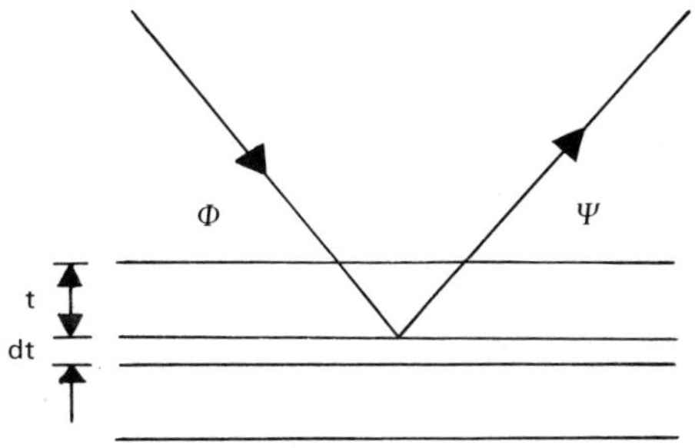

Fig. 3-8. Geometry of secondary excitation in a specimen.

impinging on the specimen. The emerging intensity:

$$\begin{aligned} \mathrm{d}I_{i,\lambda_J} &= \mathrm{d}I_{t,\lambda_J} \exp - (\mu^*_{\lambda_J} \rho t \operatorname{cosec} \psi) \\ &= Q_i \rho_i I_{0,\lambda_\mathrm{p}} \exp \rho t (\mu^*_{\lambda_\mathrm{p}} \operatorname{cosec} \phi + \mu^*_{\lambda_J} \operatorname{cosec} \psi)\, \mathrm{d}t \end{aligned} \tag{7}$$

Since cosec ϕ, cosec ψ, $\mu^*_{\lambda_\mathrm{p}}$ and $\mu^*_{\lambda_J}$ are constant for a given instrument and specimen we may set:

$$\bar{\mu} = (\mu^*_{\lambda_\mathrm{p}} \operatorname{cosec} \phi + \mu^*_{\lambda_J} \operatorname{cosec} \psi)$$

Equation (7) becomes:

$$\mathrm{d}I_{i,\lambda_J} = Q_i \rho_i I_{0,\lambda_\mathrm{p}} \exp - (\bar{\mu} \rho t)\, \mathrm{d}t \tag{8}$$

Integrating equation (8):

$$I_{i,\lambda_J} = \frac{Q_i \rho_i I_{0,\lambda_\mathrm{p}}}{\bar{\mu} \rho} [1 - (\exp - \bar{\mu} \rho t)].$$

If we permit t to become very large, that is, an infinitely thick sample, then:

$$I_{i,\lambda_J} = (Q_i \rho_i I_{0,\lambda_\mathrm{p}}) / \bar{\mu} \rho \tag{9}$$

For an infinitely thick sample, which for most elements is less than a few millimeters the intensity of a beam scattered from specimen is given by equation (9).

We may proceed from equation (9) along several paths to produce correction methods for quantitative analysis. We look at two, each of which is typical of correction procedures suitable for exploration analysis.

Linear correction method

Allowing $\bar{\mu}$ of equation (9) to be defined by a sum of linear coefficients with each term representing the effect of absorption and secondary fluorescence of element j on element i:

$$\bar{\mu} = \Sigma a_{ij} \tag{10}$$

Since $\rho = \Sigma\rho_i$ where ρ_i is the density of element i in the sample. Then $Q_i\rho_i I_{0,\lambda_p} = \bar{\mu}\rho = \Sigma a_{ij}\rho_i$ and $a_{ii} = 1$, so that for pure i:

$Q_i\rho_i I_0/I_{100i} = a_{ii}\rho_i$.

Therefore:

$$a_{i1}\rho_1 + a_{i2}\rho_2 + \ldots + [1 - I_{100i}/I_i]\rho_i \ldots + a_{in}\rho_n = 0 \tag{11}$$

where I_i and I_{100i} are analyte line intensities of element i from the specimen and pure element i respectively.

Since $\rho_i/\rho = w_i$, equation (11) becomes:

$$\sum_{j=1}^{n} a_{ij}w_i + (1 - R_i)w_i = 0, \qquad i \neq j \tag{12}$$

with $I_{100i}/I_i = R_i$.

Equation (12) gives the basis for quantitative absorption and secondary fluorescence corrections. If ten elements are to be analyzed in a sample, at least ten standards must have ten elements in differing proportion to determine the coefficients a_{ij}. All elements in a specimen except trace or very minor elements must be included in the equation for accurate results.

The method outlined, due to Lucas-Tooth and Pyne (1964), although quite simple and often powerful, has a fundamental weakness. Equation (10) is but an approximation because the preceding equations do not explicitly account for secondary fluorescence. In geological specimens, secondary fluorescence can be appreciable, especially in metal ore materials. For example, when analyzing for titanium in the presence of iron and calcium, the error in the amount of titanium reported can be as great as 50%, without secondary fluorescence corrections. Analysts avoid such errors by requiring a limited range of compositions of the analyte specimen and having the standards closely approximate the composition of the analytes. Almost all correction methods useful for exploration samples proposed before 1970 are similar to the Lucas-Tooth and Pyne method and have similar weaknesses. Rasberry and Heinrich (1974) proposed equations to explicitly correct for secondary fluorescence:

$$\frac{w_i}{R_i} = 1 + \sum_{j=1}^{n} a_{ij}w_i + \sum_{j=1}^{n} \frac{a_{ij}w_i}{1 + w_i}, \qquad i \neq j \tag{13}$$

The fundamental difference between previous methods and the Rasberry-Heinrich method is the second summation in equation (13) which approximately fits the form of enhancement of an analyte line by secondary fluorescence. This improvement should give better accuracy and wider compositional range of applicability than simpler methods.

Fundamental parameters method

We may make corrections for absorption and secondary fluorescence using fundamental parameters to define excitation, absorption and secondary fluorescence (Criss and Birks, 1968). If we derive a set of equations in a similar manner to equations (6) through (9), but now assume a distribution of X-ray wavelengths from the tube similar to the distribution in Fig. 3-2a, taking into account enhancement of an X-ray line of element i by other elements in the sample, we have the following equation:

$$I_{\lambda_L} = g_i w_i \sum_{\lambda_p} \left(\frac{\delta_{i,\lambda_p} \mu^*_{i,\lambda_p} I_{\lambda_p} \Delta\lambda_p}{\mu^*_{M,\lambda_p} \operatorname{cosec} \phi + \mu^*_{M,\lambda_L} \operatorname{cosec} \psi} \right.$$

$$\times \left\{ 1 + \frac{1}{2\,\mu^*_{i,\lambda_p}} \sum_{\lambda_J} (\delta_{i,\lambda_p} w_j K_j \mu^*_{i,\lambda_J} \mu^*_{j,\lambda_p}) \right. \tag{14}$$

$$\left. \left. \times \left[\frac{1}{\mu^*_{M,\lambda_p} \operatorname{cosec} \phi} \ln\left(1 + \frac{\mu^*_{M,\lambda_p} \operatorname{cosec} \phi}{\mu^*_{M,\lambda_J}}\right) + \frac{1}{\mu^*_{M,\lambda_L} \operatorname{cosec} \psi} \ln\left(1 + \frac{\mu^*_{M,\lambda_L} \operatorname{cosec} \psi}{\mu^*_{M,\lambda_J}}\right) \right] \right\} \right)$$

The variables in equation (14) are defined as follows:

w_i	=	weight fraction of analyte element i
w_j	=	weight fraction of matrix element j
δ_i, δ_j		has value 1 if λ_p is short enough to excite λ_L or λ_J and value 0 for longer wavelengths
g_i	=	function of absolute intensity from element i
K_j	=	$[1 - 1/r]\ \omega$ for spectral line of element j that excites λ_L of element i
I_{λ_i}	=	intensity of analyte line of element i
I_{λ_p}	=	intensity of the primary beam in the interval $\Delta\lambda_p$
i, j	=	analyte and matrix element respectively
M	=	specimen matrix including i
r	=	absorption edge jump ratio, the ratio of the local high to the adjacent low in Fig. 3-4
$\Delta\lambda_p$	=	increment of primary spectrum
λ_J	=	wavelength of a spectral line from element j capable of exciting λ_i
λ_i	=	wavelength at analyte line of element i
λ_p	=	one wavelength of primary beam
μ^*_i, μ^*_j, μ^*_M	=	mass absorption coefficients of analyte i, matrix element j and matrix respectively
ϕ, ψ	=	angles of Fig. 3-8
ω	=	fluorescent yield, related to Q_i of equation (7).

The fundamental parameters method requires knowing the distribution of radiation from the tube, mass absorption coefficients as a function of wavelength and the fluorescent yield. From tables of these and related values we can determine all values in the previous list and equation (14). Equation (14) cannot be solved for w_i explicitly and interative converging procedures must find the final values of w_i. The method, while very powerful, requires a computer of larger size than for the previously discussed methods.

The fundamental parameters method has several advantages in exploration analysis. Pure element standards or a single multi-element standard is sufficient for many types of analyses. For practical rock analyses, perhaps 20—30 standards are necessary for analyzing any type of rock, whereas 200 or more standards are required for the linear equation methods to cover the same elemental ranges. Although one multi-element standard is all that is necessary in the fundamental parameters method, the mass absorption coefficients and fluorescent yields for the lighter elements are not accurate and multiple standards must cover a range of elemental concentrations. For example, to analyze a shale-sandstone sequence, four standards might prove adequate; a granite, a tonolite, a basalt and a granite spiked with quartz. The fundamental parameters method allows inclusion in the equations of elements determined by other methods. Carbon may be analyzed by thermogravimetric analysis and sodium and magnesium analyzed by atomic absorption. The results of carbon, sodium and magnesium analysis can be included in the correction equations for the elements actually analyzed for by X-ray spectrometry. The linear equation method does not have the same flexibility. Like the linear equation method all elements in a specimen, except one element by difference and trace elements, must be determined for accurate analytical values.

Heavy matrix method

This method is useful for measuring one or a few elements in exploration specimens. The method may also be used when no computer is available for absorption and fluorescence corrections.

We may express the efficiency of excitation of an analyte line of a specimen as:

$$\frac{I_{s,i}}{I_{0,\lambda_p}} = \frac{P_{s,i}\, w_i \mu^*_{i,\lambda_p}}{\Sigma\, w_j(\mu^*_{j,\lambda_p} + A\mu^*_{j,\lambda_p})} \tag{15}$$

with

$I_{s,i}$ = analyte line intensity of the ith element of the specimen
I_{0,λ_p} = incident beam intensity
w_i, w_j = analyte and jth component weight fraction respectively
μ^*_{i,λ_p} = mass absorption coefficient of ith element for primary beam
μ^*_{j,λ_p}, μ^*_{j,λ_s} = mass absorption coefficient of jth element for primary beam and analyte X-ray line respectively
$P_{s,i}$ = constant for ith element related to Q_i of equation (7)
A = constant geometric function.

We can expand equation (15) to:

$$\frac{I_{s,i}}{I_{0,\lambda_p}} = \frac{P_{s,i}\, w_i \mu^*_{i,\lambda_p}}{w_h(\mu^*_{h,\lambda_p} + A\mu^*_{h,\lambda_s}) + w_d(\mu^*_{d,\lambda_p} + A\mu^*_{d,\lambda_s}] + \Sigma w_j(\mu^*_{j,\lambda_p} + A\mu^*_{j,\lambda_s}]} \quad (16)$$

where the subscript h refers to a highly absorbing additive and d refers to a diluent.

If we make either the heavy absorber or the diluent term in equation (16) large by adding say one gram of heavy absorber and ten grams of diluent to one gram of specimen, a change in the matrix elements of the sample affects the denominator of equation (16) very little. $I_{s,i}$ becomes proportional to w_i. This is the simplest of methods that have proved effective with geological exploration samples.

Evaluation of correction methods

Both the fundamental parameters method and the linear equation methods and their near relatives are the most suitable for total rock analysis. Both methods require analysis of all but one non-trace element in each specimen. The fundamental parameters method is proving the best for exploration analysis. It has great flexibility and requires few standards. One variation of the fundamental parameters method called CORSET (Stephenson, 1971) has been proved in several laboratories on both scanning and energy dispersive spectrometers. The linear equation method is fast and accurate in its corrections over a limited range of compositions. For mineral exploration analysis, the large number of standards and standard measurements partially cripples the linear equation method. The heavy matrix method is good when analyzing rocks for a few elements. However, the heavy matrix method causes significant loss of sensitivity and an increase in measurement time to achieve desired statistical accuracy. This method also requires more costly and time consuming sample preparation with at least three weighings per specimen. The heavy-matrix method is perhaps the best rock analysis method when no computer is available. We can combine features of the heavy matrix method to increase the correctable range of compositions and the accuracy of the fundamental parameters and linear equation methods. Mixing some diluent with the specimen material reduces differential absorption and secondary fluorescence effects in the specimen making correction more accurate over wider composition ranges. The price for more accurate correction is, of course, loss of sensitivity and analytical speed.

Sample preparation (quantitative analysis)

To obtain intermediate- to high-accuracy analyses by X-ray fluorescence analysis, samples must be prepared with attention to detail. Poor sample preparation abrogates the quality of present-day spectrometers and correction methods. The nature of geological specimens complicates sample preparation. Differences in shape, composition and size of rock particles effect analytical results to such an extent that every effort must be made to minimize these effects. Geological samples usually require the measurement of light elements where absorption prevents light-element radiation from all but the surface of the specimen. Good mixing and smooth sample surfaces are necessary.

Exploration samples fall into two categories: samples where only a few elements are of interest and samples where a total rock analysis is necessary. For both types of specimens, economic restrictions demand fast inexpensive methods capable of analysis of large numbers of samples with relatively high accuracy. Samples should be processed and analyzed for from \$ 5—\$ 25 each depending on the number of elements determined and desired accuracy. Sample preparation procedure, therefore, must be relatively simple. The procedure must apply to all types of rock samples.

Fused pellets

Fusion is the best method of preparation of silicate rock materials for accurate analysis. Ground specimens with a constant proportion of a mixture of lithium tetraborate and lithium carbonate flux with, say, 1 part specimen to 2—8 parts of flux are fused in a carbon or platinum/gold crucible. If the matrix dilution method is used, a heavy absorber such as lanthanum oxide can be added. The resulting glass buttons may be polished or used directly. Fusion realizes several necessary or desirable specimen attributes. Fusion removes particle size and shape effects which reduce accuracy. Fusion affects good sample mixing and presents a smooth specimen surface necessary for light element measurement. Fusion also introduces a diluent making absorption and secondary fluorescence corrections less consequential. Fusion eliminates carbonates and water from the specimen, which must be measured by other methods when doing total rock analysis. Finally, fusion accomplished with automatic fusion devices, is actually faster and cheaper than other methods suitable for high-accuracy quantitative analysis.

Briquettes

Briquettes are suitable for sulfide ores and other rocks where fusion may not be appropriate. The specimen is ground to pass a 325-mesh sieve and mixed with a binder in constant proportion. The ratio of binder to sample may vary from 0.15 to 10. A press capable of 10,000—25,000 psi makes a coherent pellet of the specimen-binder mix. Binder serves several purposes besides holding the sample together. Binder makes the mass absorption coefficient of the matrix more uniform from specimen to specimen requiring less drastic absorption and enhancement corrections and aids forming a smooth specimen surface. The binder helps eliminate density, particle shape and particle size differences among specimens and standards. For the matrix dilution method we might use proportions of 1 : 1 : 10 of specimen, heavy binder (e.g., cesium iodide), light binder (boric acid). For total rock analysis we eliminate the heavy binder when possible and maximize the ratio of specimen to binder.

For rock analysis, the briquette method with mixed binder is sometimes discouraging. Mixing is the major problem. Mixing of rock samples to get a uniform specimen is often difficult because of the hygroscopic nature of many rock forming minerals. Another disadvantage of the briquette method is that non-measurable elements such as carbon and hydrogen remain in the sample and must be accounted for in the correction programs.

Sometimes, when specimens and desired accuracy permit, briquettes may be made with the binder as a backing for the sample. Boric acid is poured into an aluminum cup with sample poured over it and pressed. This method, requiring no weighing, gives maximum sensitivity, but poorer accuracy than the fusion method. It is rapid and inexpensive. The method is a particularly appropriate sample preparation technique for exploration drill cuttings where high accuracy is not necessary because of sample contamination.

X-RAY DIFFRACTION ANALYSIS

X-ray diffraction analysis determines the mineral composition of rocks. A monochromatic beam of X-rays from an X-ray tube impinges on a flat sample surface of ground rock or mineral powder. The minerals in the powder comprise well-ordered arrays of atoms whose arrangement describe families of planes in each mineral particle. The atomic arrangements characterize the spacing of the planes, which in turn characterize the mineral. "Reflection" of impinging X-rays from the planes occurs as an angular function of the interplanar spacing. The number of photons reflecting from a plane and the reflection angle identify and quantify each mineral in the specimen.

In the hands of a competent exploration analyst, X-ray diffraction can be one of the most powerful tools for mineral exploration. X-ray diffraction methods allow the mineralogist to identify minerals and quantitatively determine amounts of minerals in rocks and ores. The mineralogist can identify and quantify minerals quickly and inexpensively whether the rocks are fine- or coarse-grained.

X-ray diffraction theory

If the spectrum from an X-ray tube with an appropriate target, say copper, is passed through nickel foil the output of the tube is a nearly monochromatic beam of wavelength 1.54 Å corresponding to copper-K_α radiation. If now a geometry of a spectrometer and a collimated X-ray beam is constructed as in Fig. 3-9, X-ray diffraction can occur. X-ray diffraction occurs because X-ray photons cause an *electron* in an atom of a crystal to act as a secondary source, emitting photons of the same wavelength as the incident photons. All electrons in all atoms emit the same way. Each electron acts as a source with waves emanating from it spherically.

Every crystal has families of planes whose atomic environment is the same around each plane. Fig. 3-10 shows two such families of planes. There are an infinite number of families of planes in each crystal. Each family of planes in a particular crystal or mineral has a definite distance between the members of the family. When X-rays impinge on electrons, diffraction will occur if the path difference between spherical waves from electrons associated with successive planes is an integral number of wavelengths.

Since the environment about all planes in a family is identical, we can assume that the planes themselves reflect (diffract) the X-rays as in Fig. 3-9b. Referring to the geometry of Fig. 3-9b, X-rays impinge with the wave front $\overline{S_1S_2}$ and are reflected from the planes to give the diffracted wave front $\overline{P_1P_2}$. Diffraction occurs only if the path difference between $\overline{S_1OP_1}$ and $\overline{S_2AP_2}$ is

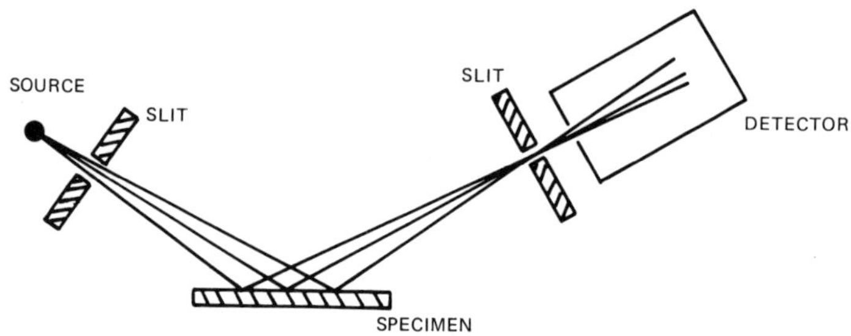

a

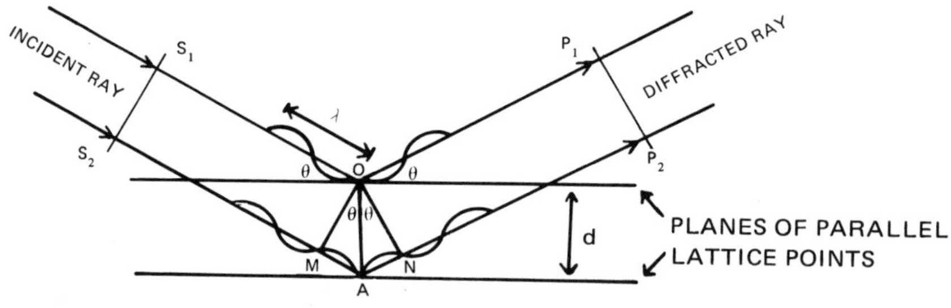

b

Fig. 3-9. X-ray diffraction geometry. (a) Diffractometer instrument geometry. (b) Geometry of diffraction from one family of planes from one specimen particle.

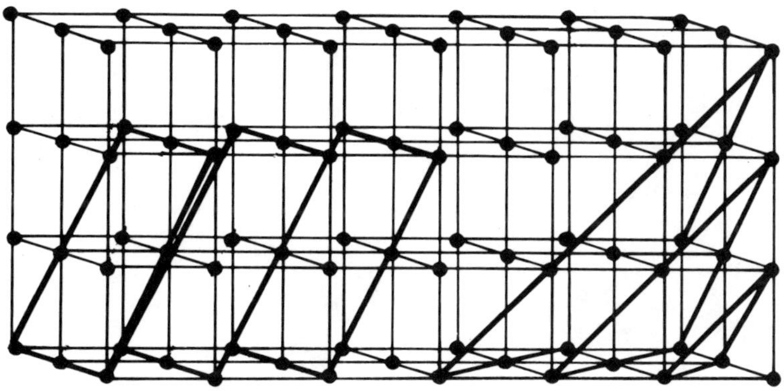

Fig. 3-10. Representation of a crystal showing two families of planes.

one X-ray wavelength so that:

$$\lambda = \overline{S_2AP_2} - \overline{S_1OP_1} = \overline{NA} + \overline{AM} = 2d \sin \theta$$

Thus diffraction occurs when the following equation is satisfied:

$$\lambda = 2d \sin \theta \qquad (17)$$

where d is the spacing between a particular set of planes. Equation (17), the Bragg equation, is the basis for all X-ray diffraction analysis.

Qualitative analysis

Each mineral has a different chemical composition and arrangement of atoms in its structure. Differences in atomic arrangement or composition cause the families of planes to be spaced differently, causing diffraction to occur at different angles for each mineral. Each mineral will have diffraction occurring at angles characteristic of that mineral. Since each grain of each mineral diffracts individually, the pattern of diffraction from a mixture of two minerals would appear as in Fig. 3-11, a diffraction pattern of calcite ($CaCO_3$) and quartz (SiO_2).

The different compositions and arrangements of atoms in minerals also produce different amounts of diffraction from each plane. Each mineral therefore not only has its own set of characteristic spacings between planes and consequent diffraction angles, but also has a characteristic intensity for each line, giving us two sets of criteria for mineral identification: d-spacing and line intensity.

A file maintained originally by American Society for Testing and Materials and now maintained by J.C.P.D.S. (Joint Committee for Powder Diffraction Standards) (Berry et al., 1971), catalogue the characteristic spacings and relative intensities of diffraction lines of over 20,000 inorganic, organic and

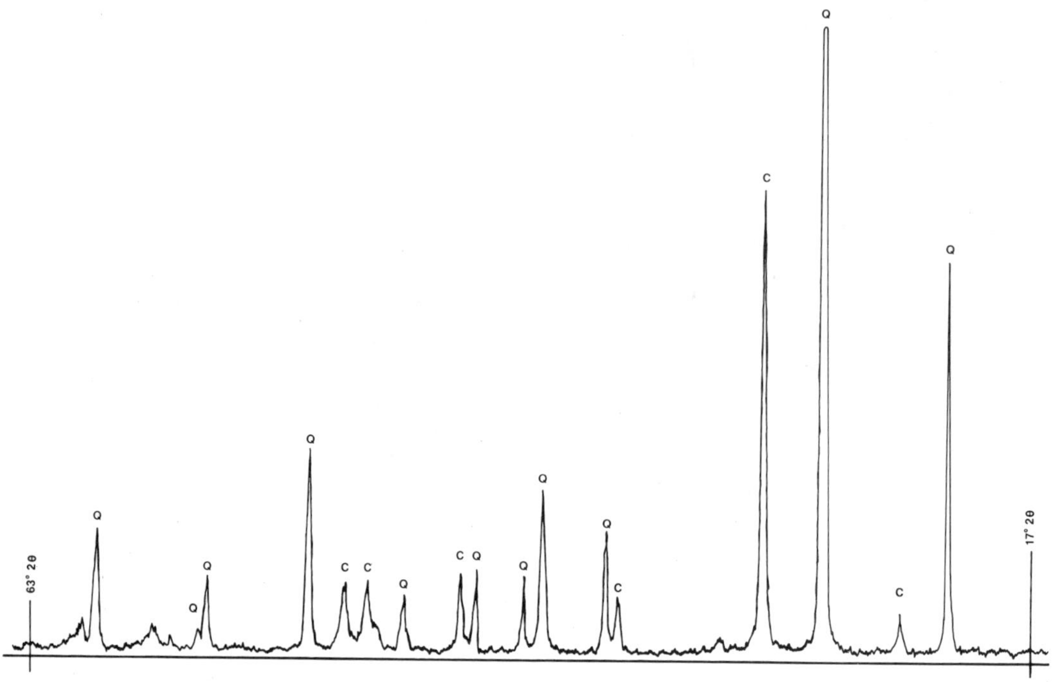

Fig. 3-11. Diffraction spectrum of quartz and calcite mixture.

mineral compounds. This indexed and cross-referenced file allows identification of most mineral species.

Quantitative analysis

Quantitative mineral analysis by X-ray diffraction is simple and relatively straightforward in principle. Suppose a mineral is ground to a powder and placed in the geometry of Fig. 3-9. Suppose the specimen in Fig. 3-9 is rotated about its axis at a velocity ω and the detector is rotated about the same axis at 2 ω. The beam, specimen, and detector will always have the same angular geometry with respect to beam, sample surface and detector as the specimen in Fig. 3-9. Any particle, one of whose planes has the orientation of Fig. 3-9, will diffract.

The intensity of radiation from a given diffraction line is proportional to the volume fraction of the mineral in the sample if there is not differential absorption of X-rays. The intensity of any diffraction line will be proportional to the volume of particles oriented to diffract, that is, oriented to satisfy equation (17). If we reduce the volume of a mineral's diffracting particles by adding a second component, the intensity will be reduced proportionally to the reduction in volume of the component.

Absorption and intensity of diffraction

The only effect on the intensity of an X-ray line diffracted from a well ground powder, other than changing the volume of particles of a given mineral, is absorption of X-rays, usually different for each different mineral in the sample. We can express the mass absorption coefficient of a mineral as:

$$\mu_j^* = \sum_{i=1}^{n} w_i \mu_i^*$$

where w_i is the weight fraction and μ_i^* is the mass absorption coefficient of the ith element in the mineral. The mass absorption coefficient of a sample of N mineral components is:

$$\bar{\mu}^* = \sum_{j=1}^{N} x_j \mu_j^*$$

where x_j is the weight fraction of the jth mineral.

The intensity of a given line from the jth mineral is proportional to the volume fraction f. The intensity decreases as the linear absorption $\bar{\mu} = (\bar{\mu}^*/\rho)$ of the sample increases:

$$I_j = \frac{K_j f_j}{\bar{\mu}}$$

The volume fraction f is simply:

$$f_j = \frac{x_j/\rho_j}{\Sigma x_j/\rho_j}$$

where ρ_j is the density of the jth component, and:

$$\bar{\mu}^* = \frac{\Sigma x_j \mu_j^*}{\Sigma x_j/\rho_j}$$

giving:

$$I_j = K_j \frac{x_j/\rho_j}{\Sigma x_j \mu_j^*} \tag{18}$$

We can use equation (18) for an N-component specimen by regarding the sample as consisting of just two components, the analyzed component, $j = 1$, and the sum of all other components, the matrix, M. The mass absorption coefficient of the matrix is:

$$\mu_M^* = \frac{\sum_{j=2}^{N} \mu_j^* x_j}{1 - x_1}$$

Equation (18) becomes:

$$I_1 = \frac{K_1 x_1}{\rho_1[x_1(\mu_1^* - \mu_M^*) + \mu_M^*]} \tag{19}$$

Equation (19) relates relative intensity of one component to absorption by that component and the matrix. Fig. 3-5 illustrates equation (19) for the conditions $\mu_1^* < \mu_M^*$, $\mu_1^* = \mu_M^*$ and $\mu_1^* > \mu_M^*$, represented by lower, middle and upper lines respectively with percent component 1 as the abscissa.

All-component method

If we have pure component 1, whose intensity is:

$$I_{1_0} = \frac{K_1}{\rho_1 \mu_1^*}$$

we may ratio the intensity of a component 1 in an unknown sample:

$$\frac{I_1}{I_{1_0}} = \frac{x_1 \mu_1^*}{x_1(\mu_1^* - \mu_M^*) + \mu_M^*} \tag{20}$$

Equation (20) is the working equation for quantitative analysis of minerals in an N-component system. One pure standard must be measured for each component. Particle shape, size and orientation effects generally require that I_0 be determined by extrapolation from standards made from the unknowns. If we use equation (20) to determine x_1 in an N-component specimen, all N components must be measured and determined. Since we cannot solve for x_1 explicitly, we must determine it by iterative computational procedures. The method outlined by equation (20), although requiring a computer, is fast, simple and requires no introduction of standard materials into the sample.

Internal-standard method

Suppose an internal standard mineral(s) is added in known and constant amount to a sample with fraction f_j of component j, then:

$$I_j = K_j f_j / \bar{\mu}, \qquad I_s = K_s f_s / \bar{\mu}$$

and

$$\frac{I_j}{I_s} = \frac{K_j' x_j' \rho_s}{K_s x_s \rho_j}$$

Solving for x_j', the weight fraction of the jth component including the internal standard as part of the specimen:

$$x_j' = K_j' \frac{I_j}{I_s}$$

$$x_j = \frac{x_j'}{1 - x_s}$$

so that:

$$x_j = K \frac{I_j}{1 - x_s} \tag{21}$$

From equation (21) we see that if an internal standard is added in constant amount to a series of standard samples with known amounts of the jth mineral, a linear curve always results when I_j/I_s is plotted against x_j.

This method is simple, fast and effective when one or more components are to be determined.

Sample preparation

Samples are usually prepared for qualitative analysis by X-ray diffraction as loose powders or oriented-grain mounts. Samples should be ground to —325 mesh to insure minimum particle orientation and produce maximum intensity and reproducibility. For consistent results, geological materials require that each sample be put into the sample holders in the same way, preferably by the same person. Clay minerals in exploration samples often require special preparation to enhance diffraction lines from planes parallel to their platy shape. Exploration analysis, however, often precludes elaborate sample handling procedures. One simple procedure to orient clay minerals and micas for qualitative X-ray diffraction uses a small amount of the sample to make a slurry with water. The preparer then smears a glass slide with a thin layer of the slurry. As the slide dries, flat particles orient parallel to the slide.

Good quality quantitative analyses require specimen mounts with uniform density, packing, and usually require rotation in the X-ray beam to average sample inhomogeneities. We have found that the best procedure for preparing almost all types of exploration samples for quantitative mineral analysis is the following. We grind the sample to —325 mesh. Add internal standard and mix if the internal standard method is used. Press the sample at 10,000—25,000 psi into an X-ray fluorescence analysis aluminum cup to produce a $1\frac{1}{4}$-inch diameter specimen. We find that for suites of exploration samples the pressed mount produces consistent and reproducible grain orientation, intensities and sensitivities. The pressed specimens may be kept as long as necessary because they are in their own holders.

Accuracy and sensitivity

The lowest detection limit for minerals determined by X-ray diffraction analysis ranges from 1 to 5 vol. % in loose powder samples and 0.2—2 vol. % for pressed samples. A highly absorbing matrix increases the detection limit. For example, copper-K_α radiation impinging on an iron-rich matrix results in absorption by the iron to decrease detection limits to ten volume percent. In such instances, radiation such as cobalt-K_{α} must be used to minimize absorption.

With careful specimen preparation, accuracies will range around ±10% of the amount of a major mineral present in the specimen when using a loose powder. Pressed pellets can give accuracies of about ±4% of the amount of a mineral present in the sample. Optimum accuracy is about ±2%.

INSTRUMENTATION

X-ray fluorescence

Scanning spectrometers

X-ray fluorescence instruments have four essential components, a generator and X-ray tube, the spectrometer, X-ray detection electronics, display and control equipment.

The generator transforms line voltage and current to produce a high voltage and a low current to power the X-ray tube located in the spectrometer. The generator and the X-ray tube constitute the X-ray source. Fig. 3-12 shows a typical X-ray generator. X-ray generators have controls to adjust voltage and current between the X-ray tube's cathode and anode. Generators for high-precision fluorescence analysis provide internally stabilized, constant-potential voltage and stabilized current. Switches allow resetting voltage and current precisely each time the unit is turned on.

Modern X-ray spectrometers (Fig. 3-13), although having the essential geometry depicted in Fig. 6a, have many more features to measure all possible wavelengths with precision and sensitivity. Crystal changers with up to six crystals allow changing wavelength ranges. Collimators with different slit widths allow the operator to choose between resolution and intensity. Two detectors, a scintillation counter and a gas-flow proportional counter permit detection of higher- and lower-energy X-rays respectively. A heater heats the spectrometer to a constant temperature to keep analyzing crystals spacings and other geometries constant. Vacuum systems evacuate the spectrometers, to eliminate absorption of long wavelengths. Some spectrometers such as the ones shown in Fig. 3-13 have sample changers holding many samples and capable of either automatic or manual operation.

Detection electronics comprises several components. The detectors, a scintillation counter for short wavelengths and a flow-proportional counter for longer wavelengths, and their preamplifiers produce pulses whose voltage is roughly proportional to the energy of the X-ray photons entering them. Pulses, one for each photon detected, pass into an amplifier which amplifies and shapes them. The shaped signal passes into a pulse-height selector which acts as a pass-band filter to eliminate unwanted energies. From the pulse-height selector, the signal may be used in two different ways. A pulse

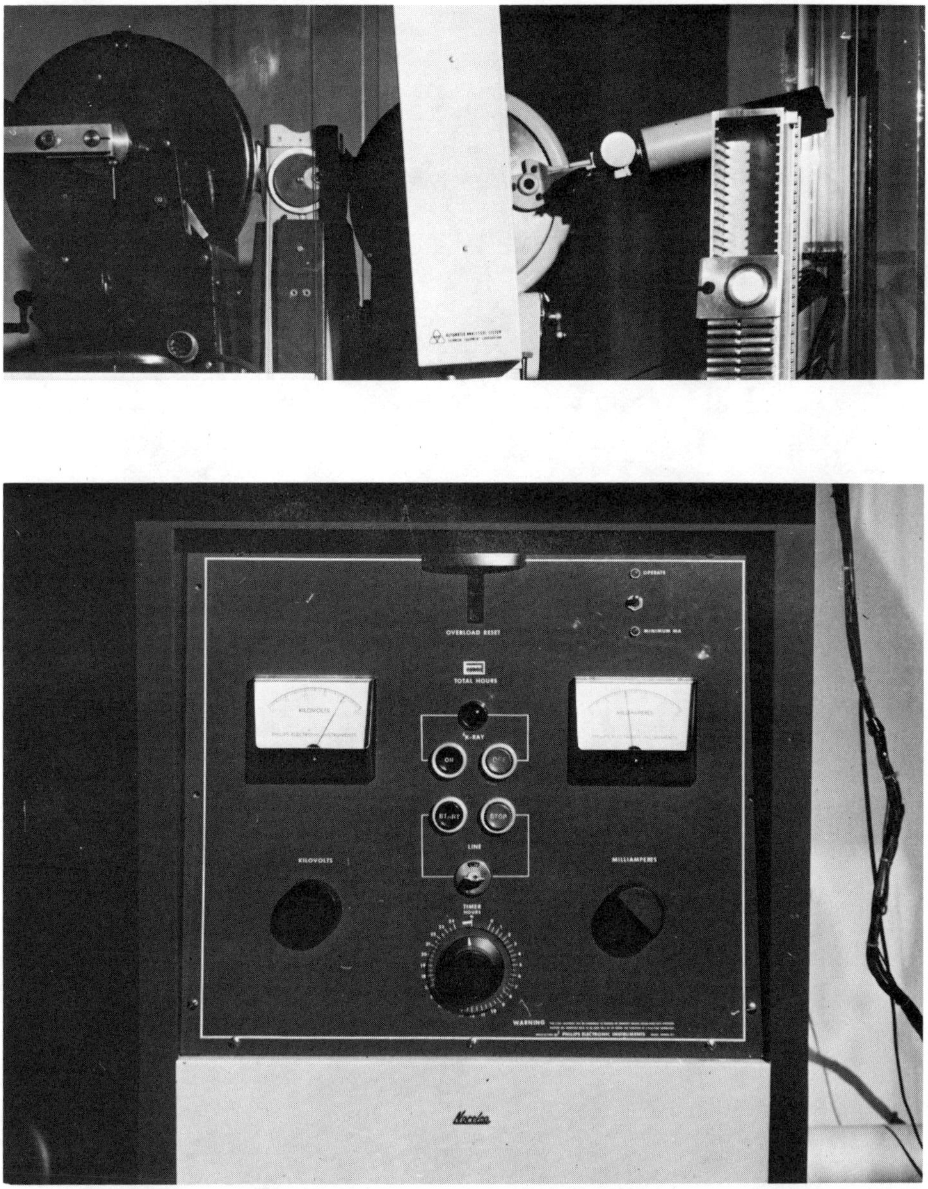

Fig. 3-12. Typical X-ray generator containing high-voltage transformer, inside voltage controls (lower left), current controls (lower right), switching (lower center) and X-ray tube housing (vertical housing in upper left).

counter accumulates individual pulses. The pulses may also pass into an RC time-constant circuit which averages them and converts the averages to an analog signal for display on a recorder.

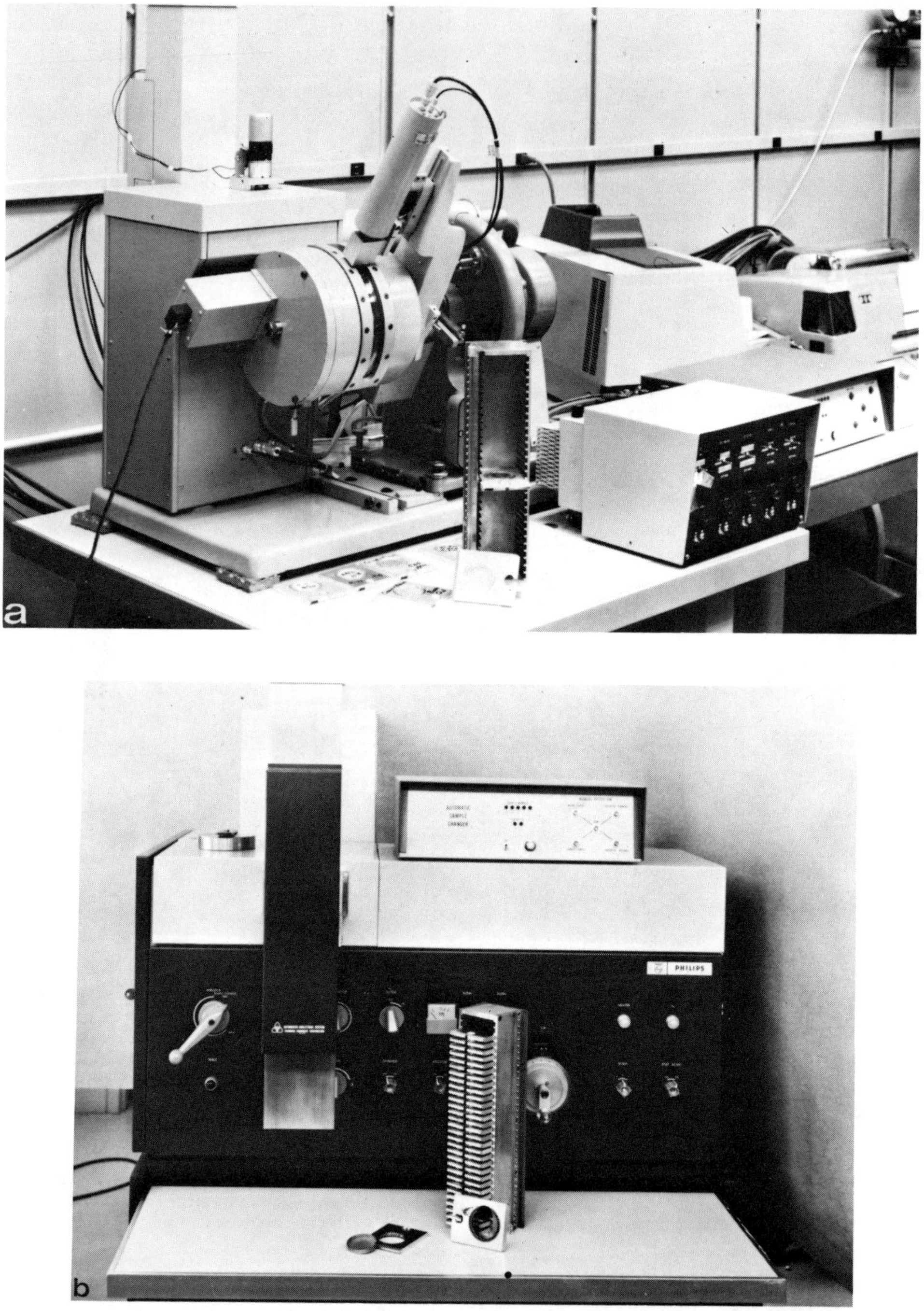

Fig. 3-13. Scanning X-ray spectrometers. (a) Spectrometer showing scintillation counters on scanning arm. Drum with window contains analyzing crystals and gas flow proportional counter. Sample cassette with specimen holders (front center). (b) Modern automated spectrometer in enclosure; geometry of spectrometer similar to (a).

Fig. 3-14 shows an electronic panel of an X-ray spectrometer. The spectrometer components described are common to both automatic and manual spectrometers. Automatic spectrometers also have a computer, interfacing electronics and peripheral devices such as a teletype, card reader or disk drive

Fig. 3-14. Electronic control and signal detection panel. Right side contains detection electronics. Left side contains computer system.

for subsidiary data and program storage. Fig. 3-14 also shows automation components of an automatic spectrometer system.

Exploration analysis requires high sample throughput at low cost. Cost and high sample analysis rates make automatic fluorescence spectrometers necessary or very desirable to service extensive exploration programs. Automated X-ray fluorescence analysis can deliver up to fifty high-quality total rock analyses per day on a scanning spectrometer. If, however, lower-quality analyses are acceptable or few elements are required, sample throughput rate can be as great as one hundred samples per day.

Complete automatic analysis by X-ray fluorescence uses a computer to insert a specimen then to slew the spectrometer to each element's emission line, determine proper measurement time, find the peak, position the line, measure the line, measure nearby backgrounds and store the measured data. Before and after measurement, the computer may calibrate the spectrometer and check the calibration. When all lines are measured the computer then processes the data with quantitative data correction programs and prints the analyses in whatever form required by the operator. The computer then inserts the next specimen and continues with analysis. Both spectrometers shown in Fig. 3-13 are automatic units. The control system for these units is shown in Fig. 3-14.

Energy dispersive spectrometers

The X-ray generator and X-ray tube are similar in concept, if not physical form, to those of the scanning spectrometer. The function of the energy dispersive spectrometer has been described earlier. Fig. 3-15 shows an energy dispersive spectrometer. The system as illustrated can do qualitative and quantitative analysis. A small computer allows data collection and data processing by programs which separate overlapping spectral lines and calculate photon counts under a peak. Other programs correct data and present results of analyses as in the scanning spectrometer.

X-ray diffractometry

X-ray diffraction instruments are simpler than X-ray fluorescence instruments. The same or a slightly simpler X-ray generator used for fluorescence analysis powers an X-ray diffraction tube. Ordinarily the generator has few stabilizing circuits and frequently does not operate at constant potential but with full-wave rectification.

The X-rays emanate from an X-ray tube, similar in function to the fluorescence tube, but different in the geometry of construction. The X-rays pass from the tube through a filter which monochromatizes the beam, filtering most non-K_α radiation. A sealed proportional counter or a scintillation counter detects the diffracted signal. Signal detection and display electronics are the same as those of a scanning X-ray fluorescence spectrometer. Fig.

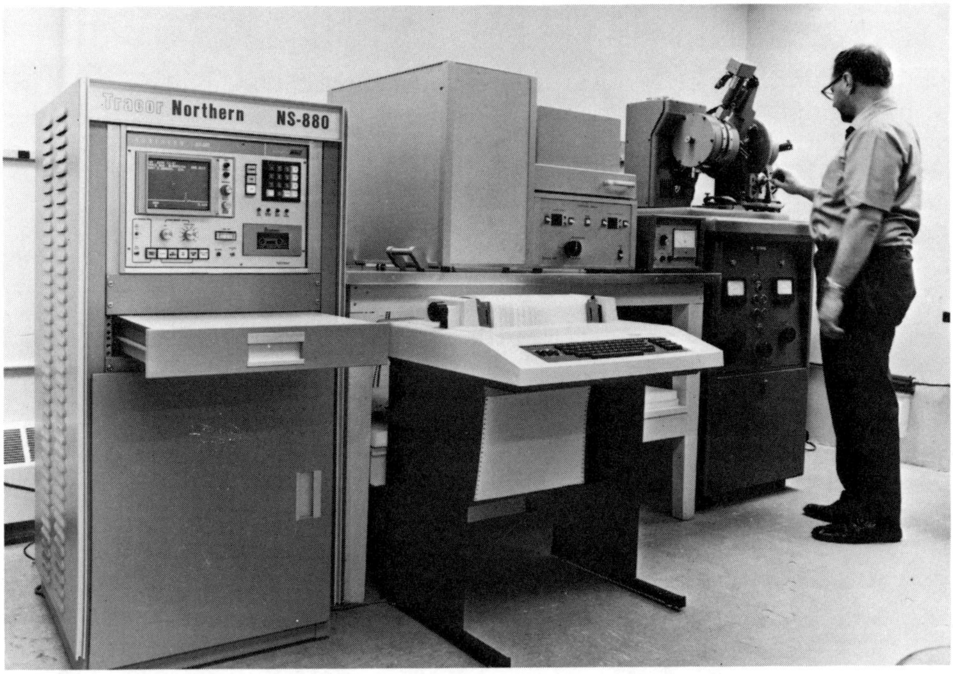

Fig. 3-15. Energy dispersive spectrometer. Console on left contains multichannel analyzer in upper portion with computer hidden below. X-ray chamber is on table. (Courtesy Colorado School of Mines Research Institute.)

3-16 shows a typical X-ray diffractometer with tube and generator. There are no portable field X-ray diffractometers but there is a small 120-V X-ray diffractometer that could be mounted in a van for field use. This small instrument has limited flexibility but could be automated for quantitative analysis.

Automated diffractometers use the same generator, goniometer and counting electronics as manual models. Automatic diffractometers have stepping motors replacing the synchronous drive motors in the goniometer. An automatic sample changer and a control computer completes the automatic hardware. The electronic/computer panel shown in Fig. 3-14 serves an automatic diffractometer as well as an automatic X-ray spectrometer. Standard-sized automatic as well as manual diffractometers may constitute instruments in a mobile exploration laboratory as depicted in Fig. 3-17.

For some types of exploration analysis, manual diffractometers suffice for quantitative mineral analysis, while for other types an automatic diffractometer is desirable. We define quantitative analysis in two ways. In the conventional sense, we define quantitative analysis as the quantitative determination of one or more mineral constituents in a sample. We also define

Fig. 3-16. X-ray diffractometers; manual (left), automatic (right).

quantitative analysis as the quantitative measurement of any diffraction parameter representing the attribute of one or more minerals in a specimen such as width of a diffraction peak.

Special instrumentation

Mobile exploration laboratories

The object of X-ray fluorescence analysis, X-ray diffraction analysis and other types of analysis is to give the exploration geologist information he cannot otherwise obtain. The competent and experienced geologist can estimate visually many parameters of potential ore rocks. Exploration analysis must, therefore, give him qualitative and quantitative definition and accuracy beyond his capability. Such definition and accuracy demands proper procedures and equipment. At present, this implies use of a conventional laboratory.

The U.S. Geological Survey in Denver, Colorado, has successfully mounted standard laboratory direct-reading emission spectrometers in vans. The vans go to the field site, where the analyst performs analyses as the geologist collects samples. Other groups (Australian Mining, 1970; Kinson and Belcher, 1970; Lynch and Durham, 1966; Holman and Durham, 1966) have built and operated mobile exploration laboratories for several years. Most have been emission spectrographic laboratories and all seem to have worked well after solving problems associated with mobility.

Mineral exploration programs cost much time and money. They will cost more in the future as exploration drilling becomes deeper and the predominant exploration method. A simple concept of moving analytical exploration services to the field with the geologists, if executed properly, should substantially shorten exploration time and save exploration drilling and large-team exploration costs. Along with field analytical services, the geologists should have very fast interactive mineral logging and chemical mapping during surface and drilling phases of minerals exploration.

Typically, the geologist collects exploration samples or executes a drilling program and must wait weeks or months for analytical results with which to make exploration decisions. To save costs the geologist could have such data available to him during exploration. Frequently he must revisit the exploration site with a drilling rig for additional data. Often he must collect and analyze more samples or drill more holes than necessary to insure that there will be enough information for future decisions. The samples the geologist has already collected or the holes he has drilled should tell him how he should collect the most significant new samples or drill the next holes. During the drilling program the geologist frequently has limited knowledge of trends, gradients and quality of ore until analyses can be completed at later dates.

A good rock analysis system with special data presentation and synthesis capabilities brought to or near the exploration site would add to exploration and profit the exploring group in time and money. The geologist would have almost immediate analyses to guide his next collections or drilling. He could collect more efficiently, spend more time looking for trends or indications and collect fewer but more significant samples. He would minimize recollection. With immediate analysis of samples the geologist could determine the best drilling strategy. Moreover, if the service laboratory accumulated data and plotted maps, sections, ore pointers and preliminary orebody models, rapid exploration decisions would be expedited.

We have examined two economic deposits and found, in retrospect, that an on-site mobile laboratory could have significantly reduced exploration drilling.

A clay deposit constituted the first study. The area was drilled on a close grid. No analyses were done on cores at the time of drilling. Later, maps of mineralogical, chemical and ceramic tests defined the deposits. The important tests were ceramic and mineralogical properties. Correlating results with core spacing and drilling depth, we found that on-site analysis could allow initial drilling at twice the grid spacing, that is, one-fourth the number of holes of the actual drilling program. To present clay distribution to the accuracy available from the original cores, only about one-half the holes drilled would have been necessary. About 50% of the drilling and subsequent sample analysis could have been saved.

The second study was also after the fact. We analyzed cores from lead deposits. We analyzed the cores chemically by several methods and mineralogically by X-ray diffraction and binocular microscope. The deposit was irregularly shaped with narrow elongate mineralized zones. We estimate that on-site analysis would have saved about 25% of the coring and associated analyses. A part of this saving was the result of a discovery during analysis that the magnesium content of the carbonate wall rock "pointed" to the ore.

Speeding mineral exploration with immediate on-site analytical service would yield other, less quantifiable benefits. Land costs would be reduced. Faster exploration would give more continuity to an exploration program, minimizing effects of changing personnel, short field seasons, and short-term changes in economic climate.

A mobile laboratory needs to meet just a few criteria to be extremely useful. The instruments must have the capability to analyze for major, minor and trace elements with both speed and accuracy. The mobile laboratory must have the capability to analyze mineral content of rocks both qualitatively and quantitatively. Few instruments must do the analyses and each instrument must be automated to attain the necessary speed and accuracy.

Fig. 3-17 illustrates one possible version of a mobile laboratory. Except for automation, each instrument conforms to conventional laboratory-based specifications. The X-ray fluorescence instrument does major and minor element analysis of rocks. X-ray diffraction does mineral analysis. A plasma-source emission spectrometer analyzes trace elements and some minor elements. Thermogravimetric analyses determine CO_2 and H_2O. The computer controls the instruments, determines elemental concentration and collates data from each instrument. The computer system, in addition to instrument control and data processing, integrates analyses geographically and stratigraphically, presenting the data as stratigraphic sections and three-dimensional contour maps.

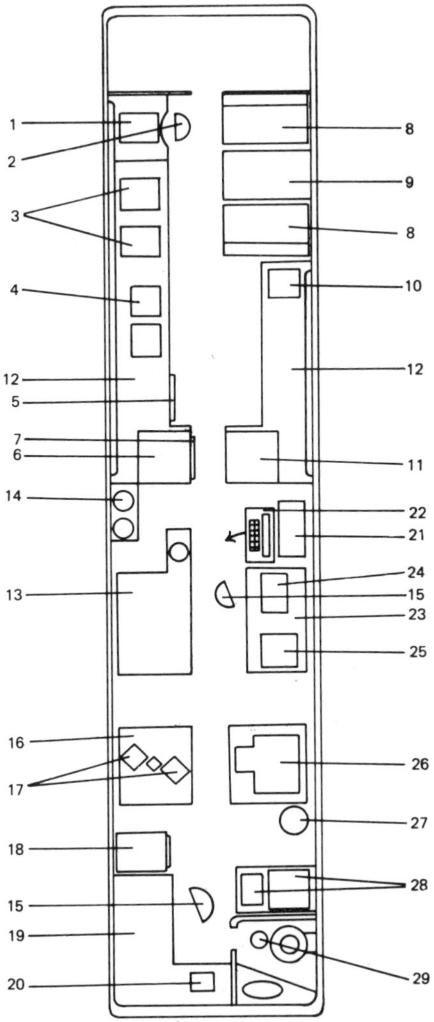

Fig. 3-17. Mobile exploration laboratory. Preparation room: *1* = balance, *2* = chair, *3* = grinder, *4* = sink, *5* = refrigerator, *6* = drying oven, *7* = furnace, *8* = seats, *9* = table, *10* = press, *11* = hood, *12* = work benches. Instrument room: *13* = plasma source direct-reading emission spectrometer, *14* = argon tanks for plasma, *15* = chair, *16* = X-ray generator, *17* = diffractometers (automatic), *18* = card file, *19* = desk, *20* = typewriter/microscope, *21* = waste, *22* = teletype, *23* = electronics and computer system, *24* = card reader, *25* = plotter, *26* = X-ray spectrometer, *27* = P10 gas tank for spectrometer, *28* = thermogravimetric analyzer, *29* = bath. Utilities and power in underneath compartments.

Portable field spectrometers

X-ray fluorescence analysis instruments have been designed for field use. Some instruments are very portable and others are transportable in a vehicle. One of the most portable of such instruments measures about $12 \times 9 \times 8$

inches and weighs a few pounds. For strictly portable instruments, an isotope button such as ^{238}Pu produces gamma-rays which excite X-ray fluorescence in a specimen. The specimen may be a normally prepared sample or an unprepared sample such as a rock or soil in-situ. The fluorescent X-rays from the specimen pass through a pair of thin foil filters, eliminating some unwanted radiation by absorption. A scintillation counter detects the filtered X-rays. Scintillation counters produce electrical pulses of voltage approximately proportional to photon energy allowing a pulse-height selector to further eliminate unwanted energies. A pulse-counter counts and displays photons from the analyte line.

Portable units usually measure one or two elements, but can measure more if their energies are sufficiently well separated. Absorption of X-rays by air, the inefficiency of the scintillation counter at long wavelengths, the absorption of X-rays by the filter and the high energy of the gamma-ray source permits measurement of elements of atomic number higher than chromium.

A second type of portable X-ray fluorescence unit, on the market for some years, has a more nearly conventional spectrometer configuration. A single crystal-spectrometer measures fluorescent X-rays from a specimen. X-rays are produced by a small low-power generator and tube. This type of spectrometer has the advantage of good photon energy resolution but is more bulky than the radioactive-source scintillation-counter type of unit. Fig. 3-18 illustrates a portable unit.

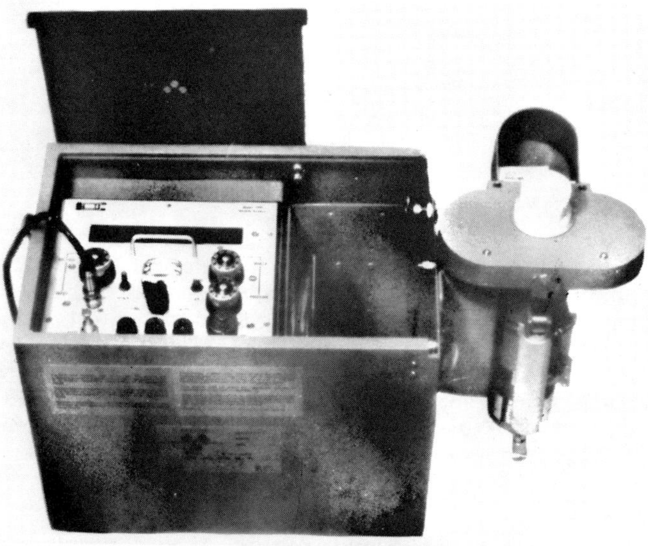

Fig. 3-18. Portable X-ray spectrometer box measures 12 inches across. Source and detector in detachable assembly on right. (Courtesy Columbia Scientific Industries Corp.)

There are a few exploration applications where the portable analyzer as described above can serve usefully. A portable unit gave a reasonably good estimate of iron content in Precambrian taconite beds in Wyoming. The estimates were better than visual (hand lens) estimates because of the difficulty of visually quantifying amounts of particles in differently textured dark matrixes. Application of a portable fluorescence unit proved useful in exploration of copper-bearing black shales. Wollenberg et al. (1971) very effectively used a portable X-ray spectrometer to explore for niobium, zirconium and lanthanium plus cerium in Greenland. Fig. 3-19 illustrates part of their analytical results across five rock units. Rhodes (1971, 1974) summarized the status of portable X-ray analyzer applications.

Before the exploration geologist rushes to buy a portable unit for use on the outcrop or on a drill-core sample, he should realize that usually a few accurate analyses of selected materials will give him comparative information to visually estimate elemental content adequately for early exploration needs. A portable unit may serve usefully where visual discrimination fails.

Downhole spectrometers

If we use the concept of the first-described portable X-ray fluorescence spectrometer with a radioactive isotope source, place the measuring device and source in a cylinder with appropriate cabling, we have a primitive downhole X-ray spectrometer.

The downhole spectrometer becomes a simple logging device. The oper-

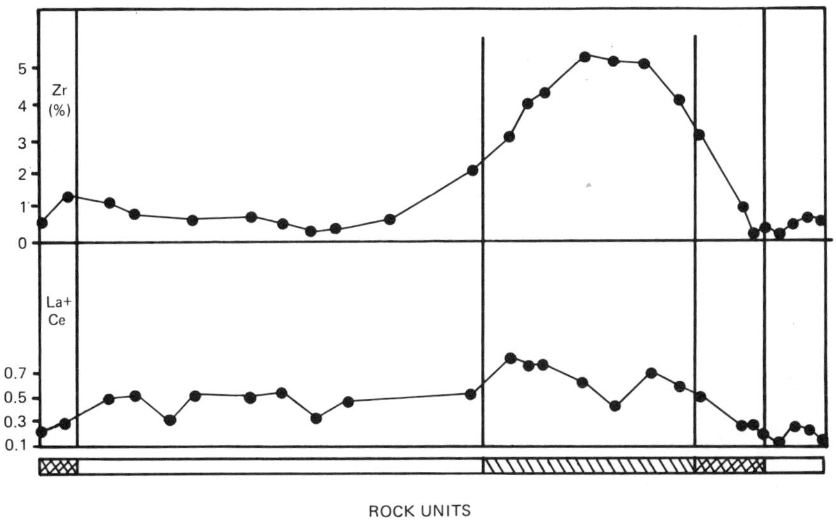

Fig. 3-19. Profiles of zirconium and lanthanum plus cerium measured by portable X-ray fluorescence analyzer, across four rock unit boundaries (vertical lines). (After Wollenberg et al., 1971.)

ator lowers it into a drill hole, logging the spectrum while the spectrometer measures at fixed intervals down the hole. The spectrometer can, if configured with one or more scintillation counters, measure a few elements at a time.

Since the downhole spectrometer requires a van to carry the spectrometer, electronics, cables, etc., we may make the spectrometer more sophisticated than a portable X-ray unit. The scintillation counter pulses may pass into a multichannel analyzer as do the pulses from energy dispersive detectors. A computer may then mathematically strip the energy spectrum stored in the multichannel analyzer allowing detection and some quantification of several elements simultaneously.

Although still in the experimental stages the downhole spectrometer, like the portable X-ray fluorescence spectrometers described above for surface use, does have potential utility. The downhole spectrometer gives information to the geologist he could not get otherwise because he cannot go down the hole to look at the rocks. If a downhole spectrometer can give good information in the particular exploration situation, the amount of drilling and expense can be reduced.

Future trends

Future trends in X-ray instrumentation as used by mineral exploration are difficult to predict. One trend that is certain is the evolution to more automation in both fluorescence and diffraction. Automation speeds the analytical process making it economically feasible to use X-ray methods routinely. Instrumentation seems to evolve from manual to automatic in predictable stages with the manual instrument increasing in complexity with automatic functions making a primitive debut. As automation proceeds, the instrument becomes simpler while the computer does more and more decision making and work.

The final stage is an automatic instrument from which a moderately well-trained operator can extract useful analyses easily. X-ray instruments are now about in the middle of the evolutionary stage. When evolution completes, X-ray analysis will be able to accomplish almost all the quantitative analyses of solids in mineral exploration except low levels of concentration of trace elements.

As automatic instruments evolve, becoming more simple and more compact, it will be possible to move them into the field. The mobile laboratory concept as shown in Fig. 3-17 uses presently available equipment. However, a compact suitably monitored high-frequency generator could substitute for the traditional generator. Using a fully automated energy dispersive X-ray analyzer would reduce size and weight of X-ray diffraction and fluorescence instruments to less than one-half that of Fig. 3-17 allowing them to become the hand-maidens of field exploration.

APPLICATION OF X-RAY ANALYSIS

Reporting of systematic applications of X-ray analysis to mineral exploration seldom appears in the literature for several economic reasons. Further, except for diffraction, most modern X-ray instrumentation and methods discussed in this chapter date from 1964, or later precluding extensive reporting from any but the most recent literature. Finally, the application of X-ray analysis to mineral exploration is broad enough to preclude comprehensive review. Therefore, this section elucidates some of the kinds of applications rather than reporting systematic studies.

Non-metallic minerals

Limestone and dolomite

The carbonate producing industry, including limestone for cement manufacture, steel production and other uses, requires knowledge of the content of Ca/Mg ratios, quartz content and other contaminating impurities such as clay minerals, etc. Both X-ray fluorescence and X-ray diffraction analysis can provide much of the information needed to determine quality of deposit. Quantitative X-ray diffraction gives very good estimates of calcite ($CaCO_3$), dolomite [$(Ca,Mg)CO_3$] and quartz content in unmetamorphosed and metamorphosed deposits. Silicate impurities can also lend themselves to quantitative determination in marbles and to a lesser extent in unmetamorphosed limestone and dolomite. X-ray fluorescence gives accurate estimates of Ca, Mg, Fe, Si and Al.

Both the linear equation and the fundamental parameters methods of correction give good results since CO_2 dominates the matrix. It is possible to define any economic carbonate deposit using only X-ray fluorescence and X-ray diffraction coupled with supplementary use of other analytical techniques.

Bauxite and laterite ores

Bauxite and laterite ores constitute the chief source of aluminum and one of the main sources of nickel. Ores of this type are particularly amenable to exploration analysis by X-ray fluorescence and to a lesser extent by X-ray diffraction.

X-ray fluorescence gives good analytical values for Al, Si, Fe, Ti and Ni in bauxites and laterites. Exploration analysis is fast and relatively accurate. When coupled with weight-loss-on-ignition, extractable Al_2O_3 and SiO_2 by wet extraction, X-ray fluorescence analysis almost completes the chemical data needed for exploration decisions. Furthermore, it is not difficult to analyze 100 samples per day with a scanning X-ray spectrometer, allowing analysis to coincide with exploration.

Quantitative X-ray diffraction can provide good estimates of mineral and chemical content of bauxite ores. Frequently, the aluminum-bearing minerals, gibbsite [$Al(OH)_3$] and related species have well-defined X-ray diffraction lines and the amounts of each mineral can be estimated accurately. The iron-bearing minerals of bauxite, although hydrated, tend also to have well-defined diffraction peaks and estimation of amount is not difficult. On the other hand, residual quartz has its peak masked by gibbsite and its measurement is more difficult.

Employing X-ray diffraction to outline nickelferous laterite and other laterites can be very difficult. Frequently the iron minerals are variously hydrated and do not show well-defined diffraction lines, rendering quantitative diffraction analyses impossible.

Clays

Ceramic clay minerals are primarily kaolinite and halloysite ($Al_2O_3 \cdot 2\ SiO_2 \cdot 2\ H_2O$), two minerals of nearly the same composition but slightly different structure. Bauxitic clay deposits may also contain hydrated aluminum oxide minerals such as gibbsite [($Al(OH)_3$] and related minerals boehmite, bayerite, etc. Toward the opposite compositional type, clay deposits may contain illite (a muscovite-like clay mineral) admixed with the kaolinite. Quartz, anatase (TiO_2) and other minerals are often present as contaminants.

Although standard ceramic tests provide necessary information on ceramic behavior, a combination of both X-ray diffraction and X-ray fluorescence analysis yield important exploration information. For most deposits diffraction analysis is the most comprehensive method of identifying and quantifying the clays mineralogically. With special standard and specimen preparation techniques, mineralogical analysis can give quantities of the various types of clays and other minerals with an error about ±7—10% of the amount present. Special characteristics of the clay minerals such as crystalline disorder which often areally indicates changes in composition of the clay minerals and amounts of fluxing elements may be quantified by X-ray diffraction.

X-ray fluorescence analysis gives good estimates of aluminum, silicon, and iron oxides in clay deposits. Perhaps as important as determining the more abundant elements, X-ray fluorescence can give good analyses of the fluxing elements Na, K, Mg, and Ca. Where the latter group of elements are the only elements of interest, other instrumental methods would be superior to X-ray fluorescence analysis.

Other non-metallic minerals

Exploration for other non-metallic materials such as glass sands, bentonite, zeolites, phosphates, gypsum, etc., effectively use the combination of X-ray diffraction and X-ray fluorescence analysis to define the quality and useful limits of a deposit.

Energy materials

Coal

The use of X-ray diffraction and X-ray fluorescence in coal exploration is minimal. There is some potential for use but the potential remains undeveloped. Coals comprise organic material and subsidiary mineral contaminants. Kaolinite, illite, pyrite, small quantities of quartz, a few resistant detrital minerals and gypsum comprise the subsidiary suite of minerals that eventually becomes ash. Sulfur, an environmental contaminant when the coal burns, occurs in the organic fraction, in sulphates and in sulfides (mainly pyrite). B.T.U., ash, sulfur and water contents constitute the most important non-geological or industrial parameters to determine coal quality.

X-ray diffraction analysis is unable to determine any of the industrial parameters directly. Diffraction analysis can determine mineral, sulphate and pyrite content if the coal is ashed at low temperatures. The value of low-temperature ashing followed by diffraction analysis on routine exploration samples is questionable. Eastern U.S. coals, however, frequently have sufficient pyrite and occasionally sufficient gypsum to allow inorganic sulfur determination in the raw coals by diffraction methods.

Research by the U.S. Bureau of Mines (Berman and Ergun, 1968) has shown that X-ray fluorescence analysis can determine total sulfur and total ash content of coals very accurately. X-ray fluorescence analysis determines total sulfur content directly. Determination of ash content by X-ray fluorescence requires correlation of chemical content with mineral content thence with ash content. Although an indirect method of finding ash content may seem less useful than direct methods, there are advantages. Both sulfur and ash are determined simultaneously, ash materials can be deduced accurately, sample preparation is simple and measurement is rapid.

Other factors limit the value of X-ray analysis of coals. The cost of X-ray analysis equipment prohibits investment by the smaller companies unless the equipment can also serve quality control in an operating mine. Low-sulfur, low-ash coals of the western U.S. can be analyzed by less expensive instrumental methods than X-ray diffraction and X-ray fluorescence.

Petroleum

The petroleum industry has made the most sophisticated exploration use of X-ray analysis of all industries. They have used X-ray methods in spite of the fact that X-ray methods do not usually lead directly to finding oil or defining reserves and give only supplementary data to the information bank necessary for comprehensive exploration. The petroleum industry research groups have done most of the exploration analysis partly because they have adequate laboratory facilities and partly because they also have the special competence required. Three types of X-ray analysis dominate petroleum exploration and correlation studies: stratigraphic definition, facies analysis and depositional environment definition. Most exploration studies have centered around X-ray diffraction analysis because diffraction analysis can differentiate fine-grained sediments and carbonates. X-ray fluorescence has had limited use in petroleum exploration.

Unfortunately much of the excellent work using X-ray diffraction analysis as an exploratory tool remains unpublished and in company files. Two examples of exploration analysis by X-ray diffraction were conducted some years ago by W.T. Caneer.

Preliminary exploration drilling off the coast of Santa Barbara County, California, presented a thick sequence of sediments from several formations. The formations, comprising sandy siltstones and mudstones, were very difficult to differentiate using drill cuttings. Faulting complicated the nearly impossible task of stratigraphic correlation.

X-ray diffraction analysis allowed differentiation of the formations and approximate stratigraphic correlation. Table 3-II summarizes the results. The example is interesting because the rocks were so fine-grained that only X-ray diffraction could provide diagnostic mineral determinations.

In another oil exploration program done by Caneer, exploration analysis of a shale and carbonate sequence of southern Canada found that the minor mineral chlorite as defined by X-ray diffraction became less iron-rich near ancient oil-bearing reefs. Two measured parameters, the ratio of the (001) to (002) diffraction peaks and the X-ray diffraction behavior of the chlorite after heating to 600°C indicated the relative distance from the reefs. Examination of samples from exploration drilling over a geographic area located direction and distance to isolated reefs. Closer to reefs, changes in

TABLE 3-II

Summary of mineralogical analyses by X-ray diffraction, Ventura Basin, California

Formation and member	Illite/kaolin ratio	Diagnostic minerals
Pico and Repetto	1.2—3.2	microcline
	0.9—1.1	microcline
Upper Sisquoc	0.5—0.9	—
	1.0	—
	1.1—1.3	—
	1.6	—
Lower Sisquoc	1.1—1.4	cristobalite
	1.6—2.5	cristobalite
	1.7	cristobalite + microcline
Basal Sisquoc	0.7	—
(laminated)	1.0—2.4	plagioclase + cristobalite
Monterey	5.5	cristobalite
	mica, no clay	cristobalite

the dolomite/calcite ratio as determined by X-ray diffraction further defined the reef edges. Wells drilled as a result of the X-ray analysis produced oil.

X-ray fluorescence analysis has much more limited value than X-ray diffraction for petroleum exploration. The reason is simple, most sedimentary rocks are less distinctive chemically than they are mineralogically or texturally. Only special chemical characteristics of rocks make fluorescence analysis worth while in petroleum exploration. For example, some black shales or zones in black shales have unusually high copper or zinc concentrations. X-ray fluorescence easily quantifies such anomalies, which can then serve as key stratigraphic horizons. In carbonate sequences, X-ray fluorescence analysis can define trends in such elements as barium and strontium which, like the chlorite example cited above, might help locate porous oil-bearing zones.

Uranium

About 95% of economic uranium deposits and reserves are in sedimentary formations and most are in sandstones (Saum and Link, 1969). Most sedimentary deposits result from leaching of a sandstone by groundwater with reconcentration in economic mineral quantities. ^{238}U decay leads to many radioactive daughter products among them Th, Bi, Pb, Rn and K. Each of these daughter products emits gamma radiation of characteristic energy measurable radiometrically. Frequently Mo, Cu, Ag, Se, Co and V migrate with the uranium to the ore site. Quick-test methods can detect most of these emigrant elements.

Radiometrics, tests for "pointer" elements and geologic features of uranium deposits make systematic analysis by X-ray diffraction and fluorescence usually unnecessary until the drilling phase of exploration.

The drilling program begins when samples of subsurface materials are needed. These samples will be of one of two types. The most expensive and most useful samples are taken by the core drill. The samples are long cylinders of rock and the geologist is able to examine the lithology, mineral content, etc., in detail. Samples taken with the core drill are much more reliable than those acquired in other ways because the core is enclosed within the steel core barrel in the drill hole and no contamination occurs. Usually core drilling is used in the final phase of an exploration program and continues during mine development. The most common and inexpensive drilling methods cut chips from the bottom of the drill hole and use air or mud to carry these chips to the surface. Although samples obtained this way are of less value to the geologist, the cost is only one-tenth that of core-drilling.

Drilling programs fall into three categories: (1) information or reconnaissance drilling in which the hole is put down to determine geologic information about the area, and hopefully, to make a uranium discovery; (2) outline drilling which is used to determine the outline of a discovery and extent of mineralization; and (3) sample or core drilling to obtain information on exactly how much ore of what grade can be reliably counted on for planning a mine.

Reconnaissance drilling on random centers spaced more than a mile apart is common in uranium exploration in the sedimentary basins. When uranium or altered host rocks are located by a reconnaissance drilling program, outline drilling is begun. Outline drilling delineates lithologic units, buried stream channels, and other favorable hosts for the ore material. Emphasis is placed on obtaining information on the nature of the ore deposit rather than attempting to detail the entire deposit.

At the outset of drilling, X-ray fluorescence and diffraction can serve usefully in the exploration process. Suppose exploration for uranium with the conditions stated above, that is, reconnaissance drilling for the "roll" type of deposit schematically depicted in Fig. 3-20. During the information or reconnaissance drilling phase, holes *A* through *D* of Fig. 3-20 might be drilled. Table 3-III lists some of the features found in each drill hole. In Table 3-III, the letters F for fluorescence and D for diffraction indicate the application of each tool for defining particular parameters. Interestingly X-ray diffraction powerfully addresses determination of several useful parameters during the exploration drilling, but is little used except by companies who had used diffraction to find other types of energy sources.

Fluorescence analysis has good sensitivity for the pointer elements because the matrix (sandstone) has low absorption coefficients. Outside the orebody, fluorescence often has insufficient sensitivity to measure uranium except with use of special preparation methods. In the orebody uranium content is easily measured.

When core drilling begins, X-ray methods supply systematic mineral and chemical data to define the new discovery.

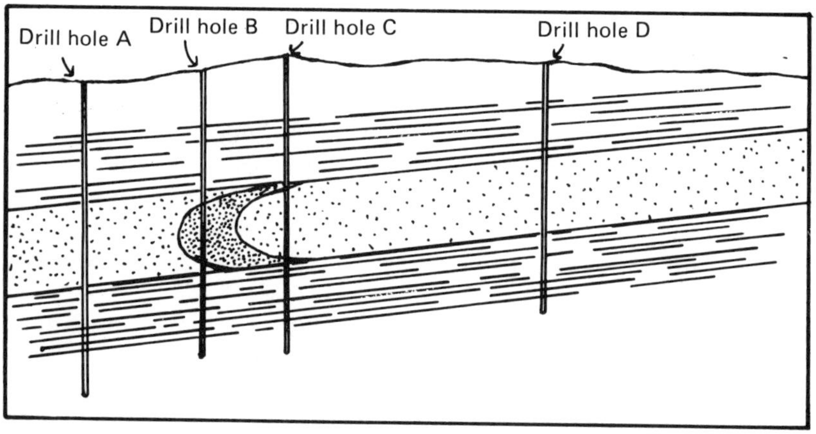

Fig. 3-20. Typical uranium roll ore deposit in sandstone. (After Adler and Sharp, 1967.)

Metals

Both X-ray diffraction and X-ray fluorescence analysis have great power as tools in metals exploration. As with uranium exploration, the usefulness of both methods occurs mainly during the later stages or ore finding when the systematic definition and use of physical chemical parameters might indicate the presence of ores or the direction to ores and the definition of quality and quantity of ore after the body is found.

The metals mining industry has for some years used X-ray fluorescence analysis very effectively for quality control in mines. Some mine operations have used modern computer controlled systems to give good analyses of ore and gangue elements to monitor mine operation. They use X-ray fluorescence analysis to systematically map the mining operation as it proceeds and make projections for future mixing of ores for beneficiation. Some companies use continuously operating "on-stream" X-ray fluorescence analyzers to help control the entire beneficiation process. Most mining operations use X-ray diffraction less than X-ray fluorescence.

Although X-ray methods receive wide use in mine and mill, for a plethora of reasons ranging from legitimate financial and technical considerations to considerations based on less appealing reasons, X-ray fluorescence and X-ray diffraction analysis have not contributed their full potential in metals exploration even after the metals prospecting phase is over and during systematic exploration of a local area. The speed and comprehensive chemical and mineralogical analytical capability of both methods can characterize an ore body easily and inexpensively relative to other costs of exploration.

Because X-ray fluorescence and diffraction analysis, when properly executed, can be so universally applicable in metals exploration we elucidate examples of the weakness of each method in metals exploration rather than dwelling on strengths.

X-ray fluorescence analysis frequently gives poor analytical results on some by-product elements. For example, X-ray fluorescence analysis indicated no or incorrect amounts of Co in Pb ore from southeast Missouri lead district. However, electron microprobe analysis showed a Co-rich phase concentrated around ore-mineral grains. Calculation based on the premise that the electron microprobe samples were representative indicated Co could be extracted economically as a by-product. During X-ray fluorescence analysis, most of the cobalt radiation was absorbed by the highly absorbing Pb-Cu-Zn matrix. Cd, Au, Mo and Ag frequently are not always analyzable when in by-product quantities. Se and Ge, by-products of Cu and Zn processing, are also not easily quantified in the raw ores. In general, minute-quantity by-products are not easily analyzed by fluorescence analysis.

Native metals such as Au and Cu are difficult to analyze without special treatment. X-ray fluorescence also proves too insensitive for most metals in the water samples frequently collected for analysis during prospecting or preliminary exploration.

TABLE 3-III

Mineralogical and chemical characteristics of uranium ore rolls of Fig. 3-20 *

	Drill hole *A* Eh— unoxidized zone unaltered sand	Drill hole *B* Eh— ore roll (unoxidized) ore and "protore"	Drill hole *C* Eh+ oxidized zone barren interior altered sand	Drill hole *D* Eh+ oxidized zone barren altered sand
Color:	light gray	drak gray	light gray, greenish yellow, bleached, brownish reddish	light gray, greenish yellow, bleached brownish reddish
Pyrite:	present, local	abundant; subhedral aggregates or euhedral	present, scarce	absent or sparse; corroded where scarse; untarnished euhedral where sparse
Uranium:	trace	F concentrated ore	absent or trace	absent or trace
Molybdenum:	F low (<10 ppm)	F high (100 ppm)	F very high (1000—10,000 ppm)	low (<10 ppm)
Vanadium:	F low (1—10 ppm)	F moderate to high (500—2000 ppm)	F moderate (1000 ppm)	F low (10—20 ppm)

Copper:	F variable (<20 ppm)	F low (20 ppm)	F low (10 ppm)	F low (<1—2 ppm)
Selenium:	F low (<0.5 ppm)	F moderate to high (2—200 ppm)	F very high (200—1200 ppm)	F moderate (10—60 ppm)
Fe^{2+}/Fe^{3+}:	2	>2	<2	1
Heavy opaques:	present	absent or scarce	scarce	scarce
Calcite:	D scarce	D abundant	D sparse	D absent or sparse
Clay:	D low-iron mont. ** kaolinite/mont. sharp (00*l*) lines	D low-iron mont. kaolinite/mont. sharp (00*l*) lines	D high-iron mont. D kaolinite/mont. broad (00*l*) diffraction lines	D high-iron mont. kaolinite/mont. broad (00*l*) diffraction lines
Limonite:	D absent	D usually absent	D present	D sometimes abundant
Hematite:	D absent in reduzate facies	D usually absent or localized in ore	D present sometimes abundant	D sometimes predominant
Zeolites:	D present	D present	D absent	D absent
Sulfate:	moderately abundant	high	low	very low

* Based only on available data, primarily applicable to rolls in Tertiary sandstones; selected to indicate promising exploratory guides from Harshman (1972); Adler and Sharp (1967) and W.T. Caneer (personal communication); Granger et al. (1961).
** Mont. = montmorillonite.

X-ray diffraction analysis, although a very powerful exploration tool, occasionally suffers similar deficiencies to X-ray fluorescence analysis. X-ray diffraction has a sensitivity limit ranging upward from about 0.25% for a given mineral. A sensitivity of 1% is typical and with a highly absorbing matrix, as many ores might have, the analyst can find a sensitivity limit as high as 4%. Poor sensitivity means that economically valuable mineral phases could be missed or at best, poorly quantified.

Lead

We present one example of the application of X-ray analysis to exploration of metals. In southeast Missouri Pb occurs in mineralized zones in the Cambrian Bonneterre Formation, a dolomite. Fluorescence and diffraction analysis, easily able to quantify the elemental and mineral richness of the mineralized zones, could also serve to indicate direction to ore during exploration drilling. Within the sensitivity range of X-ray fluorescence, elemental analysis of Bonneterre dolomite showed little relation between such elements as Cu, Fe, Zn, Mn and Cd in the country rock and nearness to the mineralized zones.

However, the Bonneterre Formation, although regionally dolomatized,

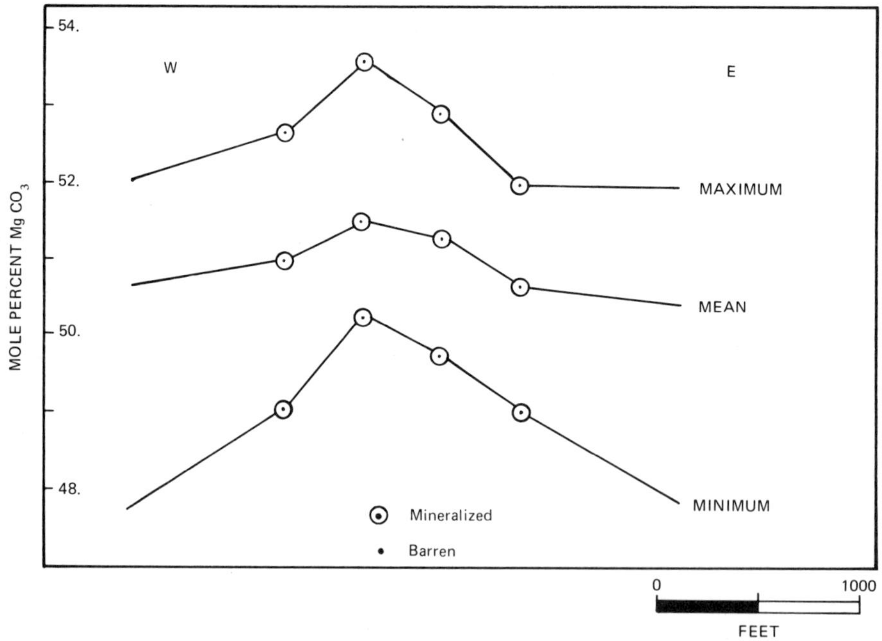

Fig. 3-21. Plot of maximum, mean and minimum mole percent $MgCO_3$ content of core samples along an east-west traverse across a mineralized zone in the Bonneterre Formation, Missouri (After Briceno, 1975.)

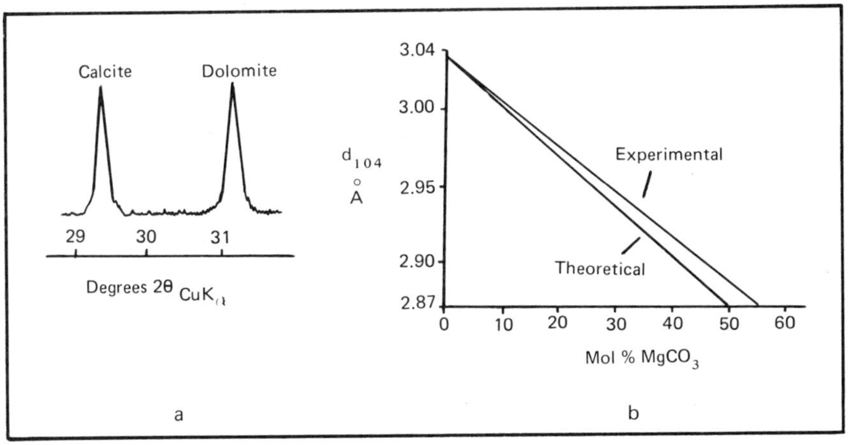

Fig. 3-22. (a) Diffractogram showing the positions for the $d(104)$ reflections for calcite ($CaCO_3$) and dolomite $(Ca,Mg)CO_3$. (b) Plot for $d(104)$ versus mole percent $MgCO_3$ for Ca-Mg carbonates. (After Goldsmith and Graf, 1958.)

contained higher concentrations of Mg near the ores, the relative Mg enrichment being partly dependent on the original lithologic character of the rock.

Fig. 3-21 illustrates the variation of Mg with respect to Ca in the wall rock zones, showing how the element concentration can be used to indicate ore zones. Variation between maximum and minimum can be related partly to textural parameters of individual facies of the original biogenic limestone. Diffraction gave the results of Fig. 3-21, although fluorescence analysis would have yielded the same results. Fig. 3-22 illustrates the method of determination of the mole percent $MgCO_3$ of Fig. 3-21. Similar relationships between mole percent $MgCO_3$ and mineralized zones were found by Klugman et al. (1962) in Colorado.

CONCLUSIONS

Quantitative analysis is the primary function of both X-ray fluorescence and diffraction in exploration analysis. Each method has strengths and weaknesses. For multi-element total rock analysis at high sample rates and low cost per sample, X-ray fluorescence is perhaps the single most powerful instrumental method for exploration analysis. Fluorescence analysis is particularly good when Al, Si, Ti and Fe constitute analyte elements. Fluorescence has sufficient sensitivity for metals of the first-row transition elements for almost all exploration needs.

On the other hand, X-ray fluorescence analysis with standard instrument configurations shows analytical weakness, as do other types of instruments for individual elements, if there are overlapping lines such as shown by uranium in the presence of strontium. Special instrument configurations can often abrogate specific analytical weakness. For example, a special X-ray tube and vacuum system allows measurement of certain low-energy lines of uranium rather than the higher-energy lines, reducing overlap and background. However, special instrument configurations or modifications are expensive and require special care. Fluorescence analysis is inadequate for routine trace-element analysis below 1—50 ppm for metals and below higher levels for lighter elements. Routine fluorescence analysis is incapable of analyzing elements below Na.

X-ray diffraction and microscopic analysis give the best results for general mineral analysis of any methods now in use. Both methods yield good results for most coarse-grained rocks. For fine-grained rocks, X-ray diffraction yields better values than microscopic analysis. The character of the analyte specimen generally limits accuracy.

Quantitative analysis by X-ray diffraction is usually fast and inexpensive. Quantitative diffraction techniques are very good for analyzing carbonate rocks, igneous and metamorphic rocks and sandstones when peaks do not overlap. Diffraction provides the best method for analyzing shales and mudstones. However, clay minerals, fine-grained iron and aluminum oxides, certain carbonates and a few other rock-forming minerals, have poorly defined structures and compositions, creating difficulties in characterizing them in standards and specimens. To analyze these minerals quantitatively often require special competence and experience.

X-ray fluorescence and X-ray diffraction analysis as used by the mineral industry for exploration has had a history marked with both success and failure. Two factors have accounted for success or failure. Technological reasons account for some failures. Until about eight years ago fluorescence equipment did not have sufficient sophistication to meet the requirements of flexibility and speed to satisfactorily serve an exploration program. Earlier equipment had insufficient stability, too few analyzing crystals in the spectrometers to cover the wavelength range necessary for general exploration analysis, etc. Operators had to attend instruments at all times during measurement and do primitive corrections by hand calculation or use large remote computers. If the operator wished to avoid corrections, sample preparation became more complicated. In short, primitive equipment precluded extensive use of fluorescence analysis for general mineral exploration. Where fluorescence analysis had success it was brought about by competent and determined analysts.

The use of X-ray fluorescence analysis in mineral processing and material production caused development of better equipment coincident with advent of improved transistor and integrated circuit electronics and inexpensive control computers. Equipment is now good enough for exploration requirements.

Personnel competence has caused spotty success for exploration analysis using X-ray fluorescence. Many companies have charged untrained personnel with their X-ray fluorescence analysis programs. For example, one company brought a young man, recently graduated from high school, from one of its mines to operate an X-ray fluorescence and diffraction analysis program. The analysis program failed and the equipment sat idle. On the other hand, a certain very competent commercial analyst, using ancient cast-off equipment, obtains enviable exploration analyses accurately, rapidly and inexpensively.

The history of X-ray diffraction analysis has suffered similarly but less profoundly than X-ray fluorescence analysis. Diffraction analysis has proved itself in many specific exploration programs. Except in the hands of a few specialists, diffraction has served primarily as a qualitative and semiquantitative analysis tool. Time, expense, low accuracy of quantitative mineral determinations have precluded wide-spread use of diffraction as a quantitative exploration method. Many exploration problems such as locating and characterizing orebodies call for special experience and competence. Some companies have been willing to commission the special competence necessary to make effective use of X-ray diffraction, many have not.

Unlike X-ray fluorescence equipment, basic diffraction equipment, has changed little in more than twenty years except for improvements in counting electronics. Present equipment with additions for automation, if desired, is adequate to serve exploration analysis. The major problem associated with measurement remains what is has been, the character and preparation of samples. Usually experience and initiative can surmount difficulties with sample preparation and analytical consistency and reproduceability.

Modern X-ray spectrometry and diffraction, primarily due to recently improved equipment and automation, are taking their place as major tools of mineral exploration. The trend will continue and probably accelerate over the next twenty years. Sophisticated automatic control and data processing coupled with relatively simple hardware will deliver large quantities of accurate analyses and aleviate the need for highly skilled analysts. The drawback of low sensitivity for some elements and compounds is not likely to be overcome in the near future, preventing X-ray analysis from becoming the analytical panacea for mineral exploration. Despite this drawback, X-ray analysis is an extremely powerful tool and should always be considered as one of the potential analytical methods in any mineral exploration program.

REFERENCES

Adler, H.H. and Sharp, B.J., 1967. Uranium ore rolls-occurrence genesis and physical and chemical characteristics. *Utah Geol. Surv., Guidebook to the Geology of Utah*, 21: 53—77.

Australian Mining, 1970. Mobile laboratory for isolated exploration areas. *Aust. Min.*, 62: 69.

Berman, M. and Ergun, S., 1968. Analysis of mineral matter in coals by X-ray fluorescence. *U.S. Bur. Mines, Rep. Invest.*, 7124, 20 pp.
Berry, L.C., Post, B., Weissman, S. and McMurdie, H.F. (Editors), 1971. *X-Ray Powder Diffraction File.* Joint Committee on Powder Diffraction Standards, Swarthmore, Pa.
Bertin, E.P., 1975. *Principles and Practice of X-Ray Spectrometric Analysis.* Plenum Press, New York, N.Y., 2nd ed., 1079 pp.
Briceno, H., 1975. *Geology and Geochemistry of the Magmont Mine, Iron County, Missouri.* Masters Thesis, Colorado School of Mines.
Criss, J.W. and Birks, L.S., 1968. Calculation methods for fluorescent X-ray spectrometry — empirical coefficients vs. fundamental parameters. *Anal. Chem.*, 40: 1080—1086.
Goldsmith, J.R. and Graf, D.L., 1958. Relation between lattice constants and composition of the Ca-Mg carbonates. *Am. Mineral.*, 43: 84—101.
Granger, H.C., Santos, E.S., Dean, B.C. and Moore, F.B., 1961. Sandstone-type uranium deposits at Ambrosia Lake, New Mexico — an interim report. *Econ. Geol.*, 56: 1179—1209.
Harshman, E.M., 1972. Geology and uranium deposits Shirley Basin, Wyoming. *U.S. Geol. Surv. Prof. Paper*, 745, 82 pp.
Holman, R.H.C. and Durham, C.C., 1966. A mobile spectrographic laboratory. *Geol. Surv. Can. Paper*, 66-35, 15 pp.
Jenkins, R., 1974. *An introduction to X-Ray Spectrometry.* Heyden and Son Ltd., New York, N.Y., 163 pp.
Kinson, K. and Belcher, C.B., 1970. A mobile laboratory for nickel prospecting and drilling operations. *Australas. Inst. Min. Metall., Proc.*, 234: 59—65.
Klug, H.P. and Alexander, L.E., 1974. *X-Ray Diffraction Procedures for Polycrystalline and Amorphous Materials.* John Wiley and Sons, New York, N.Y., 2nd ed., 965 pp.
Klugman, M., Bloom, D. and Stevens, D., 1962. Mg/Ca ratios in carbonate wallrock. *Natl. Meet. AIME, New York, N.Y.*, Reprint.
Lucas-Tooth, H.J. and Pyne, E.C., 1964. Accurate determination of major constituents by X-ray fluorescence analysis in the presence of large interelement effects. *Adv. X-Ray Anal.*, 7: 523—541.
Lynch, J.J. and Durham, C.C., 1966. Field laboratories for Operation Bathhurst, a combined chemical and spectrographic analytical program. *Geol. Surv. Can. Paper*, 66-54: 188—194.
Rasberry, S.D. and Heinrich, K.F.J., 1974. Calibration for interelement effects in X-ray fluorescence analysis. *Anal. Chem.*, 46: 81—89.
Rhodes, J.R., 1971. Design and application of X-ray emission analyzers using radioisotope X-ray or gamma-ray sources. *Am. Soc. Test. Mater., Spec. Tech. Publ.*, 485: 243—285.
Rhodes, J.R., 1974. Portable X-ray and on-line nuclear instrumentation — present status of application in the USA mining industry. *Panel Meet., Nuclear Techniques in Geochemistry and Geophysics, IAEA, Vienna.*
Saum, N.M. and Link, G.M., 1969. Exploration for uranium, I. *Min. Ind. Bull.*, 12(4).
Stephenson, D.A., 1971. Theoretical analysis of quantitative X-ray emission data: glasses, rocks, and metals. *Anal. Chem.*, 43: 1761—1764.
White, E.W. and Johnson, G., 1970. *X-Ray and Absorption Wavelengths and Two-Theta Tables.* American Society for Testing and Materials, Washington, D.C., Data Series D.F.S. 37A, 2nd ed., 293 pp.
Wollenberg, H., Kunzendorf, H. and Rose-Hansen, J., 1971. Isotope-excited X-ray fluorescence analysis for Nb, Zr, and La + Ce on outcrops in the Ilimaussage intrusion, South Greenland. *Econ. Geol.*, 66: 1048—1060.

Chapter 4

X-RAY FLUORESCENCE GEOCHEMICAL ANALYSIS ON THE SURFACE OF MARS

BENTON C. CLARK, A.K. BAIRD, PRIESTLEY TOULMIN, III, HARRY J. ROSE, Jr. and KLAUS KEIL

INTRODUCTION

With the advent of the space age, it has become technically feasible to begin the geochemical exploration of the planets and satellites of our solar system, and techniques employing nuclear or atomic radiation have already contributed prominently to that exploration. The first in-situ analysis of the inorganic chemical composition of Martian surface material employs a highly miniaturized, energy dispersive X-ray fluorescence spectrometer. This experiment is on two Viking spacecraft launched by Titan III-C rockets late in the summer of 1975 and landed on the surface of Mars in July and September of 1976. In the sections which follow, we discuss in detail the X-ray experiment we have sponsored for the Viking mission and provide brief reviews of past usages of spaceborne nuclear geochemical techniques.

THE VIKING MISSION TO MARS

After 10 month journeys of some 700 million kilometers through space, the two Viking spacecraft reached the vicinity of Mars, retro-rockets fired and the spacecraft were injected into elliptical orbits about the planet. At this point in the mission, the scientific study of Mars began, using techniques of remote sensing. Dual telescopic vidicon cameras photograph the surface with overlapping frames to provide contiguous coverage with a resolution at closest approach of 50 m per picture element. A multichannel infrared radiometer maps reflected solar radiation and thermal emissions from areas as small as 5 km in diameter. From these data can be deduced the surface thermal inertia, emissivity, and albedo, which are functions of not only grain size distribution and bulk density, but composition. With a visible-light channel and five channels covering the range 6—35 μm (Kieffer et al., 1972), it may be possible to characterize those areas with physiochemical similarities and recognize geographical dissimilarities more subtle than morphological differences. In addition, a fixed-grating infrared spectrometer tuned to the 1.4-μm absorption band of water molecules is mapping the water vapor content of the Martian atmosphere.

When sufficient data had been accumulated by the orbiter science experiments to allow final selection of optimum landing sites, a signal from ground controllers on earth initiated the complex series of steps which led to separation of the Viking-1 Lander from the Viking-1 Orbiter. Only after the landing and performance of primary mission objectives of Lander-1 was a similar command signal sent to Viking-2.

The Viking Lander (Fig. 4-1) has been described as the most sophisticated and complex automated spacecraft ever built. There are many reasons for this, among which the following are some of the more important:

(1) Since the landing from orbit took place within a 5- to 10-min period, a time short compared to the radio link travel time to earth (up to 20 min), the landings had to be fully automatic.

(2) The atmosphere of Mars was not completely known (pressure; argon content; winds) and there was an uncertainty in the height of the terrain at the landing site, which required an adaptive landing flight.

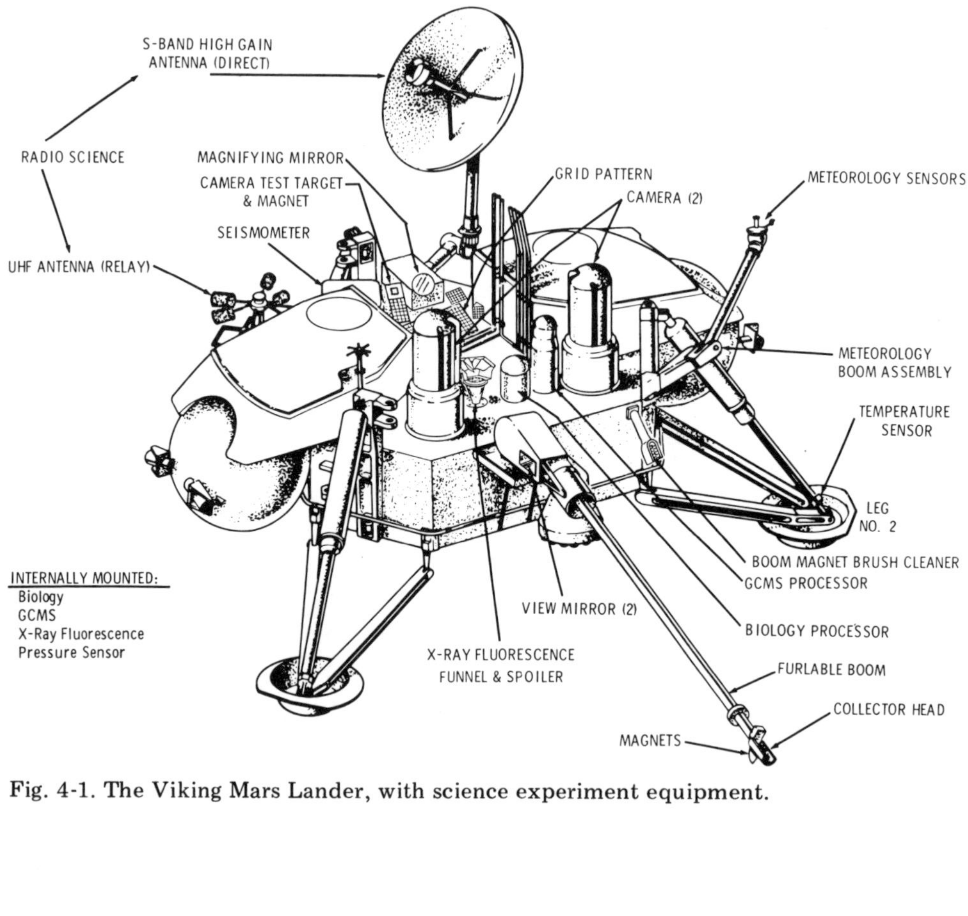

Fig. 4-1. The Viking Mars Lander, with science experiment equipment.

(3) The science complement included 13 separate experiments, only one of which has been flown previously; two experiments (biology and organic chemistry) most likely surpass in sophistication and technical complexity all past space science hardware.

(4) In accordance with an international treaty limiting the contamination of Mars with viable organisms from earth, both lander spacecraft were sterilized by heating in a dry nitrogen atmosphere at 112°C for 40 h prior to launch.

A central computer inside the Lander controlled all aspects of the landing descent as well as all operations of equipment after the landing was completed. To initiate descent, separation nuts were fired, retrorockets were powered on, and pulsed-burns were made to maintain correct orientation for entry. An ablative aeroshell protected the lander against the heat of entry. An inertial reference unit, containing gyros and accelerometers and two separate radar systems provided the computer with the information required to control the descent profile. Final deceleration was provided in two increments — first by a parachute opened at 6.4 km and jettisoned at 1.2 km, and finally by three terminal propulsion retro-rockets. The rockets utilized unique 18-nozzle engines fueled by ultrapure hydrazine, which provided for minimal alteration of the physical and chemical properties of the landing site.

Once landed, Viking began automatic execution of a sequence of engineering functions and scientific experiments. The sequence is modified as desired by updating the program stored in the central computer; this is accomplished by transmitting from earth to the on-board S-band communications system. Scientists are able to interact in an adaptive manner with their experiments, changing sequences and instrument operating parameters. Scientific experiments conducted on the landers include the search for life forms (biology investigation), multispectral imagery of the landscape by dual facsimile cameras, collection of meteorological data (temperature, wind velocity, and pressure), seismic monitoring for Marsquakes, analysis of the soil for magnetic components and physical properties, assay for organic compounds by combined gas chromatography/mass spectrometry, and lastly, analysis of soil and pebble-sized fragments by X-ray fluorescence spectrometry using the instrument described below. For additional information concerning Viking, its mission and scientific results, the reader is referred to the following issues of *Science*: Vol. 193, No. 4255; Vol. 194, No. 4260; and Vol. 194, No. 4271 (1976).

DESIGNING AN INSTRUMENT FOR SPACE

With the extreme constraints imposed upon the instrument designer, it is not surprising that virtually none of the spaceborne geochemical experiments are simply derivatives of already existing instruments. The first and foremost objectives in design of an instrument for space are miniaturization, ruggedness, and reliability. Obviously, these three objectives cannot necessarily be optimized independently, and tradeoffs must be made. For the designer who is also the scientist, as usually is the case, a greater conflict arises: how to obtain the greatest quantity and quality of scientific information, yet satisfy the applicable engineering requirements. This, indeed, is the justification and the necessity for deep involvement by the scientist during all phases of design.

Planetary missions provide even more constraints than usual, for the equipment must face not only the harsh environments of deep space but the entry and landed environments as well. Table 4-I summarizes the environments encountered in a Mars mission and lists some of the characteristics of each environment which must be considered. The specifications matrix resulting from crossing the right- and left-hand columns of Table 4-I is, of course, quite formidable.

In designing the X-ray fluorescence experiment for Mars (Clark and Baird, 1973a) we were confronted with a design requirement unique to Viking, viz., that all components be heat sterilizable. This rendered impractical our first choice for the X-ray sensor, the high-resolution Si solid-state detector, because of its susceptibility to degradation by any contaminants released from its housing during heating and because of the difficulty in providing the requisite cryogenic temperature for low-noise operation. The next best sensor, in terms of resolution of fluorescent X-rays from low-atomic-number atoms is the gas-filled proportional counter. When properly designed and constructed the performance of this counter is unaffected by heat sterilization.

Reliability considerations also greatly influenced the design. Originally, the concept for sample handling was placing the soil in a shallow cup, smoothing with a knife edge, and then moving the cup into position for analysis. It was also planned to keep the soil inlet covered with a solenoid-

TABLE 4-I

Environments for a Mars mission

Types of environments	Environment components
Transportation, handling	Shock, vibration, acceleration
Prelaunch	Vacuum/pressure
Launch and boost	Temperature (including heat sterilization)
Interplanetary cruise	Humidity
Mars orbit	Electromagnetic interface
Deorbit and entry	Acoustics
Landing	Atmospheric composition, wind, dust
Martian surface	Solar flare and cosmic radiation

actuated lid to minimize convection-current heat losses. Both of these ideas were rejected, however, on the basis that a failure in either mechanism could ruin the entire experiment. The alternatives selected were to (1) provide a fixed sample-holding cavity with thin plastic windows on two of its vertical walls, and a trap door at the bottom (see "Instrument description" below), and (2) eliminate the inlet-lid and tolerate the relatively minor additional temperature excursions.

Since our experiment was added to Viking more than a year after the other experiments were selected and the spacecraft detailed design was well underway, the instrument had to be designed to fit within an available "gap" between the life-detection experiment and data processor unit, resulting in a packaging arrangement far from that which would be optimum for ease of assembly and maintainability.

Electronic design was governed by several factors, including the selection of components from only a narrow range of approved devices with proven reliability; a prescribed all-digital interface with the data system; and requirement to minimize power drain by power switching certain portions of circuitry. This latter requirement is of critical importance since the entire power source for the Viking lander is only 70 watts, i.e., one light-bulb equivalent of power to operate the computer, mechanisms, communications systems, and all science experiments on-board the spacecraft. The power sources are two thermoelectric converter units with the heat provided by SNAP-19 units containing plutonium oxide fuel.

In the preceding discussion, we have emphasized only a few of the many technical complexities that confronted the instrument design team. A more complete list is provided in Table 4-II. Before describing the instrument as it

TABLE 4-II

Design criteria for flight instruments

Scientific performance versus	
Weight and volume available	Age-sensitive elastomeric compounds
Type and amount of power available	Malfunction isolation and test points
Sterilization heat soak	Quality assurance factors
Radiation safety	Cleanliness control
Reliability	RTG radiation environment
Failure detection	Operating and useful life
Interchangeability	Emissions, outgassing, and vapor pressure
Attachment points	Workmanship implementation
Electrical connectors	Mandatory parts, materials, and processes
Operating safety	Identification, standardization and traceability
Dissimilar metals	Experiment's radioisotope environment
Stress corrosion	Electrical ground system
Electromagnetic compatibility	Packaging against amplification
Thermal dissipation	

was finally developed, the context of this experiment in the history of exploration of planetary surfaces will be briefly reviewed.

PREVIOUS NUCLEAR GEOCHEMICAL MEASUREMENTS OF PLANETARY SURFACE

The first proposal for a spaceborne nuclear technique for geochemical exploration is that by Arnold (1958) a scant two years after the opening of the space age by the first successful launch of an artificial satellite, Sputnik I. Arnold proposed passive gamma-ray spectroscopy of the lunar surface to detect emissions of natural radioactivity in the soil, including ^{40}K, Th, and U, and radioactive nuclides of other elements produced by nuclear reactions with the incident cosmic ray flux. In 1962, three spacecraft, Ranger 3, 4, and 5, carried 32-channel gamma-ray spectrometers to perform this proposed experiment. Each spectrometer consisted of a 7-cm × 7-cm CsI scintillator crystal surrounded by a plastic scintillator, both viewed by the same photomultiplier, whose output pulses were analyzed by the "phoswich" pulse-rise-time discrimination technique to reject cosmic ray events originating in both scintillators and to accept gamma-ray events in the CsI crystal. Although the Ranger craft did not pass sufficiently close to the moon to allow detection of lunar gamma rays, an important result was the experimental demonstration of the high background of secondary gamma rays produced in the spacecraft structure by cosmic rays, and that its interference was reduced to a negligible value by deployment of the spectrometer away from the spacecraft by a 2-m boom (Arnold et al., 1962).

In 1966, gamma-ray spectrometers were flown on the Soviet lunar orbiters, Luna 10 and 11. These spectrometers were also of the phoswich type, like the Ranger instruments, but the central detector was a 3-cm × 4-cm NaI(Tl) crystal, and the experiment was mounted inside the spacecraft rather than on a boom (Adler and Trombka, 1970). Although the emissions were not correlated to levels of specific elements, Vinogradov et al. (1967) concluded that the natural radioactivity of the lunar maria corresponds to that of terrestrial basalts.

The landmark experiment in remote geochemical analysis was performed by the alpha-particle instrument mounted on the Surveyor 5 spacecraft which softlanded on the moon in September, 1967. This experiment, as well as the basic method of chemical analysis by spectrometry of Rutherford scattering of alpha particles, was developed by Turkevich (1961). In this technique, the sample is irradiated at normal incidence by a tightly collimated beam of 6.11-MeV alpha particles from six ^{242}Cm sources. Alpha particles scattered directly backwards (180° scattering angle) suffer an energy loss which is greater for scattering by light nuclei than heavy nuclei. Each element produces a characteristic energy loss, so that for a thick sample of a pure element the energy spectrum of backscattered alphas is virtually constant up to the critical energy, then decreases to zero. A sample containing many elements produces a staircase spectrum, with each step indicative of the presence and relative concentration of some element. In addition, some elements, such as N, F, Na, Mg, Al, and Si undergo nuclear reactions with alpha particles to produce protons having distinctly characteristic spectra which provides additional identification. Silicon solid-state detectors, both Li-drifted and surface barrier types, were used to measure the resulting alpha and proton spectra. Elements detected during the Surveyor missions were O, Na, Mg, Al, Si, (K + Ca), Ti, and Fe (Franzgrote et al., 1970). The outstanding achievement of the alpha-backscatter experiment in providing the first chemical analyses of the lunar surface was capped by its success in showing that the composition was very similar at sites on two lunar maria (Surveyors 5 and 6), but distinctly different from that at a highlands site (Surveyor 7). Further, analyses of returned samples (Apollo and Luna programs) verified the chemical conclusions reached from the alpha-backscatter experiment.

In late 1970, the Soviet automated moon rover Lunokhod-1 carried "RIFMA", an X-ray fluorescence spectrometer experiment mounted 30 cm above ground level, and accumulated spectral data from the local terrain both while the rover was stationary and during traverses. Kocharov et al. (1971) report that the experiment employed two tritium-containing X-ray sources, several side-window proportional counters, and selective filters. A 64-level pulse-height digitizer provided the spectral analysis. Several hundred spectra were obtained during the course of rover operations. Concentrations for Mg, Al, Si, Ca, and Fe have been given, and upper limits have been placed on K and Ti.

In September, 1971 and April, 1972, a special integrated geochemical experiment (Metzger et al., 1973) was flown as part of the Apollo 15 and 16 astronaut missions to the moon. This package remained with the orbiting command service module of Apollo and thus was able to view the lunar surface from an altitude of 110 km. One instrument in the experiment was an improved version of previous gamma ray spectrometers, this time including a 7-cm × 7-cm NaI crystal and an optically isolated plastic scintillator equipped with its own photomultiplier tube. The plastic was operated in anticoincidence with the crystal, rather than in the phoswich mode, for rejection of cosmic ray-charged particles. An extended boom placed the spectrometer 7.6 m from the spacecraft. High-resolution 512-channel spectra were transmitted in real time directly to earth. Collimation was impractical, so the mapping distance was defined by the spacecraft altitude, with one-half of the detected gamma rays originating within a 240-km diameter circle on the lunar surface. The experiment observed gamma-ray lines from the naturally radioactive elements (K, U, Th) and from inelastic scattering events of neutrons (cosmic ray secondaries) with nuclei of constituents of the lunar regolith.

Comparing the spectra obtained over areas for which "ground truth" is available from Apollo and Luna sampling missions, it has been possible for the experimenters to obtain empirical calibration factors and thus deduce concentrations of K, (Th + U), Fe, Ti, Si, and O over the lunar surface (Trombka et al., 1975). Data analysis still in progress should yield maps of U also, and possibly some information concerning Ca and Al distribution.

A second experiment in the Apollo geochemical package was an X-ray fluorescence analyzer experiment based upon the very imaginative concept of searching for the fluorescent X-rays emitted from the lunar surface as a result of excitation by natural X-ray emissions of the sun. Instrumentation included three large-area, collimated, beryllium-window proportional counters, two of which had selective filters (Mg and Al foil, respectively). A fourth detector viewed away from the lunar surface toward the sun to monitor the sometimes-variable solar X-ray flux. An 8-channel pulse-height analyzer produced spectra every 8 seconds. When integrated over a 60-second time period to improve statistics, these covered a ground area approximately 110 km by 150 km. (Adler et al., 1972.) Although an earlier attempt at a similar measurement on Luna 12 obtained little data on composition, Adler et al. (1973) have now mapped the Mg/Si and Al/Si ratios over nearly 25% of the lunar surface. Elements of atomic number higher than Si are not efficiently excited because of the relative paucity of solar X-rays at higher energies.

The only successful measurement to date pertaining to the chemical composition of a planetary surface other than the moon is that of Vinogradov et al. (1973), from the Venera 8 and 9 spacecraft which landed on the surface of Venus in July 1972 and October 1975, respectively. The instrument, consisting of a 6.3-cm × 6.3-cm NaI(Tl) crystal with a 60-channel pulse-height analyzer, was mounted within the temperature-controlled interior of the spacecraft. On Venera 8, the experiment accumulated two spectra over a total period of 42 min before the spacecraft succumbed to the extremely hostile (500°C, 100 atmospheres pressure) environment. Relatively high levels of K, U, and Th were detected quite similar to average granites on earth. The K/U ratio was also close to values typical of terrestrial crustal materials. Venera 9 reportedly measured lower values of K, U, and Th at its landing site.

INSTRUMENT DESCRIPTION

The Viking X-ray fluorescence spectrometer (XRFS) is constructed in two parts, which are assembled into the final configuration during installation in the spacecraft. The first part is the funnel-inlet unit, consisting of a conical funnel, wind spoiler, and delivery tube. For sample delivery, the Viking surface sampler arm is positioned with its scoop nestled down into the XRFS wind spoiler, which provides a mechanical barrier to minimize sweeping away of fines by wind. Normal delivery is by vibratory sieving through 2-mm circular openings in the bottom of the sampler scoop, into the XRFS funnel, and thence by gravity feed through a 2.5-cm diameter tube into the sample analysis cavity. An alternate delivery mode is to first sieve out the fines with the scoop positioned over the local terrain and then deliver the residual fragments by inverting the scoop while over the XRFS funnel. A screen with 1.2-cm square openings is mounted over the funnel to prevent jamming of the delivery tube by large sample pieces.

The second part of the XRFS assembly is the main unit, a rectangular box 24 cm × 15 cm × 7.6 cm and weighing 2.1 kg, which contains the X-ray sources, detectors, sample chamber, and all necessary electronics. The two parts are mounted to a common equipment-mounting plate, one from above and the other from below (Fig. 4-2) with the circular delivery tube positioned directly over the sample holding chamber.

A typical sample quantity delivered by the scoop is 35 cm^3. In Fig. 4-3, the sample chamber is filled with a sample of this amount, which is approximately 40% more than that required to fill the chamber to a level above the ports through which the X-ray analysis is made. These ports (only one is depicted in Fig. 4-3; the second is perpendicular to the first) are 2.2 cm in diameter. The port windows are very thin plastic films reinforced by metal grids (Toulmin et al., 1973) which have high transmission for the excitation and fluorescent X-rays. Radioisotope sources provide the stimulus X-rays.

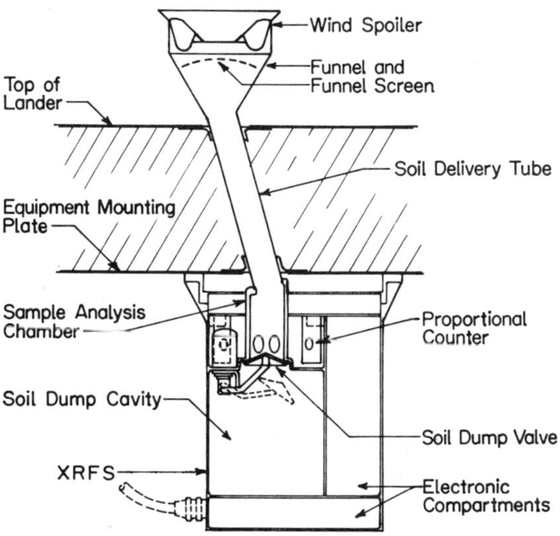

Fig. 4-2. The X-ray fluorescence spectrometer (XRFS); cutaway view from the side. The hatched area is the location of insulation material which protects the inside of the lander from the extreme cold (as low as −120°C) during the Martian night.

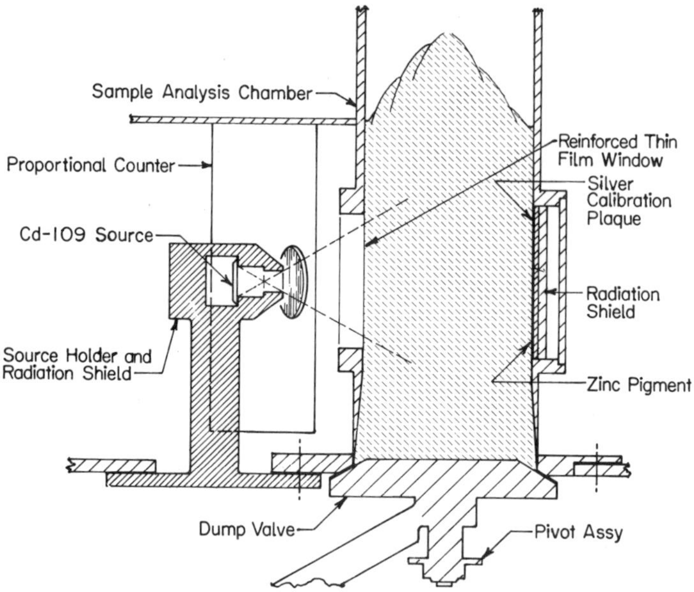

Fig. 4-3. Sample analysis chamber, showing one of the four proportional counter detectors and one of the two radioisotope X-ray sources.

Two nuclides are employed: ^{55}Fe, emitting 5.9-keV X-rays (MnK_α) and ^{109}Cd, emitting 22.2-keV X-rays (AgK_α). Half-lives of the sources are 2.60 and 1.24 years, respectively. Accelerator grade material with high specific activity and virtually no contamination by extraneous radionuclides is used for both sources. When the experiment was landed on Mars, source strengths were 270 mCi of ^{55}Fe and 55 mCi of ^{109}Cd. The ^{109}Cd source holder, shown in Fig. 4-3, is a composite of tungsten and silver layers designed for strong absorption of the 87.7-keV gamma-ray from ^{109}Cd, and for collimation of the source such that the X-rays are allowed to impinge only upon the sample window. Less-tight collimation was employed for the ^{55}Fe source in order to provide the desired level of count-rate and consequently there is some impingement of source X-rays on the sample holder. To prevent the production of extraneous Al fluorescence signals from the holder, a thin electro-coating of Ni was applied. When Ni is excited by an ^{55}Fe source, only the L lines are produced and their energies are too low for detection.

Four different proportional counter (PC) tubes are included, each optimized for a specific region of the fluorescent X-ray energy spectrum by selection of window material, gas filling, and inner liner. A description of the selected properties has been given elsewhere (Toulmin et al., 1973).

Analysis proceeds by taking pulse-height spectra from the four PC tubes. Insufficient weight and volume were available for a multi-channel analyzer, so a stepping single-channel analyzer circuit is used to scan the fluorescent

X-ray peaks. The control logic has been designed so that the scan may be initiated at any of 16 equally spaced channels within the 128-channel spectrum, with dwell times adjustable from 8 to 492 seconds per channel. A programmable high-voltage bias supply also permits the setting of system gain in increments of 5% over a dynamic range of 10,000 to 1 by command from earth.

As with all analyses by the technique of X-ray fluorescence, the observed signals for the light elements originate from only a very shallow surface layer because of strong absorption within the sample. The "half-thickness" value for an X-ray is the 50% transmission point for that X-ray energy in the material specified (Table 4-III). Alternatively stated, 50% of the X-rays will have originated in the region between the surface and a depth equal to the half-thickness value, when measured along the path of the X-ray (45° to the surface normal for our instrument). From the simple exponential absorption law, it follows that 90% of the signal originates between the surface and 3.3 times the half-value depth, and 99% from surface to 6.7 times the half-value depth. The values of Table 4-III make it clear that for soils devoid of a clay-sized fraction, the fluorescent X-rays for the light elements will originate almost exclusively from the single-grain layer adjacent to the sample chamber window.

Upon analyzing the spectra obtained on a given sample, and deciding it is sufficient and complete, a signal is sent to command a trap door to open in the bottom of the sample chamber. A cycling solenoid will open/close the trap door 4 times per second for a period of one minute to dump the sample into a "dump cavity" within the XRFS (Fig. 4-2). This cavity can contain 688 cm^3 of sample material, or from 5 to 20 samples, depending on average sample size.

TABLE 4-III

Half-thickness values for key elements in typical matrices

Element	X-ray energy (keV)	Half-value thickness (μm)				
		quartz	rhyolite	basalt	ultramafic	calcite
Mg	1.25	1.5	1.2	1.0	1.0	1.0
Al	1.49	2.4	1.9	1.5	1.2	1.7
Si	1.74	3.6	2.2	1.8	1.7	2.5
K	3.31	5.9	5.5	6.0	6.1	15.2
Ca	3.69	8.0	6.6	7.7	8.1	20.6
Ti	4.51	14.1	11.3	10.6	10.7	7.0
^{55}Fe	5.90	30.5	24.0	21.8	21.8	14.5
Fe	6.40	38.9	30.5	27.6	27.7	18.3
Sr	14.16	423.7	284.4	175.0	166.2	178.6
^{109}Cd	22.16	1415.2	930.2	555.8	526.5	569.6

METHODS OF DATA INTERPRETATION

The basic output of the XRFS instrument is spectra with numerous peaks corresponding to the presence of fluorescent X-rays from particular elements. In Fig. 4-4, the spectra of 7 pure elements as recorded in PC-2 are superimposed upon a spectrum of the backscatter peak produced by the ^{55}Fe source. As will be our convention for this paper, this plot has the ordinate (counts per channel) plotted logarithmically so that for those spectra including trace elements, both major and minor peaks will be revealed. Elements of adjacent atomic number overlap quite strongly due to the limited resolution attainable with the PC tube. Careful analysis using computer techniques is required to determine a small concentration of one element in the presence of large concentrations of its nearest-neighbor or next nearest-neighbor element.

Some spectra of geological materials for PC-2 are shown in Fig. 4-5. Each spectrum has been multiplied by an appropriate factor so that the same number of counts are in the backscatter peak. Simple inspection of the curves in this figure allows the ranking of each specimen for relative Si, S, K, and Ca content. The situation for Ti is often less clear and requires more quantitative analysis. That different rock types produce significantly different spectra is evidenced by the collection of PC-2 spectra presented in Fig. 4-6, a group which includes a sequence of igneous rocks as well as quartz and a limestone (highest Ca peak).

Not only are the major elements apparent in these spectra, but certain

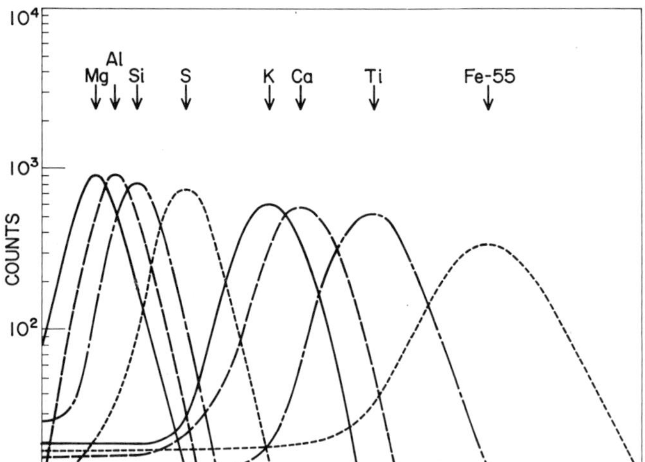

Fig. 4-4. Spectra of pure elements and a backscatter spectrum (^{55}Fe), as recorded in PC-2.

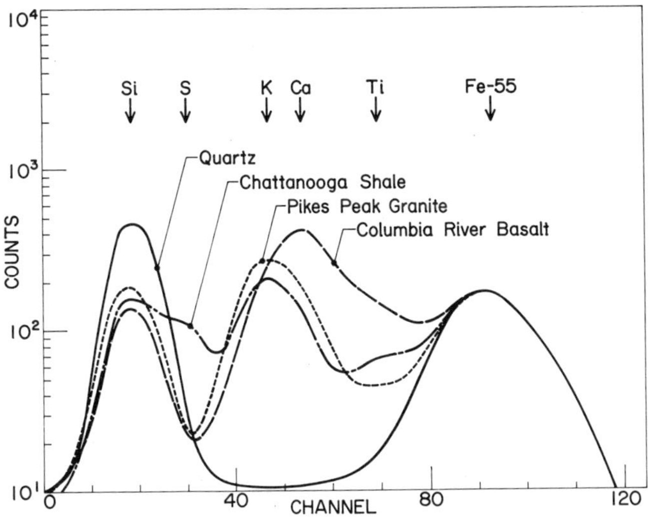

Fig. 4-5. PC-2 spectra for four distinctly different rock types. Quartz produces only two peaks, i.e., silicon and backscatter. Note the indicated levels for S, K, Ca, and Ti in the other rocks.

trace elements can be identified as well. Fig. 4-7 shows spectra from three specimens taken with the PC-3 tube and the ^{109}Cd source. Besides the ubiquitous Fe peak, peaks are clearly seen for the elements Rb, Sr, and Zr, which are present at levels of 0.1—0.3% in their respective sample. Such concentrations for these three elements are rather atypical, but a wide variety of

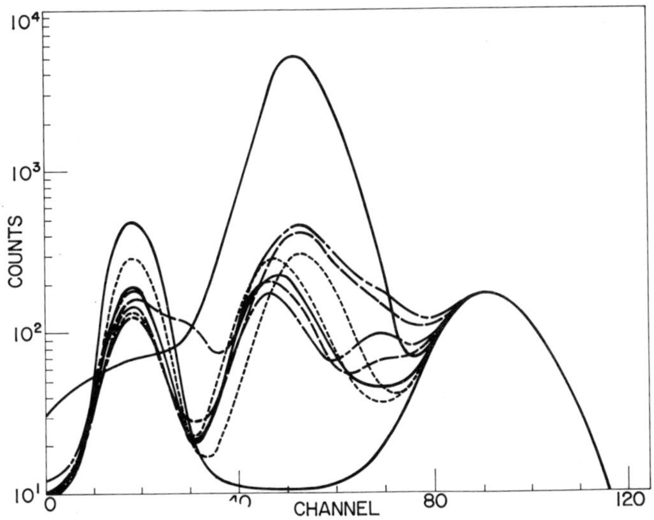

Fig. 4-6. Spectra for nine different rocks are overlaid in this graph to indicate the relative uniqueness of rock signatures.

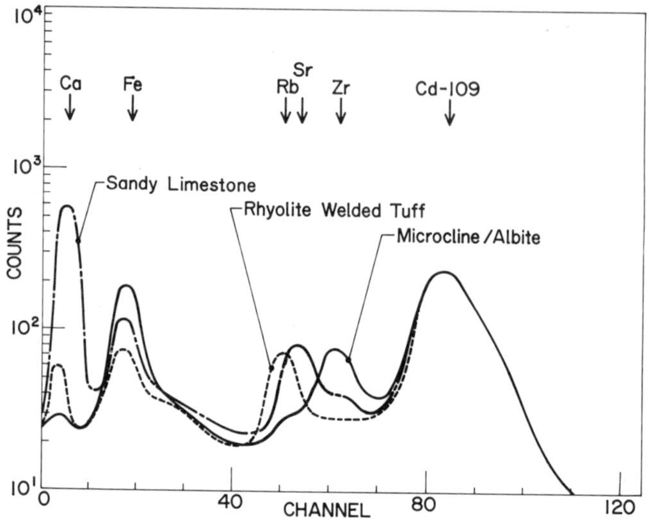

Fig. 4-7. These spectra, taken with PC-3, illustrate detection of the important trace elements, Rb, Sr, and Zr.

terrestrial rocks, and lunar samples as well, contain levels of trace elements which are readily detected. Some examples are shown in Fig. 4-8.

The foregoing discussion of spectra was intended to demonstrate the discriminatory power of the instruments. It also serves to indicate the ultimate approach to interpretation of the data received from Mars: the method of

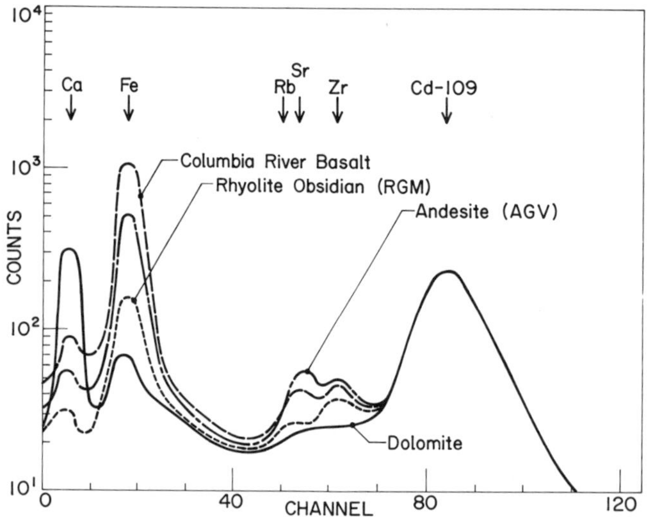

Fig. 4-8. PC-3 spectra of specimens with a range of Fe content and typical trace element concentrations.

analogues, or what is known as "signature analysis". The XRFS flight instruments have each been tested with 25 rock types as well as standards of pure elements. These spectra are stored in a computer library and compared to the spectra telemetered from the Martian surface. A series of laboratory studies using other rock standards, soils, and contrived mixtures will later be undertaken with the two remaining flight instruments. These studies will continue until an analogue is found or synthesized which matches, within statistical limits, the spectra data obtained.

Other methods of analysis were needed to supplement the signature analysis technique, particularly during the mission when decisions are required on a timeline not consistent with a systematic laboratory study. For this, a mathematical model was developed (Clark, 1974) which includes theoretical corrections for matrix absorption and enhancement effects, theoretical calculations of backscatter intensity, PC-tube response functions, and empirical correction factors obtained from laboratory tests. With this model, spectra can be calculated for rocks of any specified composition.

An algorithm employing an iterative series approach to spectrum matching (Fig. 4-9) has been programmed and successfully used to calculate element concentrations in knowns and in blind samples (Clark and Baird, 1973a, b). Although studies are still in progress, it is clear that the elements in the periodic table can be classified into eight distinct groups for which the XRFS has different analytical capabilities. A discussion of these groups follows.

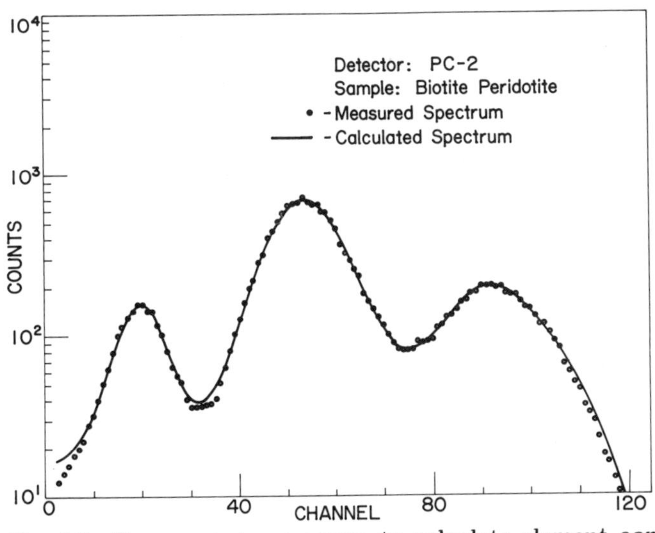

Fig. 4-9. The computer program to calculate element concentrations matches the measured spectrum with a theoretical spectrum derived from the deduced composition. This figure indicates the accuracy of match that can be achieved.

Elements 1—10 (H to Ne). The first 10 elements of the periodic table cannot be identified directly by our technique because the fluorescent X-rays they produce are of such low energy that they do not effectively penetrate the PC-tube entrance windows or soil chamber windows. However, the presence of elements within this group can be detected by the intensity of the back-scattered X-ray signal since they strongly influence the bulk absorption and scattering properties of the sample material.

Elements 11—13 (Na to Al). The elements Na, Mg, and Al emit very soft X-rays (1.04—1.49 keV) which penetrate the 5-μm aluminum window of PC-1. However, the fluorescent yields for these elements (1—3.5%) are low, and some interference occurs due to the generation of extraneous Al signal by fluorescence of the window itself by higher-energy X-rays. Our present evaluation of minimum detection limits:
Na — not detectable in igneous rocks
Mg — 2.0% by weight in ultra-basic matrix
Al — 1.0% by weight in basalt.

Elements 14—23 (Si to V). Included in this group are the important rock-forming elements Si, P, S, K, Ca, and Ti. Element 23 (V) is the highest element in the periodic table which can be efficiently excited by the ^{55}Fe source, and all elements in this group are detected in PC-2 (see Figs. 4-4 to 4-6).

The minimum detection limit for Si is of no concern for many geological samples, but is of considerable importance if the specimen is say, limonite or limestone. For these cases, the Si detection limit is slightly less than 1.0% by weight. Accuracy of Si determination when at levels of 20—30% is of the order of ±1.5% by weight at a confidence level of 75%. Analysis for P is poor in most cases because of interference by high levels of Si. The same is partially true for S, although the sulfur in certain chondrites has been detected readily, and the 4.6% S in Chattanooga Shale is obvious from the spectrum (Fig. 4-5). Sensitivity for the elements K and Ca is quite good, but they do interfere with one another. In a K-free dunite matrix, the minimum detection limit for Ca has been established as 0.02%, and for a Ca-free quartz sand containing minor K-feldspar, the detectable limit for K is also 0.02%. On the other hand, in a high-Ca basalt, the limit for K becomes 0.15%. The diagnostic geochemical ratio of K/Ca has previously been demonstrated to be measurable with useful accuracy over at least the range 0.1—10.0 (Clark and Baird, 1973a). For the element Ti, no interference occurs from the presence of large quantities of Fe because the ^{55}Fe source does not excite the latter. The ^{55}Fe backscatter peak and Ca peak interfere with Ti to some extent, but all analyses for Ti to date have been well within the accuracy desired for petrologic interpretation. The L lines from Ca do, however, overlap the Ti emission.

Elements 24—30 (Cr to Zn). The elements with atomic number 24 through 30 include Fe (26), and other geochemically significant elements such as Mn, Co, Ni, Cu, and Zn. Sensitivity for Fe is quite satisfactory (0.03%) and limited chiefly by fluorescence from the stainless-steel edges of the PC-tube entrance window. Determination of the levels of other elements in this group is usually difficult or impossible with the XRFS because of the interference by the relatively higher concentrations of Fe itself. In a granitic matrix containing 2% Fe, the detection limits for Ni, Cu, and Zn are roughly 0.8, 0.4, and 0.2% respectively.

Elements 31—46 (Ga to Pd). Excellent sensitivity for As, Rb, Sr, Zr, Mo, and other elements in this group has been achieved. Under typical conditions, 30—60 ppm of these elements can be detected. The trio Ru, Rh, Pd at the upper end can be detected less easily because of overlap with the ^{109}Cd backscatter peak.

Elements 47—57 (Ag to La). Elements such as Ag, Cd, Sn, and Ba have K-shell absorption thresholds beyond the 22.2-keV ^{109}Cd X-ray and thus can be fluoresced only in the L-shell. This is accomplished efficiently by ^{55}Fe, but the resultant L-lines (2.98—4.47 keV) are generally obscured by the K-lines from much higher concentrations of K, Ca, and Ti.

Elements 58—75 (Ce to Re). Consisting of the rare earths, and Ta, W, and Re, this group of elements also can only be fluoresced in the L-shell by the XRFS sources. In this case, their emissions (4.7—8.7 keV) are obscured by the omnipresent Fe signal.

Elements 76—100 (Os to Fm). The L-lines from this group, which includes Au, Pb, Th, and U fall in the high-sensitivity energy range 8.9—16.4 keV. Characteristic and diagnostic double peaks (L_α and L_β) are produced by each of these elements. In specimens free of interfering concentrations of Rb, Sr, etc., minimum detection limits of 50 and 80 ppm are indicated for Pb and U, respectively.

The above discussion serves to illustrate the capabilities and limitations of the Viking XRFS. Confident identification of elements through peak analysis rests, of course, upon accurate knowledge of the pulse-height spectrum energy scale. To determine this scale while the XRFS is on Mars, a number of calibration standards, including samples of the elements Al, Ca, and Zn are built into the instrument.

MARS RESULTS

The Viking-1 Lander successfully touched down on the surface of Mars on July 20, 1976 in the region Chryse Planitia. Lander-2 was deployed to the surface some 44 days later at Utopia Planitia, a more northerly site on the opposite side of the planet. With the sole exception of the seismic monitoring instrument on Viking-1, all scientific instruments have operated at or above expected performance levels and completed the primary mission objectives in what has been hailed as the most successful planetary exploration mission to date. Interim results for the XRFS experiment have been published elsewhere (Clark et al., 1976a, b; Toulmin et al., 1976; Baird et al., 1976) and can be summarized as follows:

(1) From a geochemical standpoint, the fines (<2-mm size fraction) are remarkably similar at both lander sites, in spite of a distance of 6500 km between the two landing sites (a separation of 178° in longitude and 25° in latitude).

(2) The fines are predominately composed of one or more silicate minerals (bulk SiO_2 = 45 ± 5% by weight with a high content of iron (Fe_2O_3 equivalent = 19 ± 3%).

(3) Sulfur content is unexpectedly high, approximately 100 times higher than the average for the earth's crust. A duricrust (hardpan) observed by the lander cameras contains higher sulfur than the loose fines, and may be cemented by a magnesium sulfate salt comprising approximately 10% of the total sample.

(4) Potassium is less than 0.25%, at least five times lower than the average for the earth's crust. If this result is also representative of the Mars average crustal potassium content, then the low content of ^{40}Ar observed in the Martian atmosphere may reflect the dearth of source material (^{40}K) rather than a lower degree of planetary outgassing on Mars as compared to the earth.

(5) If an iron oxide coating exists over all mineral grains (as proposed by some), such a coating must be extremely thin (<0.25 μm), or highly discontinuous.

(6) The bulk density of fines after delivery to the XRFS is 1.10 ± 0.15 g cm^{-3}, implying a bulk soil porosity of 60 ± 15%.

ACKNOWLEDGEMENTS

Translating this experiment concept into an engineering reality has largely depended upon superb management by two individuals: A.J. Castro, who directed the detailed design and fabrication of the flight instruments, and W.C. Kelliher, who directed its implementation on the Viking spacecraft. T.L. Steinborn made the spectral measurements of Figs. 4-4 to 4-9.

REFERENCES

Adler, I. and Trombka, J., 1970. *Geochemical Exploration of the Moon and Planets.* Springer-Verlag, New York, N.Y., p. 53.

Adler, I., Trombka, J., Gerard, J., Lowman, P., Schmadebeck, R., Blodget, H., Eller, E., Yin, L., Lamothe, R., Osswald, G., Gorenstein, P., Bjorkholm, P., Gursky, H. and Harris, B., 1972. Apollo 16 geochemical X-ray fluorescence experiment: preliminary report. *Science*, 177: 256—259.

Adler, I., Trombka, J., Lowman, P., Schmadebeck, R., Blodget, H., Eller, E., Yin, L., Lamothe, R., Osswald, G., Gerard, J., Gorenstein, P., Bjorkholm, P., Gursky, H., Harris, B. and Arnold, J., 1973. Apollo 15 and 16 results of the integrated geochemical experiment. *The Moon*, 7: 487—504.

Arnold, J.R., 1958. Proceedings Lunar and Planetary Exploration. *Colloquium.*

Arnold, J.R., Metzger, A.E., Anderson, E.C. and Van Dilla, M.A., 1962. Gamma rays in space. *J. Geophys. Res.*, 67: 4878.

Baird, A.K., Toulmin III, P., Clark, B.C., Rose, H.J., Keil, K., Christian, R.P., Gooding, J.L., 1976. Mineralogic and petrologic implications of Viking geochemical results from Mars: interim report. *Science*, 194: 1288—1293.

Clark, B.C., 1974. X-ray cross-sections in design and analysis of non-dispersive systems. In: C.L. Grant, C.S. Barrett, J.B. Newkirk and C.O. Rund (Editors), *Advances in X-ray Analysis.* Plenum Press, New York, N.Y., 17: 258—268.

Clark, B.C. and Baird, A.K., 1973a. Ultraminiature X-ray fluorescence spectrometer for in-situ geochemical analysis on Mars. *Earth Planet. Sci. Lett.*, 19: 359—368.

Clark, B.C. and Baird, A.K., 1973b. Martian regolith X-ray analyzer: test results of geochemical performance. *Geology*, 1: 15—18.

Clark, B.C., Toulmin III, P., Baird, A.K., Keil, K. and Rose, H.J., 1976a. Argon content of the Martian atmosphere at the Viking-1 landing site: analysis by X-ray fluorescence spectroscopy. *Science*, 193: 804—805.

Clark, B.C., Baird, A.K., Rose, H.J., Toulmin III, P., Keil, K., Castro, A.J., Kelliher, W.C., Rowe, C.D. and Evans, P.H., 1976b. Inorganic analyses of Martian surface samples at the Viking landing sites. *Science*, 194: 1283—1288.

Franzgrote, E.J., Patterson, J.H., Turkevich, A.L., Economou, T.E. and Sowinski, K.P., 1970. Chemical Composition of the lunar surface in Sinus Medii. *Science*, 167: 376—379.

Kieffer, H., Neugebauer, G., Munch, G., Chase, S.C. and Miner, E., 1972. Infrared thermal mapping experiment: the Viking Mars Orbiter. *Icarus*, 16: 47—56.

Kocharov, G.Y., Borodulin, N.F., Viktorov, S.V., Voropayev, O.M., Dzevanovskaya, A.Y., Ivznov, O.G., Kiryan, G.V., Petrov, V.V. and Sakulksy, V.A., 1971. "Lunokhod-1" — mobile lunar laboratory. In: A.P. Vinogradov (Editor), *Peredvizhnaya Laboratoriya na Lune Lunokhod-1.* Nauka, Moscow.

Metzger, A.E., Trombka, J.I., Peterson, L.E., Reedy, R.C. and Arnold, J.R., 1973. Lunar surface radioactivity: preliminary results of the Apollo 15 and Apollo 16 gamma-ray spectrometer experiments. *Science*, 179: 800—803.

Toulmin III, P., Baird, A.K., Clark, B.C., Keil, K. and Rose, H.J., 1973. Inorganic chemical investigation by X-ray fluorescence analysis: the Viking Mars Lander. *Icarus*, 20: 153—178.

Toulmin III, P., Clark, B.C., Baird, A.K., Keil, K. and Rose, H.J., 1976. Preliminary results from the Viking X-ray fluorescence experiment: the first sample from Chryse Planitia, Mars. *Science*, 194: 81—84.

Trombka, J.I., Reedy, R.C., Bielefeld, M.J., Metzger, A.E., Arnold, J.R., Dyer, C.S. and Seltzer, S.M., 1975. Analytic methods in determining elemental composition from the Apollo gamma-ray spectrometer data. *Proc. 6th Lunar Sci. Conf., Johnson Space Center, Houston, Texas.*

Turkevich, A., 1961. Chemical analysis of surfaces by use of large-angle scattering of heavy charged particles. *Science*, 134: 672—674.

Vinogradov, A.P., Surkov, Y.A., Chernov, G.M., Kirnozov, F.F., Nazarkina, G.B. and

Vernadskiy, V.I., 1967. Gamma radiation of the Moon and the composition of lunar rocks (based on the measurements taken by the cosmic station Luna-10). *Geochem. Int.*, 4: 1030—1039 (translation from *Geokhimiya*, 11: 1263—1273).
Vinogradov, A.P., Surkov, Y.A. and Kirnozov, F.F., 1973. *Soderzhaniye Urana, "Veneroy 9"*. Academy of Sciences USSR, Moscow (translation published as NASA TT F-14, 911).

Chapter 5

RADIOACTIVATION METHODS

SAM S. NARGOLWALLA, NOI D. NGUYEN and AZIZ-UR-REHMAN

INTRODUCTION

In the continuing search for energy sources such as gas and oil, test holes are drilled often down to many thousands of feet. Probes or logging tools are introduced into these holes to measure properties of the geological strata intersected by them. Recording instruments located on the surface provide a convenient method of measuring the digital or analog responses communicated from the probe sensor via the logging cable. The most common procedure for obtaining drill-hole information pertinent to the composition of the geological matrix involves the extraction of drill cores which are subsequently analysed in the laboratory. The analytical data are used to determine the viability of the orebody in the exploration stage, and grade control in the development stage of mining operations. By and large the core extraction technique constitutes a major portion of the mineral prospecting and exploration budget, since it involves time-consuming operations and the use of sophisticated drilling and extraction equipment. The cost can be considerably magnified if drilling and core sampling is done at considerable depth.

Inasmuch as the drilling of boreholes and extraction of core samples are fundamental techniques for exploration of mineral deposits, the development of new geological prospecting methods, the impact of nuclear techniques on geophysical well logging, advances in electronics, automation and computer reduction of data, and improvements in general mining technology have dictated totally new concepts for the definition of what actually can be considered to be a "profitable deposit". A graphic demonstration of the importance of defining a profitable ore deposit with serious economic overtones is given by Chapman (1970). Taking a very large low-grade copper deposit as a test case, Chapman shows that deleterious financial effects can be experienced, should the copper eventually recovered from the mining operation be 15—20% lower than that predicted from drill-core analysis data. Furthermore, the necessity for accurate grade evaluation together with the distribution of metals in the total ore mass must also be considered. Wide variations in the grade of the feed to a mill introduces complex and costly undertakings. Perhaps the principal cause of this difficult problem lies in the fact that laboratory chemical and instrumental analyses can be confidently related only to the sample analyzed, and do not necessarily represent the "true" in-situ ore distribution, resulting quite often in serious grade control problems and tonnage estimation. Since the analytical information is seldom obtained immediately on site, costly delays occur in the exploration and development programs often leading to the drilling of many unnecessary holes.

Recognizing the disadvantages of existing methods for obtaining compositional information, the mining community has repeatedly expressed a dire need for totally instrumental and in-situ methods for elemental analysis in boreholes. The availability of reliable logging data on a continuous or incremental basis on site would alleviate much of the cost and the drilling of waste drill holes. Such dependable logging techniques would permit the use of cheaper percussion or rotary drilling operations at usually a much lower cost than that for diamond-drilled holes, or other sophisticated drilling methods used to maintain the integrity of the core sample.

In contrast to the currently popular core extraction-chemical analysis process, nuclear methods of analysis have the potential for in-situ elemental determinations. Due to the inherent penetrating characteristics of nuclear radiation, and depending upon the nature of radiation utilized, these techniques can permit a more realistic evaluation of the mineral content of a given deposit.

Attempts to utilize basic nuclear physics in well logging were initiated as early as 1940 when the natural gamma-ray activity in boreholes was measured. A classic dissertation on the use of neutrons as a method for well logging was presented by the famous atomic scientist Bruno Pontecorvo in September of 1941. A considerable impetus was also given to neutron experiments by the discovery of neutron fission of the uranium nucleus. It was not until 1959 that development of nuclear techniques as applied to petroleum logging was appropriately documented by Caldwell and Mills (1959), Youmans and Zimmerman (1959), and Youmans et al. (1959). With the availability of low-cost medium-intensity neutron sources, such as those generated from sealed-tube neutron generators, and isotopic and spontaneous fission sources, considerable interest for the application of nuclear techniques, such as radioactivation, for the characterization of boreholes was expressed. Recently some excellent reviews by Keys and Boulogne (1969), and Czubek (1971) discussing the general applicability of nuclear techniques for in-situ borehole analysis demonstrate radioactivation methods for the identification and/or analysis for such elements as H, Be, B, O, F, Na, Mg, Al, Si, P, Cl, Ca, Ti, Cr, Mn, Fe, Co, Ni, Cu, Ag, Cd, In, W, and Au.

This chapter attempts to summarize only those nuclear geophysical methods that can be classified as radioactivation methods. As such, techniques which provide only lithologic information by the use of nuclear scattering reactions are excluded. On the other hand, if the nuclear scattering principle is employed for elemental determinations, e.g. gamma-gamma scattering analysis for lead and zinc, such techniques are classified among the radioactivation techniques and discussed. A concise description of radioactivation principles and methods as applied to mineral exploration are covered in some detail. Comparisons between various radioactivation methods are also drawn. A section of this chapter is devoted to theoretical concepts utilized for the evaluation of analytical methods for borehole logging. A listing and description of appropriate radiation sources applicable to mineral exploration are presented with a view to optimization of basic logging systems for single or multi-element determinations. To assist the reader in proper planning of field systems, a detailed account of the properties of useful neutron and gamma-ray sources is also given.

Application of radioactivation methods as currently practiced and ongoing research activities form an important part of this chapter. Selected operations and studies are described in some detail. A critical evaluation of field data and case histories is also made. Comparisons between radioactivation methods are made in an effort to present a proper perspective to the state-of-the-art.

Finally, the impact of radioactivation techniques on general mineral exploration operations is discussed. Extensions to these techniques to new fields of venture such as oceanographic applications are presented.

THEORETICAL CONSIDERATIONS

In general, nuclear methods employed for the characterization of geological samples can be classified under two principal groups: (1) techniques utilizing gamma-radiation measurements with or without artificial gamma-ray sources, and (2) techniques in which neutron sources serve as a basis of measurement. A discussion of these techniques in terms of their applicability to mineral exploration follows:

Gamma group

Natural gamma-ray measurement

This process is used for the detection of gamma rays from the decay of natural radioactive elements such as uranium, thorium and potassium. The application of methods based on such emissions is described in Chapter 2.

Gamma-gamma scattering method

These techniques have been extensively used for the determination of in-place density of geological strata, and with some modifications also applied to the determination of elements in certain types of ore deposits. The principle of this method is based on the measurement of scattered gamma radiation from the geologic matrix irradiated with artificial gamma-ray sources. A schematic view of a typical downhole measurement system is

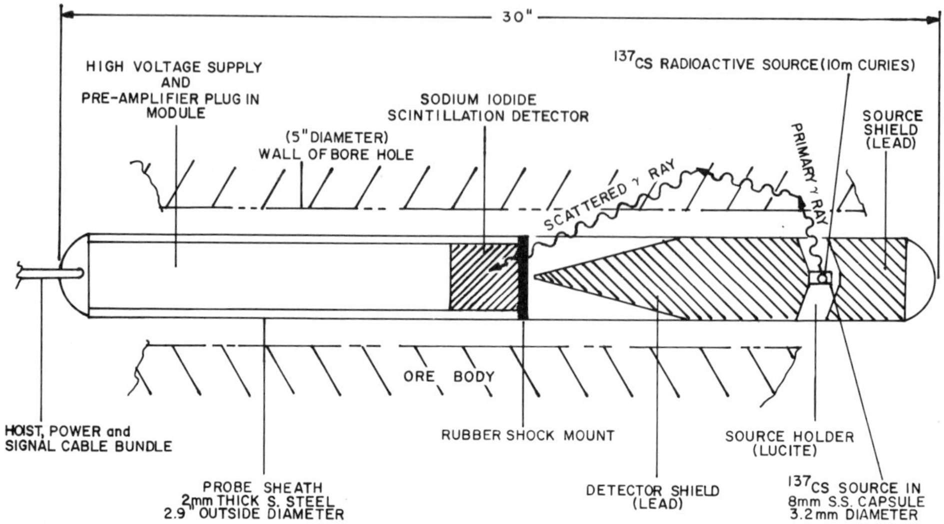

Fig. 5-1. Gamma-gamma backscattering logging system.

shown in Fig. 5-1. In the logging tool itself, the gamma-ray source is separated from a detector by a suitable shield. As a result of photoelectric absorption and Compton scattering events in the matrix, a spectrum of gamma-ray energies is measured by the detector. Two main factors influence the probe (or sonde) response. These are: (1) the in-place density of the scattering medium, and (2) the effective or equivalent atomic number of the scattering matrix. According to Czubek (1971), the equivalent atomic number, Z_{eq}, can be computed with good results for gamma-energy measurements above the K-edge, from the formula:

$$Z_{eq} = \sqrt[3.5]{\frac{\sum_i P_i(Z_i^{4.5}/A_i)}{\sum_i P_i(Z_i/A_i)}} \tag{1}$$

where P_i is the weight fraction of an element of atomic weight, A_i, and atomic number, Z_i. Except for the restrictive case of an ore containing a heavy element where one cannot establish a discrete value for Z_{eq}, for the entire photon energy spectrum, the above formula does provide satisfactory results. A simple graph (Fig. 5-2), abstracted from Czubek's (1971) review, illustrates four spectra for two different bulk densities and for two different Z_{eq}. These spectra result from the Compton scattering and photoelectric absorption of a monoenergetic gamma-ray excitation source. As can be seen, two distinct regions occur. In one energy region ($>$200 keV) the two differ-

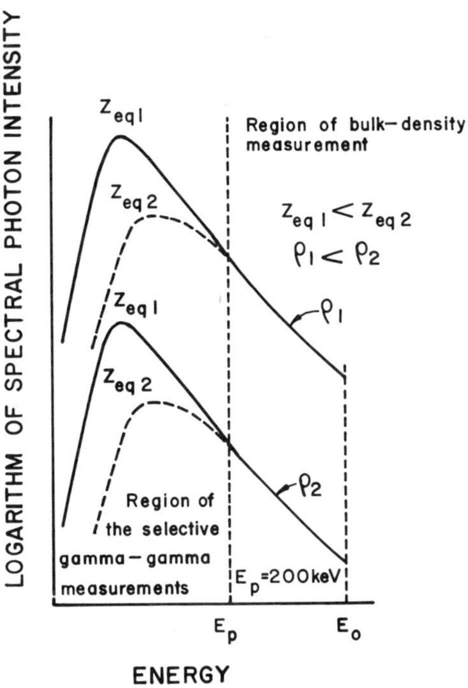

Fig. 5-2. Scattered gamma-ray spectra as a function of the density and chemical composition of the medium (Czubek, 1971).

ent densities can be clearly separated. In this region the gamma-gamma density method is applied. In the energy region below 200 keV the total response is influenced by both the density and chemical composition of the geological matrix. This region is used for the selective gamma-gamma method, usually referred to as the X-ray fluorescence method. Except for a very limited number of applications, density logging cannot be easily applied for the determination of discrete elements, and is mainly employed in oil, gas and coal exploration. On the other hand, selective gamma-gamma logging is applied to elemental determinations. However, its application till now is principally restricted to two-component systems.

In selective gamma-gamma logging, a very low-energy gamma-ray excitation source is used. The isotopes ^{75}Se (E_γ = 0.2645, 0.1359 MeV, $T_{1/2}$ = 120 days), ^{170}Tm (E_γ = 0.08426 MeV, $T_{1/2}$ = 127 days) and ^{203}Hg (E_γ = 0.27917 MeV, $T_{1/2}$ = 46.59 days) are often used. These energies are photoelectrically absorbed in the rock matrix to yield low-energy gamma responses in the detector. Since the photoelectric effect varies rapidly with the atomic number of the absorber, a differentiation can be made. The method is generally applied to the determination of heavy elements such as Hg, Pb, Bi, Sb and Ba in host rock of a reasonably constant density. Because of the very low energy of the measured radiation, the boreholes have to be relatively smooth and without the presence of drilling mud. In reality, it is essentially a surface analytical method.

As mentioned earlier, density logging in some cases can be related to the elemental abundance. Donhoffer (1968), in his concise description of the determination of zinc in a lead-zinc deposit, shows how this effect can be used for zinc determinations. His detailed study is worthy of further discussion. In this study, the principal interaction events measured are gamma energies between 0.3—2 MeV from Compton scattering in the matrix. Since for medium-Z elements between $Z = 10$ and 30, the macroscopic scattering cross-section is directly proportional to the density of the medium and the ratio Z/A is approximately constant, the macroscopic scattering cross-section is therefore a direct function of density. Thus a determination of density permits a direct relationship with the elemental content of the medium. Furthermore, the influence of the photoelectric effect, which varies as the fifth power of Z, is not significant if low ore concentrations are considered. The probe response as a function of source strength, source-detector distance, density and chemical composition of the matrix can be theoretically calculated using the photon transport theory as given by Tittman and Wahl (1965). An approximation to the solution of the differential equation for a cylindrical system is in good agreement with the semi-empirical formula:

$$I = A\rho \exp(-\mu\rho d) \tag{2}$$

as given by Bulashevich et al. (1962), where ρ is the density, μ the mass absorption coefficient of the medium, d the source-detector distance, and A is a proportionality factor which includes the source strength and the geometry. Donhoffer has evaluated this formula to obtain Fig. 5-3 for the entire density range, and for a sonde using a ^{137}Cs irradiation source separated from a detector by 30 cm. For the density range of interest, i.e. between 2.5 and 4.0 g cm^{-3}, an increase in density results in a corresponding decrease in the probe response, thereby providing a means for density determinations. Thus, assuming a two-component system, a simple relation between the in-place density and the concentration of one component can be established. This calibration curve can be used for estimating the element of interest in a similar geologic matrix from gamma-ray backscattered intensity measurements.

Resonance gamma method

Essentially this method is an extension of the Mössbauer effect as applied to the field. In a discussion of the advantages for mineral exploration, Sowerby (1973) has given an exhaustive description of this technique and suggested some possible applications for mineral exploration. The resonance scattering methods employ the use of a radioactive nuclide decaying by

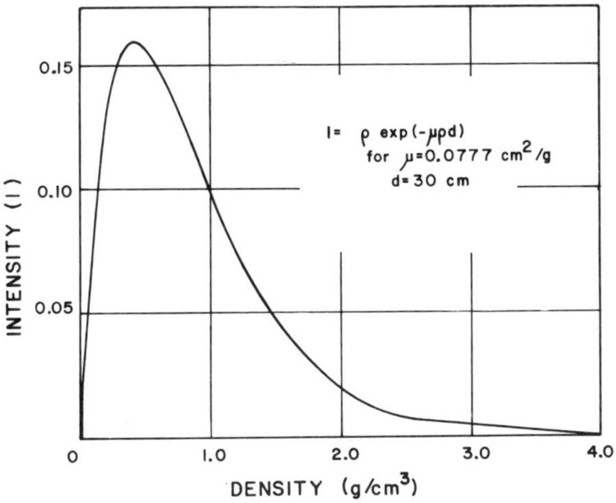

Fig. 5-3. Dependence of the intensity of scattered radiation on the density of surrounding material (Donhoffer, 1968).

gamma emission through a stable isotope of the element of interest. This gamma radiation is scattered from like nuclei in the matrix. The recoil energy loss is compensated by imparting motion to nuclei of the primary gamma source. Several methods for providing this compensating motion have been used. It is done either mechanically, thermally or by utilizing the recoil energy following radioactive decay. The reader is directed to Sowerby's dissertation for additional information regarding this technique.

Application of this technique requires a careful selection of parameters for its success in the field. In the writers' opinion a viable field application of this technique that will compete with existing exploration methods, both in economics and analytical accuracies, appears to be still in the future. However, this method exploits a rather unique approach to elemental analysis and under certain controlled conditions, such as those existing in mineral processing plants, its application for on-stream monitoring of geological materials definitely shows promise.

Photonuclear methods

This method has limited application and is almost exclusively used for the exploration of the element beryllium. If stable high-energy gamma excitation sources were available, additional elements would be amenable to this technique. Berzin et al. (1966), have described some of the basic principles governing this technique. When a low activation threshold element for incident gamma radiation is irradiated by gamma rays of energies higher than the threshold, the (γ, n) reaction takes place. A measurement of the emitted neutrons can then be related to the elemental composition. In general, the photoneutron threshold for most elements is in excess of 6 MeV, an energy not available from the highest gamma-emitting artificial radioisotope. The development of exploratory tools capable of generating gamma energies of this order for downhole applications is most difficult. If economics and other mechanical factors, such as speed and ease of operation, are concerned, this technique for elements other than beryllium becomes overly cumbersome and costly. The element beryllium differs from all other elements in the fact that the threshold energy for interaction with gamma radiation is extremely low, i.e. 1.67 MeV. Therefore, artificial radioisotope sources such as ^{24}Na and ^{124}Sb can be used to induce the (γ, n) reaction with beryllium nuclei. In particular, the isotope ^{124}Sb has been extensively used for this purpose. This isotope emits gamma energies of 1.69 MeV (46%) and 2.09 MeV (7%). The average energy of photoneutrons emitted from this interaction is of the order of 0.024 MeV. Since the activation cross-section for the (γ, n) reaction is about 1.5 mbarns at 1.7 MeV incident gamma energy, an acceptable sensitivity for the determination of this element results. The photoneutrons emitted are counted by ^{10}B-enriched BF_3 counters or scintillation counters. It is estimated that using a source strength of about 20—150 mCi of ^{124}Sb, beryllium in beryllium-bearing strata, can be determined at levels of a few hundredths of a percent.

Neutron group

This group includes techniques which perhaps have been most exploited for geological exploration. As such, the theoretical treatment of the fundamentals involved will be fairly thorough so that the reader may have an adequate basis for assessing the application of one or more of the techniques discussed herein. The specific methods within the scope of this chapter are: (1) fast neutron activation, (2) fast neutron inelastic scatter, (3) resonance neutron activation, (4) thermal neutron activation, (5) thermal neutron capture, and (6) delayed fission neutron counting. Methods related to the determination of radiation life-times, such as die-away experiments, are not included since they are generally used to obtain gross lithological information rather than to specifically determine elements of interest. In the broadest terms, all of the above methods can be generally classified as neutron activation analysis and as such, the theory of activating particles on nuclei to obtain radioactive and detectable products is applied. Several texts dealing with these fundamentals have been written by Koch (1960), Bowen and Gibbons (1963), Lyon (1964), Taylor (1964), Lenihan and Thomson (1965), Kruger (1971), De Soete et al. (1972), and Nargolwalla and Przybylowicz (1973); and the reader is referred to these for this information. Suffice it to mention that the activity induced in a nuclide, B, can be expressed as a function of the operational parameters of the technique in the following manner:

$$A = DCN_B\sigma_B\phi\left[1 - \exp\left(\frac{-0.693\, t_i}{T_{1/2}}\right)\right]\left[\exp\left(\frac{-0.693\, t_d}{T_{1/2}}\right)\right] \tag{3}$$

where

A = the measured count rate at time t_d (counts s^{-1})
D = the branching ratio of the emission to be measured
C = the detection efficiency of the counting assembly
N_B = the number of target nuclei of the bombarded element B
σ_B = the activation cross-section of the specific reaction with element, B, at the incident particle energy (cm^2)
ϕ = the particle flux of constant energy and intensity (particles $cm^{-2}\ s^{-1}$)
t_i = irradiation time
t_d = decay time from the termination of irradiation to start of count
$T_{1/2}$ = the half-life of the product nuclide.

In practice the factors, C, σ_B and ϕ are not specifically evaluated and a simple comparative method is used. In this approach a sample and a standard are simultaneously irradiated, and the element, B, in the unknown sample is determined by comparing its radioactivity with that generated in the same element in the standard. The simple relation:

$$[\text{weight B}]_{\text{unknown}} = [\text{weight B}]_{\text{known}} \frac{[\text{activity B}]_{\text{unknown}}}{[\text{activity B}]_{\text{known}}} \tag{4}$$

is used.

The five principal neutron-nuclei interactions which are commonly used for the quantitative evaluation of elements are described below.

Radiative capture (n, γ)

A nucleus which captures a neutron produces gamma rays instantaneously from the de-excitation of certain excited energy levels in the resulting compound nucleus. This reaction occurs at a significant rate only with thermal and epithermal neutrons for most elements that comprise rock or ore deposits. The gamma transition energies of interest which are most important from the aspect of spectral identification are high (2—10 MeV). It is the complexity and the lesser range of the lower-energy capture gamma rays that have so far made them not useful in the prompt gamma-ray technique. The identifying transitions for many base metals may be distinguished from those from common rock by the fact that they occur within the energy range of 6—10 MeV.

Activation process (n, γ)

Most of the nuclei formed by (n, γ) reactions are unstable and subsequently undergo radioactive decay with a characteristic half-life. The gamma rays emitted during decay are nearly all below 2 MeV in energy. Both the characteristic gamma emissions and the half-life may be used for their unambiguous analysis.

Inelastic scattering (n, n'γ)

This reaction is important only for incident neutron energies above 2 MeV; the threshold for most elements being higher than this energy. The nucleus involved in the collision process is left in an excited state and immediately returns to the ground state with the emission of a neutron and gamma rays. In some favourable conditions there is a significant cross-section for an identifying transition.

Particle reactions (n, 2n), (n, α), (n, p)

These types of reactions are initiated by high-energy neutrons only. Elemental identification may be accomplished by examining the characteristic gamma radiation produced in the subsequent decay of the product radionuclide.

Delayed neutron counting

This technique has been recently adapted for the borehole exploration of fissionable isotopes occurring naturally, i.e. isotopes of uranium and thorium. Of all fissions in ^{235}U by thermal neutron bombardment about 99.36% yield fission neutrons within 10^{-12} second and are called prompt neutrons. In the 0.64% of the fission events, a limited number (~9) of radioactive fission products with short half-life (1—55 seconds), decay by beta emission. Some of them however leave a product nucleus with energy in excess of its neutron separation energy. In these cases neutrons are emitted with half-life comparable to the preceding beta decays. These neutrons form delayed neutron groups. A measurement of such neutron energy groups offers a unique method for the exploration for ^{235}U. The longest-lived delayed neutron emitter, the 55.6-second ^{87}Br provides neutron energies of about 0.3 MeV. The basic principle involves irradiation of the geological matrix with a fast neutron isotopic or accelerator source followed by a measurement of the delayed neutron emission by a detector separated from the irradiation source by a suitable spacer.

In borehole logging for minerals, the measurement tool generally consists of a suitable isotopic (e.g. ^{241}Am-Be) or fission (e.g. ^{252}Cf) neutron source of some acceptable output separated from a gamma-ray detector, such as a NaI(Tl) scintillator or Ge(Li) solid state assembly, by a low neutron capture-high gamma-ray attenuating shield. This shield protects the detector and associated electronics from direct neutron and gamma-ray exposure from the primary neutron source. The matrix solid angle subtended at the detector depends upon such mechanical parameters such as the source-detector distance and the ratio of the shield and detector diameters. If these parameters are carefully selected, and if the materials used for the construction of the sonde are optimally chosen for their unfavourable nuclear activation characteristics, then the sonde itself will behave as a quantitative tool for the measurement of the activated geological matrix.

Ideally, in any nuclear activation method, the measurement of characteristic gamma rays can be directly related to the elemental composition. This situation is only true if such parameters as the neutron flux distribution at effective locations in the matrix, the neutron flux absorption in the matrix, the bulk density of the matrix and its moisture content, and the extent of heterogeneity of the element in the matrix, are assumed to be constant. In practice, however, all of the above parameters are subject to variation, and their effect on the sensitivity of a given element irradiated with a constant output neutron source with respect to some fixed source-detector geometry, can be severe. To understand the effect of these environmental conditions on the sensitivity, some thought is given toward a study of neutron and gamma-ray transport in a geological medium irradiated with neutrons. For reasons of simplicity, the interposing detector shadow-shield has been removed from theoretical models to be described. Furthermore, only a detector and a source have been considered without regard to perturbations which may result from the presence of other materials which comprise the probe housing, etc.

Neutron-gamma transport

In the study of neutron-gamma borehole logging, the calculation of the neutron flux (created by the radioisotopic neutron source) in both the borehole and the matrix is very important in the estimation of the total gammas produced in the matrix and measured by the detector. Moreover, the calculated neutron and gamma-ray fluxes have been used in the optimization of the experimental design parameters such as the size and configuration of the neutron-source/gamma-ray detector probe and the optimum experimental borehole size.

Using any of the fundamental theories such as diffusion, transport, or Monte Carlo, the neutron flux distribution can be obtained. To determine thermal neutron flux distribution in a borehole, three approaches can be considered. The first approach (Glasstone and Edlund, 1952) assumes a point thermal neutron source of unit strength located in an infinite medium of silica. The second approach (Glasstone and Edlund, 1952) assumes a fast neutron point source of 5-MeV energy which are slowed down according to Fermi theory to thermal energy. The third approach, as described below, gives a more realistic representation of the source-detector arrangement in the borehole logging environment. This approach provides adequately acceptable estimation of analytical parameters used in the prediction of experimental measurements. Using the two-group and two-region neutron diffusion theory to take into account the borehole diameter effects and the source-detector distance, the thermal and fast neutron fluxes can be calculated. After the thermal neutron flux in the matrix is determined, both the gamma production rate and the prompt or delayed gammas returning to the detector can be computed.

Neutron diffusion theory in its most fundamental form assumes a thermal neutron point source, and a single energy group. In the multi-group diffusion theory, a fast neutron point source is assumed and the neutrons are slowed down to thermal energies. In the fast energy region, the neutrons are classified into several energy groups depending upon the sophistication and accuracy of the results desired. It is found by Allen et al. (1967) that the assumption of a single fast group and a thermal group (two-group diffusion theory), furnishes accuracies comparable to the expected precision of the analytical measurements.

One-group diffusion theory

Tittle (1961) used one-group neutron diffusion theory to study the slowing down of neutrons from a fast neutron point source centered in an infinite continuous medium, for a single cylinder, and for concentric cylinders representing a fluid-filled borehole and the surrounding formation. Neutrons, while diffusing through the media, are both absorbed and scattered by the matrix.

The one-group neutron balance differential equation in a system described by Tittle (1961) is:

$$D \cdot \nabla^2 \phi - \frac{\phi}{\lambda_a} + S = \frac{\delta n}{\delta t} \tag{5}$$

where ϕ = neutron flux (n cm^{-2} s^{-1}); λ_a = absorption mean free path (cm); D = diffusion coefficient (cm); S = source term (n cm^{-3} s^{-1}); n = neutron density (n cm^{-3}); and t = time (s). In the case of a point source with spherical geometry, the neutron flux, ϕ, is related to the neutron density and the neutron speed, v, by:

$$\phi = nv \tag{6}$$

which is also often referred to as the flux density or fluence rate.

In a steady-state system, where the changes in neutron density are assumed to be zero, and since S is a point source, it is zero everywhere except at the origin; therefore $S = 0$ in equation (5). The equation for the epithermal neutron flux at all different points in the continuous medium is then:

$$\nabla^2 \phi - \frac{\phi}{\lambda_a D} = 0 \tag{7}$$

Since the diffusion length, L, is defined as:

$$L = \sqrt{\lambda_a D} \tag{8}$$

equation (7) reduces to:

$$\nabla^2\phi - \frac{\phi}{L^2} = 0 \tag{9}$$

Since Tittle (1961) applied one-group diffusion theory in the slowing down region, he assumed the slowing-down length, L_s, analogous to L, owing to the fact that the number of neutrons removed by slowing down is equivalent to that by absorption. Hence equation (9) is rewritten as:

$$\nabla^2\phi - \frac{\phi}{L_s^2} = 0 \tag{10}$$

The slowing-down length, L_s, may be best obtained from the experimentally derived relation from Feld (1953):

$$6\,L_s^2 = \bar{r}^2 = \frac{\int_0^\infty \phi r^4\, \mathrm{d}r}{\int_0^\infty \phi r^2\, \mathrm{d}r} \tag{11}$$

where $\bar{r}^2$ is the mean square distance of the slowing down distribution. The integrals are evaluated numerically from experimental data.

The solution of equation (10) for a point source in infinite medium is reported by Glasstone and Edlund (1952) and Feld (1953) as follows:

$$\phi = \frac{Q}{4\,\pi D r} \exp \frac{-r}{L_s} \tag{12}$$

where the epithermal flux, ϕ, is related by Hughes (1953) to the slowing-down density, q, by:

$$q = \phi\xi/\lambda_s \tag{13}$$

$$D = L_s^2\xi/\lambda_s \tag{14}$$

The relation, $D = L_s^2\xi/\lambda_s$, is needed by the normalization condition that the total number of neutrons in the medium, slowed down through the epithermal energy, E, per second, must equal the source strength, Q, since none of these neutrons are assumed lost through absorption. In equations (12) and (13), ξ and λ_s are evaluated at the final epithermal energy, E.

Tittle (1961) compared the results from equation (12) with the experimental values and the age theory, in Fig. 5-4. Curve A is the prediction of equation (12) and the experimental points obtained by measurements in a large water tank by the use of a Ra-Be neutron source and epithermal neutron detectors, i.e. cadmium-covered indium foils. The source strength, q,

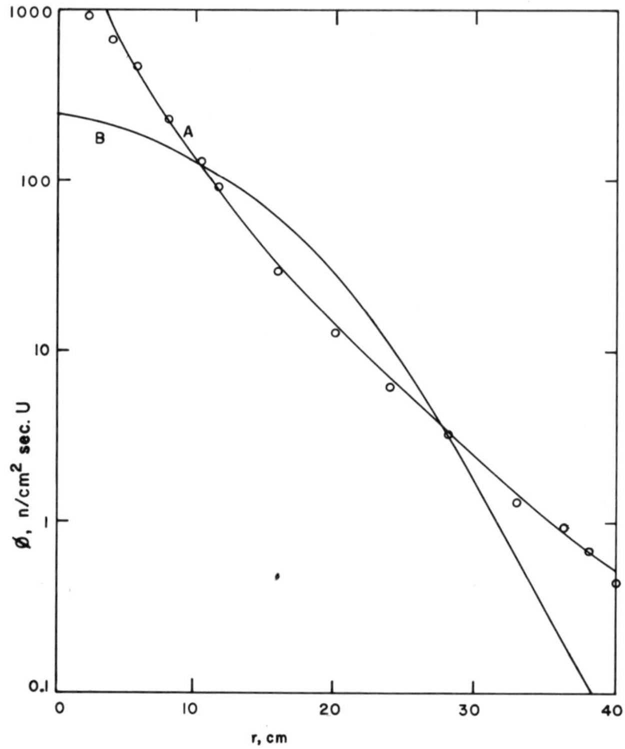

Fig. 5-4. Comparison of diffusion and age theories with experiment. The neutron source is 300 mCi of Ra-Be. Experimental points are based on measurements of the indium resonance neutron flux at 1.45 eV by Dacey et al. (1949). Curve *A*: one-group diffusion theory with $L_s = 6.8$ cm, $\xi/\lambda_s = 1.41$ cm^{-1}. Curve *B*: age theory, using same values of L_s, and ξ/λ_s. In the unit for epithermal neutron flux n cm^{-2} s^{-1} *U*, the symbol *U* stands for the neutron lethargy ($\ln E'/E_0$). It is to be interpreted as "per unit lethargy interval" in the specification of the unit of flux (Tittle, 1961).

was 4.5×10^6 n s^{-1}, L_s was 6.8 cm, and $\xi/\lambda_s = 1.41$. Curve *B* is the prediction of the age theory (Glasstone and Edlund, 1952). Agreement with experimental results is poor because of the large hydrogen content of water. However, there seems to be good agreement between experiment and the diffusion theory, except at positions near the source.

Encouraged by these results, Tittle (1961) applied this approach to a cylindrical region and also to two concentric cylinders. For an infinite cylindrical medium and assuming the condition of geometry, the neutron balance equation [equation (10)] in cylindrical co-ordinates (r, z) can be written as:

$$\frac{\delta^2\phi}{\delta r^2} + \frac{1}{r}\frac{\delta\phi}{\delta r} + \frac{\delta^2\phi}{\delta z^2} - \frac{\phi}{L_s^2} = 0 \tag{15}$$

The solution of equation (15) is:

$$\phi(r, z) = AJ_0(Cr)\exp(-Fz)$$

where A, C and F are constants and $F = (C^2 + L_s^{-2})^{1/2}$.

Then:

$$\phi_n(r, z) = A_n J_0(C_n r)\exp(-F_n z) \tag{16}$$

where the subscript n is introduced because there are several values of C, which will satisfy equation (16) and each value of C has a particular value of F, A and ϕ. The general solution of equation (15) will be:

$$\phi = \sum_n \phi_n = \sum_n A_n J_0(C_n r)\exp(-F_n z) \tag{17}$$

where:

$$C_n = \frac{x_n}{a} \tag{18}$$

and:

$$A_n = \frac{Q}{2\,\pi D F_n a^2 [J_1(x_n)]^2} \tag{19}$$

in which x_n are the values at which $J_0 = 0$ periodically.

The complete solution to calculate the epithermal neutron flux in a cylinder by one-group diffusion theory is given by:

$$\phi(r, z) = \sum_n \frac{Q}{2\,\pi D F_n a^2 [J_1(x_n)]^2} J_0\left(\frac{x_n r}{a}\right)\exp(-F_n z) \tag{20}$$

where:

$$F_n = \left[\left(\frac{x_n}{a_n}\right)^2 + \frac{1}{L_s^2}\right]^{1/2} \tag{21}$$

The flux equation (20) neglects the effect of the borehole. The application of one-group theory to the concentric cylinders is omitted in favour of the two-group diffusion theory to follow.

Two-group, two-region diffusion theory

Glauberman and Talianskyi (1958) and Kulinkovich (1965) used the two-group, two-region diffusion theory to calculate thermal and fast neutron fluxes in the borehole and the surrounding formation. Their expressions for the neutron fluxes are in the form of infinite integrals and, therefore, rather tedious to work with. Their approach is similar to that used by Allen et al. (1967). Allen et al. using the two-group, two-region diffusion theory have

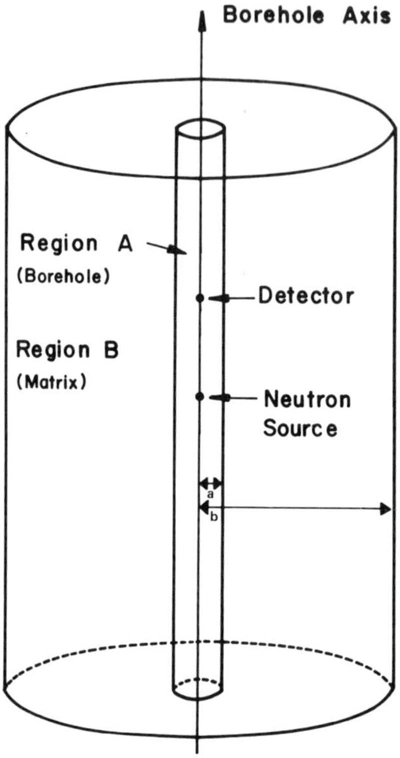

Fig. 5-5. Theoretical model for neutron flux calculations (Nargolwalla et al., 1973).

derived an expression for the thermal neutron flux in the borehole only. Their expression for the neutron flux is in the form of an infinite series and is easily adapted for computer calculations.

Nargolwalla et al. (1973) extended the method used by Allen et al. (1967) to derive expressions for thermal and fast neutron fluxes in the borehole as well as in the surrounding matrix.

Assume a point neutron source, S (i.e. ^{252}Cf with an average neutron energy of 2—3 MeV), of strength Q (10^6 n s^{-1}) is placed at the origin of the r-z coordinate axes, which is at the center of two concentric cylinders of radii a, and b (Fig. 5-5). A point detector, D, is located at a distance z directly above the source, S. The two concentric cylinders, representing the borehole of radius a, and the effective irradiated matrix region of radius b, contain 2 different media, i.e. water and the matrix (SiO_2), respectively. The height of the cylinders is assumed to be $2h$. It is further assumed that the flux in the axial direction is symmetrical.

The neutron balance equations, as shown by Glasstone and Edlund (1952), in the two regions, for fast and thermal neutrons, are expressed as follows:

$$D_{1A}\nabla^2\phi_{1A}(r,z)-\phi_{1A}(r,z)\,\Sigma_{1A}+S(r,z)=0\,,\qquad 0\leqslant r\leqslant a \tag{22}$$

$$D_{2A}\nabla^2\phi_{2A}(r,z)-\phi_{2A}(r,z)\,\Sigma_{2A}+\phi_{1A}(r,z)\,\Sigma_{1A}=0,\qquad 0\leqslant r\leqslant a \tag{23}$$

$$D_{1B}\nabla^2\phi_{1B}(r,z)-\phi_{1B}(r,z)\,\Sigma_{1B}=0\,,\qquad a\leqslant r\leqslant b \tag{24}$$

$$D_{2B}\nabla^2\phi_{2B}(r,z)-\phi_{2B}(r,z)\,\Sigma_{2B}+\phi_{1B}(r,z)\,\Sigma_{1B}=0,\qquad a\leqslant r\leqslant b \tag{25}$$

where
$S(r, z)$ = $S'\delta(r)\,\delta(z)$, where $\delta(r)$ and $\delta(z)$ are Dirac delta functions
S' = the rate of neutron density production (n cm^{-3} s^{-1})
D = diffusion coefficient (cm)
ϕ = neutron flux (n cm^{-2} s^{-1})
1 denotes fast neutrons
2 denotes thermal neutrons
∇^2 = Laplacian operator
Σ_1 = macroscopic removal cross-section for neutrons from fast to thermal energy
Σ_2 = macroscopic thermal absorption cross-section.

Σ_i can be readily calculated from:

$$\Sigma_i=\frac{\rho N_0\sigma_i}{A_0}$$

in which ρ = the density of the material (g cm^{-3}), N_0 = Avogadro's number (6.023×10^{23} atoms/g-atom), A_0 = atomic weight (g/g-atom), and σ_i = cross-section of element i (cm^2); A denotes borehole region containing water, and B, effective irradiated matrix region containing SiO_2.

The analytical solutions of equations (22) to (25), from Tittle and Allen (1966), are as follows:

$$\phi_{1A}=\frac{Q}{4\,\pi D_{1A}}\left[\frac{\exp(-\sqrt{r^2+z^2}/L_{1A})}{\sqrt{r^2+z^2}}-\frac{\exp[-\sqrt{r^2+(z-2\,h)^2}/L_{1A}]}{\sqrt{r^2+(z-2\,h)^2}}\right]$$

$$+\frac{Q}{2\,\pi D_{1A}h}\sum_n\left(\frac{A_n}{B_n}\right)I_o(\alpha_n r)\cos(\lambda_n z) \tag{26}$$

$$\phi_{2A}=\frac{Mh}{2}\left[\frac{\exp(-\sqrt{r^2+z^2}/L_{1A}}{\sqrt{r^2+z^2}}-\frac{\exp[-\sqrt{r^2+(z-2\,h)^2}/L_{1A}]}{\sqrt{r^2+(z-2\,h)^2}}\right.$$

$$\left.-\frac{\exp(-\sqrt{r^2+z^2}/L_{2A})}{\sqrt{r^2+z^2}}+\frac{\exp[-\sqrt{r^2+(z-2\,h)^2}/L_{2A}]}{\sqrt{r^2+(z-2\,h)^2}}\right]$$

$$+M\sum_n\left[\left(\frac{A_n}{B_n}\right)I_0(\alpha_n r)+\frac{C_n}{B_nS_4}I_0(\beta_n r)\right]\cos(\lambda_n r) \tag{27}$$

$$\phi_{1B} = \sum_n G_n K_0(\gamma_n r)\cos(\lambda_n z) \tag{28}$$

$$\phi_{2B} = \sum_n L_n K_0(\delta_n r)\cos(\lambda_n z) + S_3\phi_{1B} \tag{29}$$

where:

$$M = \frac{Q}{2\pi D_{2A} h}\left(\frac{L_{2A}^2}{L_{1A}^2 - L_{2A}^2}\right); \qquad B_n = \frac{Q}{2\pi D_{1A} h}$$

$$S_3 = \frac{D_{1B} L_{2B}^2}{D_{2B}(L_{1B}^2 - L_{2B}^2)}; \qquad S_4 = \frac{D_{1A} L_{2A}^2}{D_{2A}(L_{1A}^2 - L_{2A}^2)}$$

$$L_{1A} = [D_{1A}/\Sigma_{1A}]^{1/2}; \qquad L_{2A} = [D_{2A}/\Sigma_{2A}]^{1/2}$$

$$L_{1B} = [D_{1B}/\Sigma_{1B}]^{1/2}; \qquad L_{2B} = [D_{2B}/\Sigma_{2B}]^{1/2}$$

$$\gamma_n = [\lambda_n^2 + L_{1B}^{-2}]^{1/2}; \qquad \alpha_n = [\lambda_n^2 + L_{1A}^{-2}]^{1/2}$$

$$\delta_n = [\lambda_n^2 + L_{2B}^{-2}]^{1/2}; \qquad \beta_n = [\lambda_n^2 + L_{2A}^{-2}]^{1/2}$$

$$\lambda_n = \frac{n\pi}{2h} \qquad n = 1, 3, 5, \ldots$$

and:

$$A_n = \left[\frac{D_{1A}\alpha_n a K_0(\gamma_n a) K_1(\alpha_n a) - D_{1B}\gamma_n a K_1(\gamma_n a) K_0(\alpha_n a)}{D_{1A}\alpha_n a K_0(\gamma_n a) I_1(\alpha_n a) + D_{1B}\gamma_n a K_1(\gamma_n a) I_0(\alpha_n a)}\right] B_n$$

$$G_n = \left[\frac{(A_n/B_n) I_0(\alpha_n a) + K_0(\alpha_n a)}{K_0(\gamma_n a)}\right] B_n$$

$$L_n = \frac{S_4 B_n}{K_0(\delta_n a)}\left[\left(\frac{C_n}{B_n}\right)\frac{1}{S_4} I_0(\beta_n a) - K_0(\beta_n a) + (1-P) K_0(\gamma_n a)\left(\frac{G_n}{B_n}\right)\right]$$

$$C_n = \frac{R \cdot B_n \cdot S_4}{D_{2A}\beta_n a I_1(\beta_n a) K_0(\delta_n a) + D_{2B}\delta_n a K_1(\delta_n a) I_0(\beta_n a)}$$

in which:

$$\begin{aligned} R = {} & D_{2B}\delta_n a K_1(\delta_n a)\left[P\left(\frac{G_n}{B_n}\right) K_0(\gamma_n a) + K_0(\beta_n a) - K_0(\alpha_n a)\right. \\ & \left. - \left(\frac{A_n}{B_n}\right) I_0(\alpha_n a)\right] - K_0(\delta_n a)\left\{D_{2A}\gamma_n a P\left(\frac{G_n}{B_n}\right) K_1(\gamma_n a)\right. \\ & \left. + D_{2A}\left[\beta_n a K_1(\beta_n a) - \alpha_n a\left\{K_1(\alpha_n a) - \left(\frac{A_n}{B_n}\right) I_1(\alpha_n a)\right\}\right]\right\} \end{aligned}$$

$$P = \frac{D_{1A} D_{2A} L_{2B}^2 (L_{1A}^2 - L_{2A}^2)}{D_{2B} D_{1A} L_{2A}^2 (L_{1B}^2 - L_{2B}^2)}$$

The values for ϕ_{1A}, ϕ_{2A}, ϕ_{1B}, and ϕ_{2B} in equations (26) to (29) can be solved by computer. Some parameters, such as the thermal diffusion coefficient, D_2, the fast diffusion coefficient, D_1, the slowing down length, L_s, and the diffusion length, L, are given by Glasstone and Sesonske (1963 — fast group parameters for water); ANL-5800 (thermal group parameters compiled by Argonne National Laboratory, 1963); and BNL-325 (the scattering cross-section for silicon for the fast group is averaged over the energy range of interest from the cross-section energy curve reported by Hughes and Schwartz, 1958).

The computer program FLUX was used by Nargolwalla et al. (1973) to calculate the flux distribution in a simulated borehole in the laboratory. This computer program for solving equations (26) to (29) was written in a manner such that the convergence occurred when three consecutive terms in the series did not differ from each other by a fraction greater than 10^{-6}.

Theoretical model for prompt gamma rays returning to the detector

Once the thermal neutron flux $\phi_{2B}(r, z)$ from equation (29) in the matrix is evaluated, the gamma production rate, P_γ, can be easily determined.

Gamma transport — one region. Assuming that each neutron absorbed in the matrix nuclei produces a single prompt gamma photon of a specific energy, then, from Fig. 5-6, the gamma flux, P_γ, returning to the point detector can be determined by:

$$P_\gamma = \int_{-\infty}^{\infty} \int_{a}^{\infty} \frac{2\,\pi r\,dr\,dz}{(r^2 + z^2)\,4\pi}\,\phi_{2B}(r,z)\,\frac{\rho\sigma N_o}{M_o}\,\exp(-\mu\sqrt{r^2 + z^2}) \tag{30}$$

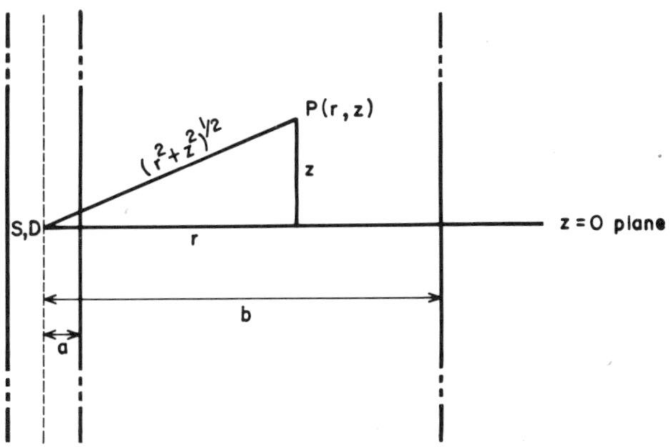

Fig. 5-6. Theoretical model for one-region gamma transport (Nargolwalla et al., 1973).

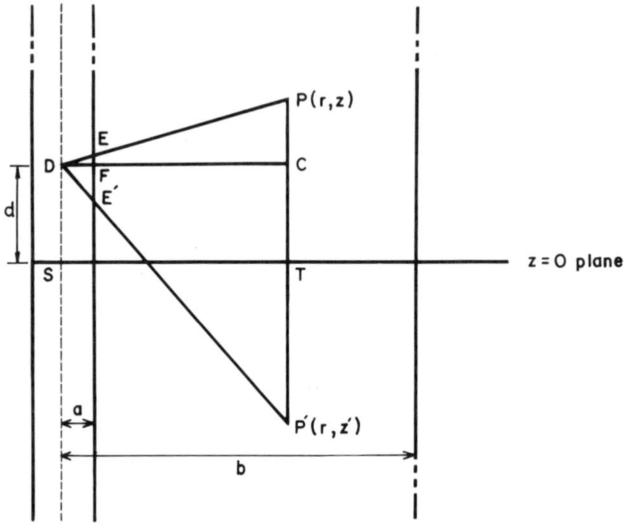

Fig. 5-7. Theoretical model for two-region gamma transport (Nargolwalla et al., 1973).

where μ is the linear absorption coefficient of some discrete gamma-ray energy, and is assumed to be the same for both the matrix and the borehole materials; ρ is the density (g cm^{-3}); M_0 the gram molecular weight of the matrix material; and N_0 the Avogadro's number. Here σ is the effective microscopic cross-section of the matrix material as a whole.

Gamma transport — two region. For practical calculation, expression (30) can be simplified. Complexity of integration increases when the source and the detector are separated by some finite distance and appropriate attenuation coefficients are adopted for matrix and borehole material (Fig. 5-7). Numerical integration is necessary in this very complex situation.

To be detected as a full energy ray by a detector at the location D, a gamma photon produced at a point $P(r, z)$ in the borehole has to traverse a distance $PE = t$ in the matrix region and a distance $DE = s$ in the borehole (water) region, in a straight path. Neglecting gamma production in the borehole, the gamma flux P_γ returning to the detector from the total matrix can be written as:

$$P_\gamma = \int_{-\infty}^{\infty} \int_{a}^{\infty} \frac{2\,\pi r\,\mathrm{d}r\,\mathrm{d}z}{4\,\pi[r^2 + (z-d)^2]}\,\phi_{2B}(r, z)\,\frac{\rho_{\mathrm{eff}}\sigma N_o}{M_0}\,\exp(-\mu_w s)\,\exp(-\mu t) \qquad (31)$$

where μ_w is the gamma absorption coefficient for water and ρ_{eff} the effective density of matrix material to account for changes in porosity. Also:

$$s = \frac{a}{r}\overline{DP} = \frac{a}{r}[r^2 + (z-d)^2]^{1/2}$$

$$t = \left(1 - \frac{a}{r}\right) [r^2 + (z - d)^2]^{1/2}$$

To evaluate the absorption coefficient for a compound or a mixture, μ can be replaced by:

$$\mu = \sum_i \left(\frac{\mu}{\rho}\right)_i \rho_i \tag{32}$$

where ρ_i = partial density of element i (g cm^{-3}), and $(\mu/\rho)_i$ = mass absorption coefficient for element i (cm^2 g^{-1}).

The values $(\mu/\rho)_i$ for elements Si, H, and O are given by Storm and Israel (1967) and Grodstein (1957) for the gamma energies investigated.

To simplify the calculations further in the case of a single element in the matrix, we can write:

$$P_\gamma = \rho_{\text{eff}} \int_{-\infty}^{\infty} \int_{a}^{\infty} \frac{2\,\pi r\,\mathrm{d}r\,\mathrm{d}z}{4\,\pi[r^2 + (z - d)^2]} \phi_{2B}(r, z) \exp(-\mu_{\text{w}} s) \exp(-\mu t) \tag{33}$$

A computer program GAMUX was written to perform this integration for the element Si, for various source detector distances, d, porosity, ρ_{eff} and gamma energy E.

It must be noted that equation (33) represents only the attenuation of gamma photons of energy E by the matrix and borehole material irradiated by the thermal neutron flux distribution, $\phi_{2B}(r, z)$. It does not imply that gammas of energy E are actually produced and subsequently detected by the detector D. For example, gamma photons of energy E = 10 MeV are not produced by neutron absorption in the silicon matrix.

Theoretical results

Neutron flux distribution. In the theoretical study described by Nargolwalla et al. (1973), the borehole is assumed to be full of water and the matrix silica sand (SiO_2) of theoretical density 2.65 g cm^{-3}, i.e. porosity = 0. The porosity is increased by the introduction of air (dry case) or water (wet case).

The calculated results (Figs. 5-8 and 5-9) show the thermal neutron flux distribution. An analogous plot for the case of a dry matrix is shown in Fig. 5-10. The effect of water in the matrix indicates an enhancement in the neutron flux close to the borehole and a sharp decrease away from the borehole. As such, the effective volume of the matrix decreases with the increase in the moisture. As can be expected for equivalent porosities, the neutron flux in the dry matrix is considerably higher.

Gamma-ray intensity returning to the detector. The number of gammas returning to the detector depends on the moisture content, borehole size and

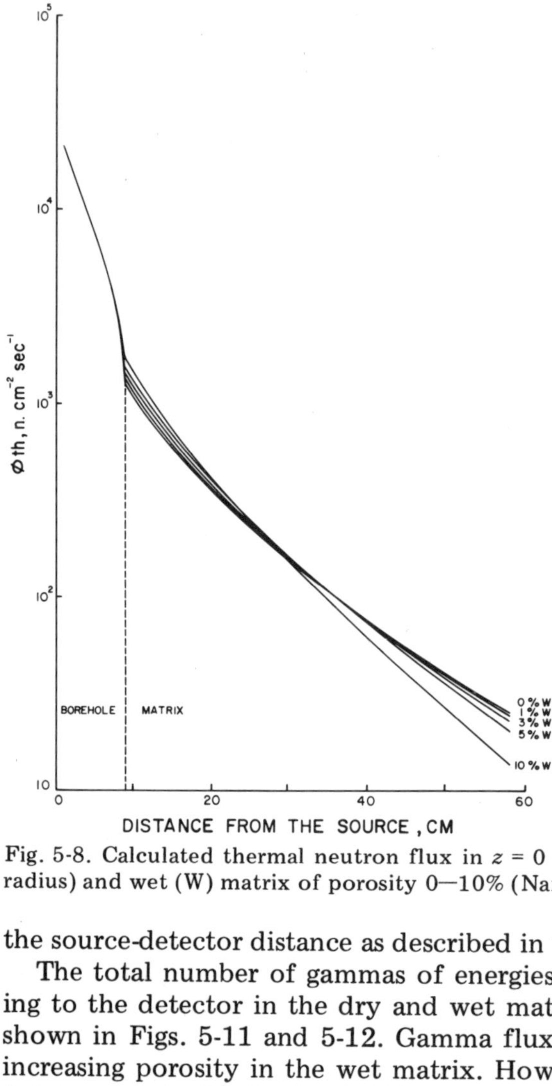

Fig. 5-8. Calculated thermal neutron flux in $z = 0$ plane for a water-filled borehole (9 cm radius) and wet (W) matrix of porosity 0—10% (Nargolwalla et al., 1973).

the source-detector distance as described in the following.

The total number of gammas of energies 2, 3, 5, 6, 8 and 10 MeV returning to the detector in the dry and wet matrix as a function of porosity are shown in Figs. 5-11 and 5-12. Gamma flux at the detector D increases with increasing porosity in the wet matrix. However, we expect all the curves to fall to zero in the limit of 100% porosity, in which case ρ_{eff} will be zero. Note the overall reduction in sensitivity with increase in the source-detector distance. From an experimental viewpoint the selection of an optimum source-detector distance can reduce the error due to the porosity effect. The effect of increasing the borehole diameter results in a decrease in the gamma-ray intensity as seen by the detector.

In interpretation of these calculated results, it must be remembered that the shield normally separating the source and detector has not been considered. In practice however, a considerable distortion in the theoretical gamma-ray pattern occurs. The design of an optimum probe depends to a

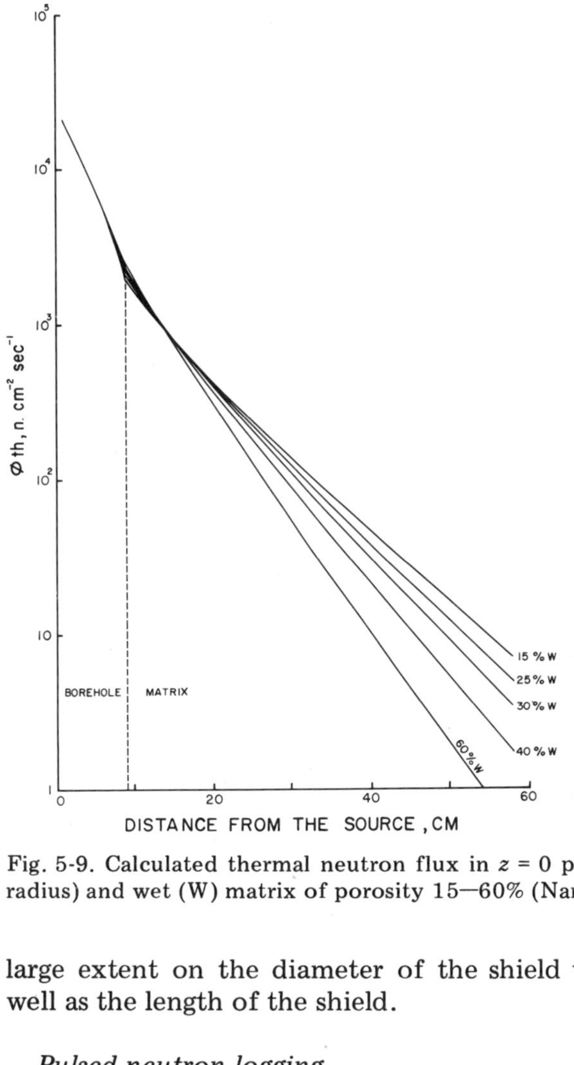

Fig. 5-9. Calculated thermal neutron flux in $z = 0$ plane for a water-filled borehole (9 cm radius) and wet (W) matrix of porosity 15—60% (Nargolwalla et al., 1973).

large extent on the diameter of the shield with respect to the detector, as well as the length of the shield.

Pulsed neutron logging

Pulsed neutron logging finds particular application for the use of inelastic scattering reactions to the determination of certain elements. Essentially, however, the object of pulsed neutron logging is the determination of fundamental neutron parameters such as, Σ_a, the macroscopic absorption cross-section; D, the diffusion coefficient and the like. In logging, the changes in such parameters from the known values can assist in the identification of elements of interest.

Denisik et al. (1963) used the Monte Carlo method for the calculation of the space-time distribution of neutrons in a SiO_2 matrix of varying porosities. The agreement between theory and experiment was excellent. Rezvanov et

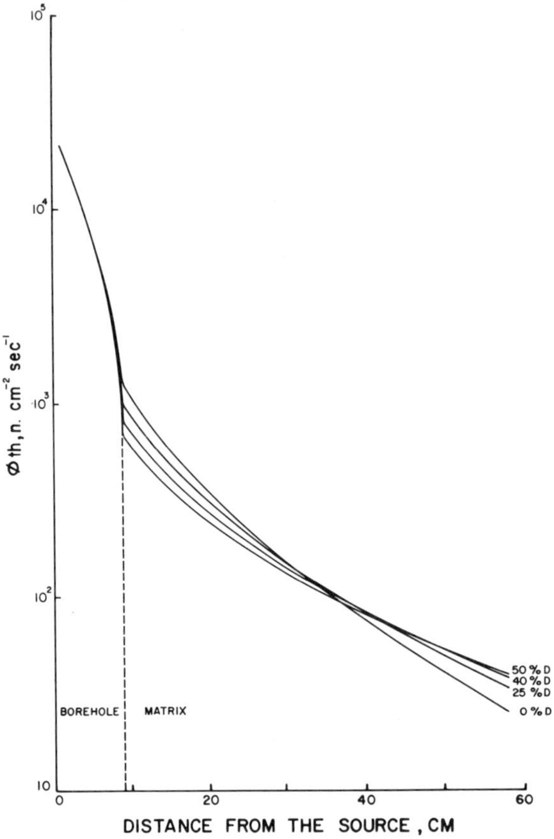

Fig. 5-10. Calculated thermal neutron flux in $z = 0$ plane for a water-filled borehole (9 cm radius) and dry (D) matrix of porosity 0—50% (Nargolwalla et al., 1973).

al. (1968) have shown results of a mathematical model of space-time thermal neutron distribution in a two-layer medium with a cylindrical interface.

Using the diffusion approximation of the neutron transport theory, Caldwell et al. (1966) calculated the neutron distribution in space and time for the borehole geometry. A computer code called "CUNLAP" (Anonymous, 1965) was used to solve the time-dependent, three-group diffusion equations. This code has the capacity of accepting up to 39 different time-step sizes. Furthermore, it is also possible for the computer code to select its own time-step size based on the rate of change of the neutron flux. The alternate direction implicit technique reported by Wachpress and Habetler (1960) is used to achieve spatial convergence of the finite difference expressions for the neutron fluxes.

To provide experimental verification for the theoretical computations, a limestone model was used. The die-away of thermal neutrons and capture

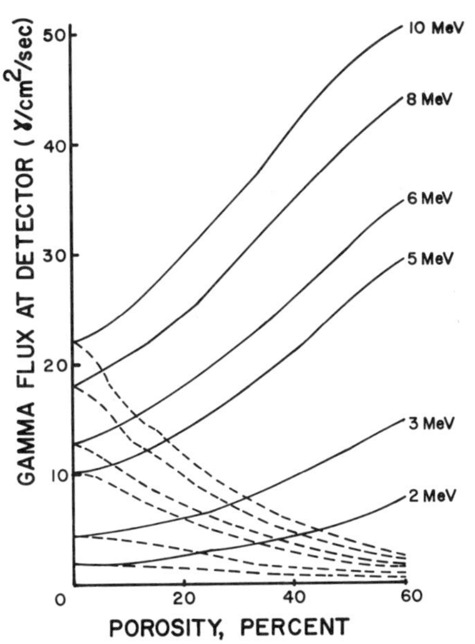

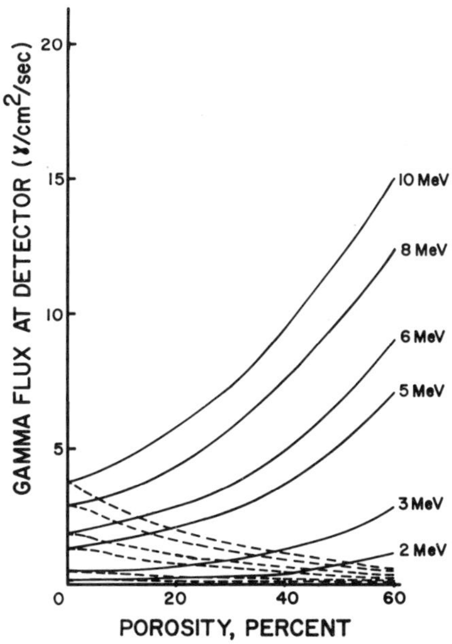

Fig. 5-11. Calculated gamma-ray flux at the detector for source-detector distance of 30 cm for a water-filled borehole (9 cm radius). Solid curves: dry matrix; dashed curves: wet matrix (Nargolwalla et al., 1973).

Fig. 5-12. Calculated gamma-ray flux at the detector for source-detector distance of 60 cm for a water-filled borehole (9 cm radius). Solid curves: dry matrix; dashed curves: wet matrix (Nargolwalla et al., 1973).

gamma rays from a pulsed 14-MeV neutron source were measured by a ^{3}He and NaI(Tl) detector, respectively. A comparison of the complete theoretical neutron and gamma-ray decay curves with experimental measurements is shown in Fig. 5-13. The experimental data was obtained for a 7-inch diameter borehole. The gamma-ray counts measured were those above 1.5 MeV energy. During each neutron burst from the pulsed neutron source, the photomultiplier tube was gated off. The data as shown have been corrected for the time analyzer dead time and for background.

As can be observed that, for the neutron curves, there is a large discrepancy between the theoretical and experimental time distribution immediately following the burst. Possible reasons for this discrepancy have been attributed to (1) the inability of the calculation to properly account for the central air column in the borehole, and (2) the effects of thermal-neutron flight time across the air column. It may also be observed that the gamma-ray curves are almost exponential over a longer time scale than the neutron curves. The theoretical gamma-ray decay curve was calculated by spatially integrating over the computed neutron distribution. Both single scattered and unattenuated radiation were included.

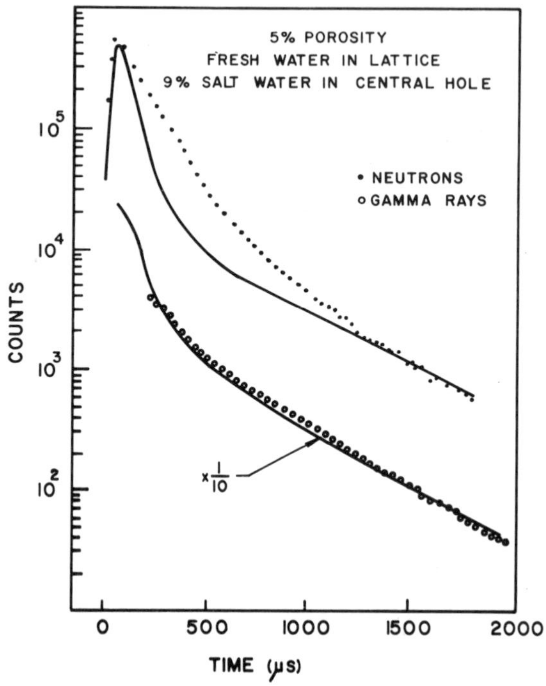

Fig. 5-13. Comparison of experimental and theoretical time distributions (Caldwell et al., 1966).

In later sections of the chapter dealing with the development of borehole exploration techniques, use is made of the theoretical concepts given above. Although the theoretical approach is a simple treatment of neutron and photon physics, a general feeling for the important parameters controlling the analytical sensitivity and possible means for the minimization of systematic errors is indicated.

RADIOACTIVATION SOURCES

Irradiation sources used in the radioactivation methods in mineral exploration can be classified into four main categories (1) radioisotopic neutron sources, (2) spontaneous fission neutron sources, (3) accelerators and neutron generators, and (4) gamma sources. Depending on the type of neutron source used, fast, epithermal or thermal neutrons can be produced. The choice of an appropriate source depends mainly on the applications to be pursued. A few criteria that must be considered in the proper selection are (a) the cost, (b) the dimension, shape and shielding required, (c) the requirement of source handling, (d) source strength, and (e) space available.

Radioisotopic neutron sources

The principal sources of neutrons are those from nuclear reactions that emit a high flux of neutrons. Some naturally occurring and some artificial alpha and gamma emitters have been coupled with a particular low-mass element having a low neutron binding energy and suitable cross section to produce neutrons. Until recently the most well-known radioisotopic neutron sources have been mainly the (α, n) and the (γ, n) sources. These consist of alpha- and gamma-emitting radionuclides mixed with a target element. The (α, n) sources utilizing alpha-emitting isotopes are mixed with a beryllium target, to produce neutrons with a wide range of energies. The principal alpha emitters used with a beryllium target are ^{226}Ra, ^{210}Po, ^{227}Ac, ^{238}Pu, ^{241}Am and ^{242}Cm. The (γ, n) sources involve a gamma emitter, such as ^{124}Sb, ^{88}Y, ^{24}Na, and ^{140}La, surrounded by a beryllium or a deuterium oxide target. Some of these neutron sources can be designed to turn the neutron output "OFF" and "ON". A survey of a few commonly used (α, n) and (γ, n) sources with their principal nuclear reactions, reaction energy Q, neutron energy and neutron yield, is shown in Table 5-I.

In practice the principal advantages of isotopic neutron sources are their stability and compact size. Their neutron outputs diminish with the decay of the gamma-emitting and alpha-emitting nuclides, depending on the half-lives of the nuclides. These neutron sources have certain disadvantages; i.e. (1) the insufficient intensity for neutron activation analysis; this shortcoming can be corrected by using large samples or by irradiating and counting repeatedly to improve the precision of the result and, (2) inability for turning off the neutron output easily.

(α, n) Sources

More than 20 principal alpha-emitting nuclides are utilized in the fabrication of these sources as described by Ansell and Hall (1971a, b).

The alpha particles emitted by the decay of a naturally occurring or artificially produced radioisotope react with the nuclei of a low-atomic-number element of suitable threshold and cross-section to produce neutrons. The alpha emitters most often used are ^{210}Po, ^{241}Am, ^{239}Pu, and ^{238}Pu (Table 5-I), and typical target materials invariably include ^{2}H and Be. In the early days the radium isotopes and their daughters mixed with a beryllium target, were about the only types of isotopic neutron source available for well logging. However, the number of (α, n) sources rapidly increased with the introduction of the smaller ^{210}Po-Be source giving more neutrons, a relatively compact ^{238}Pu-Be and other useful sources such as ^{239}Pu-Be and ^{241}Am-Be. In the selection of a suitable target material in the neutron production, beryllium is the most favorable since the number of neutrons emitted per alpha particle bombarding this target material is the largest (i.e. 77 n/10^6 alpha particles) in comparison with the yield of neutrons by other target

TABLE 5-I

(α, n) and (γ, n) neutron sources

Type	Source	Main reaction	Q (MeV)	Average neutron energy (MeV)	Neutron yield ($n\ s^{-1}\ Ci^{-1}$)	Half-life of alpha-emitting nuclide
(α, n)	^{242}Cm-Be	$^{9}Be(\alpha, n)^{12}C$	5.65	4	4×10^6	163 days
	^{244}Cm-Be	—	5.65	4	3×10^6	18.1 yr
	^{241}Am-Be mixture	$^{9}Be(\alpha, n)^{12}C$	5.65	4	2.2×10^6	433 yr
	^{226}Ra-Be mixture	$^{9}Be(\alpha, n)^{12}C$	5.65	3.6	1.5×10^7	1602 yr
	^{210}Po-Be mixture	$^{9}Be(\alpha, n)^{12}C$	5.65	4.3	2.5×10^6	138 days
	^{239}Pu-Be mixture	$^{9}Be(\alpha, n)^{12}C$	5.65	4.5	2×10^6	24,360 yr
	^{227}Ac-Be	$^{9}Be(\alpha, n)^{12}C$	5.65	4	2.4×10^7	21.8 yr
	^{228}Th-Be	—	5.65	4	2.8×10^7	1.9 yr
	^{238}Pu-Be	—	5.65	4	2.8×10^6	89 yr
(γ, n)*	^{24}Na-Be	$^{9}Be(\gamma, n)^{8}Be$	−1.67	0.2	1.4×10^5	15 h
	$^{24}Na\text{-}D_2O$	$^{2}H(\gamma, n)^{1}H$	−2.23	0.8	2.9×10^5	15 h
	^{88}Y-Be	$^{9}Be(\gamma, n)^{8}Be$	−1.67	0.16	1×10^5	108 days
	$^{88}Y\text{-}D_2O$	$^{2}H(\gamma, n)^{1}H$	−2.23	0.3	3×10^3	108 days
	^{124}Sb-Be	$^{9}Be(\gamma, n)^{8}Be$	−1.67	0.02	1.9×10^5	60 days
	^{140}La-Be	$^{9}Be(\gamma, n)^{8}Be$	−1.67	0.6	2×10^3	40.2 h
	$^{140}La\text{-}D_2O$	$^{2}H(\gamma, n)^{1}H$	−2.23	0.15	7×10^3	40.2 h

* Neutron yields are those for a 1-g Be or heavy water target at 1 cm from the gamma source of 1 Ci.

materials under the same conditions (Strain and Lyon, 1965, pp. 245—258). However, B, Li, and F are employed as target material whenever the activation by a thermal neutron flux is specifically required.

Owing to the short range of alpha particles, achievement of the highest possible neutron yield requires the alpha-emitting isotope to be intimately mixed with the beryllium target to form a powder mixture, or an alloy, depending on the intended application which demand particular shapes and proper encapsulation or shielding. However, neutron source alloys (e.g. ^{239}Pu-13Be, ^{242}Cm-13Be and ^{241}Am-13Be) have a higher neutron yield than that produced by powder mixtures (e.g. PuO_2 + metallic Be).

In the selection of a neutron source to be used for a specific purpose, the size and shape of the encapsulating container, the neutron output, and the gamma radiation emitted by the alpha-emitting radioactive nuclide are the main criteria that must be considered. Although many alpha-emitting isotopes, such as ^{226}Ra, have long half-lives compared with those of other isotopes, such as ^{210}Po, and have more intensive neutron outputs from (α, n) reactions; the gamma-ray intensity emitted by long-lived nuclides (e.g. ^{226}Ra and ^{228}Th) is so high that the gamma interference in the gamma-ray spectrometric measurements and the need for shielding makes the use of other relatively shorter half-lived nuclides (e.g. ^{238}Pu-Be and ^{241}Am-Be) more attractive. Comparing a few (α, n) sources in Table 5-I, ^{244}Cm-Be is not easily available and ^{239}Pu-Be has some user restrictions since ^{239}Pu is a fissile

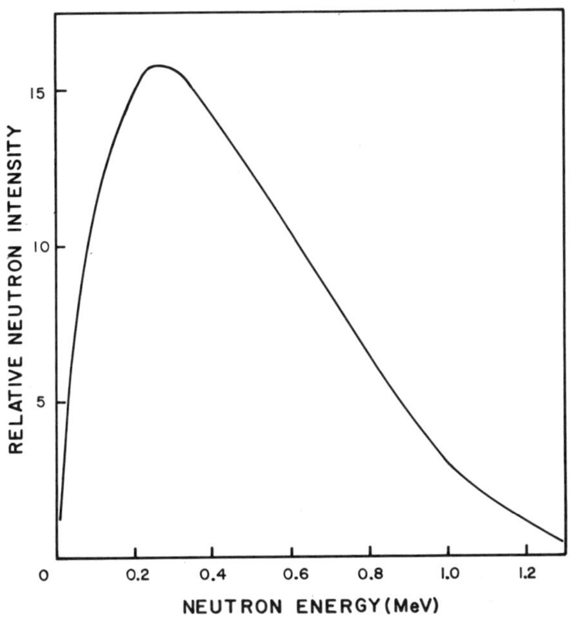

Fig. 5-14. Neutron spectrum of ^{210}Po-Li (α, n) source.

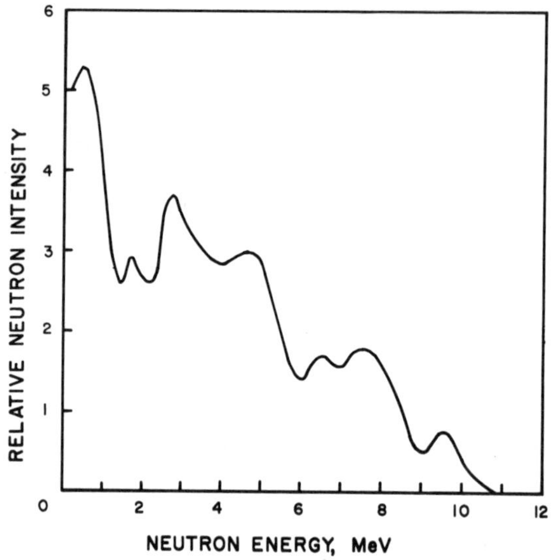

Fig. 5-15. Neutron spectrum of ^{239}Pu-Be (α, n) source.

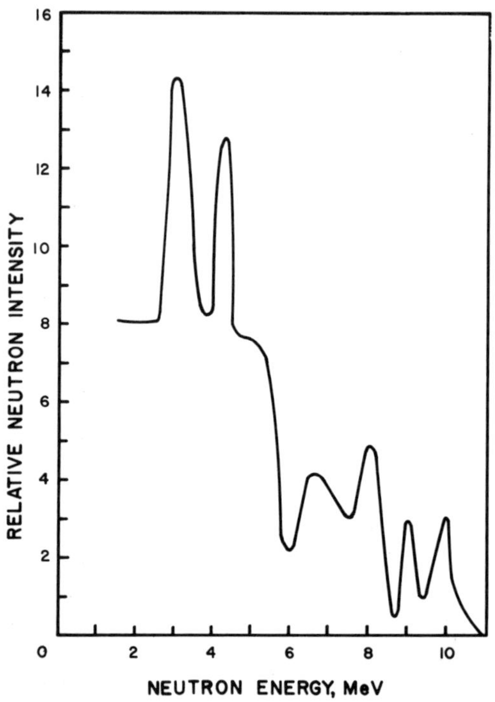

Fig. 5-16. Neutron spectrum of ^{241}Am-Be (α, n) source.

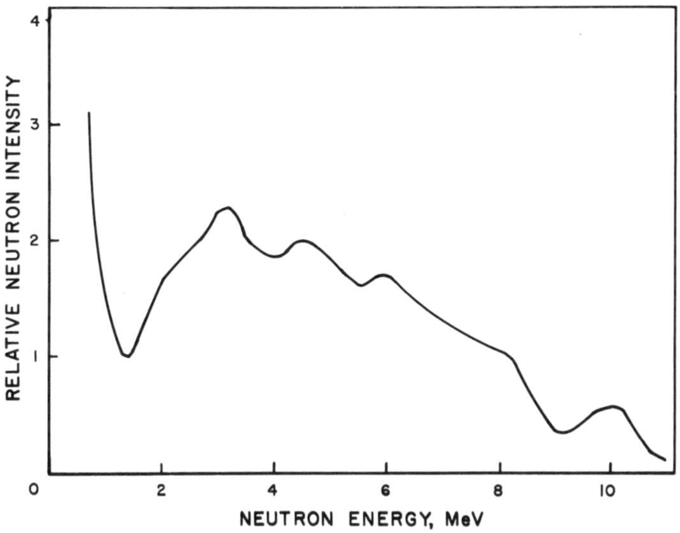

Fig. 5-17. Neutron spectrum of Ra-Be (α, n) source.

isotope. The neutron energy spectra of some of the more useful (α, n) sources are given in Figs. 5-14 to 5-17. The neutron yield of ^{241}Am-Be can be improved significantly when irradiated for 30 days at a neutron flux of 2×10^{14} n cm^{-2} s^{-1} in a nuclear reactor. In mineral analysis this source is one of the most useful low-intensity neutron producers. Other long-lived sources, i.e. ^{228}Th-Be and ^{242}Cm-Be, prepared by the irradiation of ^{226}Ra and ^{241}Am oxide, respectively, at a flux of 10^{14} n cm^{-2} s^{-1} have a relatively high neutron output of about 10^8 n s^{-1} Ci^{-1}. Other sources, i.e. ^{244}Cm-Be, ^{227}Ac-Be and ^{238}Pu-Be, incorporate alpha emitters of very high specific activity, which are principally made in reactors. The 22-year half-life ^{227}Ac has a high neutron output per unit volume and a low gamma-ray output from the source combination ^{227}Ac-Be. A by-product from the reprocessing of nuclear fuel is the 18-year half-life ^{244}Cm which, when combined with Be, has been used for high sensitivity analysis of small samples.

In practice the (α, n) sources are used for fast and thermal neutron activation analysis and are quite popular for the analysis of boreholes. Neutron activation data using these sources, such as (1) the detection sensitivities of elements irradiated with a fast neutron flux (Wing and Wahlgren, 1967, pp. 85—89), and (2) the gamma-ray spectra (Wing and Wahlgren, 1966) are available. However, detection sensitivity data for various elements irradiated by thermal neutrons can be determined from published data by Anders (1960), Wing and Wahlgren (1965), and Yule (1965). The sources described can be used for two principal types of analyses, i.e. thermal neutron activation-delayed gamma-ray measurement and thermal neutron-capture gamma-ray analysis. The latter in most cases is interference-free from the high-energy gammas of 4.43 and 2.23 MeV resulting from the use of such sources due to the Be target and the hydrogen content of moderators, respectively. Concentrations of elements such as Fe, Cu, Sr, Ti, Ni, and many others can be determined with these sources as described by Christell and Ljunggren (1965), and Gijbels (1973).

Gamma-neutron sources

These sources use two target nuclei, Be and D_2O, and the bombarding gamma rays are obtained from the decay of nuclides such as ^{24}Na, ^{54}Mn, ^{72}Ga, ^{88}Y, ^{124}Sb, ^{140}La, ^{226}Ra and ^{228}Th. Important characteristics of some of these sources are given in Table 5-I. The (γ, n) threshold for Be and deuterium are 1.67 and 2.23 MeV, respectively. To yield high neutron yields, these sources must have very thick targets.

Generally, the (α, n) sources are more widely used for mineral exploration for the following reasons: (1) the neutron yield per curie is higher than for (γ, n) sources, and (2) the (γ, n) sources produce more interfering radiation from the gamma-emitting nuclides used in the source and also they have short half-lives (see Table 5-I). However, (γ, n) sources have the advantage of being smaller in size, but are more difficult to handle.

Perhaps the most useful (γ, n) source is the ^{124}Sb-Be, because of its low cost, its ability of being regenerated by irradiating it in a high flux reactor, and its relatively high output of 2×10^6 n s^{-1} Ci^{-1}. However, owing to its relatively short half-life and low neutron output energy of about 24 keV, its use has been on the down trend. A neutron spectrum from this source is shown in Fig. 5-18. For most (γ, n) sources, the highest-energy neutrons emitted are still below normal threshold for inducing reactions such as (n, α),

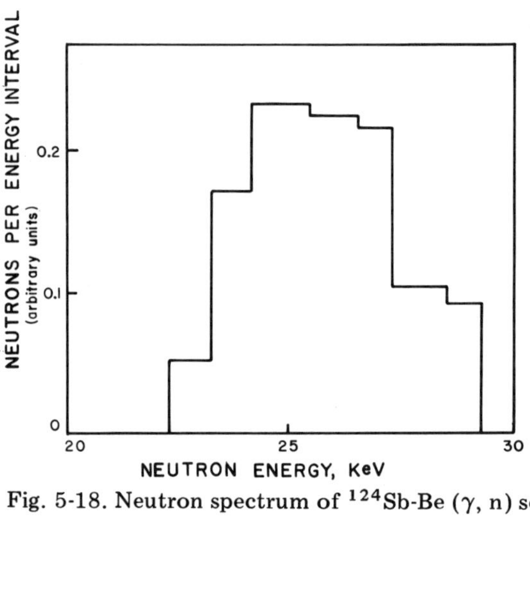

Fig. 5-18. Neutron spectrum of ^{124}Sb-Be (γ, n) source.

(n, 2n) and (n, p). In practice a thermal neutron shield, such as ^{235}U fission foil, in front of the source is used to shield out the thermal component.

Isotopic spontaneous fission neutron source

In the isotopic spontaneous fission process, neutrons are emitted. Some transuranic elements, disintegrating to a significant degree by spontaneous fission, are copious sources of energetic neutrons. Some of these sources and their emission characteristics are shown in Table 5-II abstracted from Kruger's treatise (1971). Currently however, most of the heavy elements listed in Table 5-II are produced in minute quantities from the irradiation of fuel elements in very high flux reactors for a number of years. Their availability till very recent times was extremely limited. In the future these heavy elements may be produced as by-products from controlled thermo-nuclear explosions. These events produce heavy radioelements by successive neutron capture processes to generate beta-unstable products. After several beta decays the unstable nuclide decays by spontaneous fission and alpha decay, to produce high fluxes of neutrons with a distribution comparable to that obtained from the thermal neutron fission of ^{235}U. If problems, such as heat dissipation from the sources are solved, these spontaneous fission sources will be extremely useful for the analysis of minerals. Among these sources ^{252}Cf is the most popular and useful. It is of rugged construction, small in size, provides a stable flux for a reasonable period of time (half-life 2.646 years) and has flexibility in terms of encapsulation. In Tables 5-III to 5-V and Figs. 5-19 and 5-20 are given data pertinent to this source. The source is generally encapsulated in stainless steel, platinum-rhodium alloy, zircaloy or certain types of ceramics. The source is manufactured by the U.S. Atomic Energy Commission (now ERDA) and costs $ 10 per microgram. The encap-

TABLE 5-II

Spontaneous fission neutron sources

Nuclide	Alpha decay $T_{1/2}$ (yr)	Fission $T_{1/2}$ (yr)	Fissions per 10^6 α's	Neutrons per fission
^{238}U	4.5×10^9	6.5×10^{15}	5.6×10^{-1}	—
^{236}Pu	2.85	3.5×10^9	8.1×10^{-4}	1.89 ± 0.20
^{238}Pu	86.4	4.9×10^{10}	2.3×10^{-3}	2.04 ± 0.13
^{240}Pu	65.8×10^2	1.3×10^{11}	5.4×10^{-2}	2.09 ± 0.11
^{242}Pu	3.8×10^5	7.1×10^{10}	5.3	2.32 ± 0.16
^{244}Pu	7.6×10^7	2.5×10^{10}	3.0×10^{-3}	—
^{242}Cm	0.45	7.2×10^6	6.2×10^{-2}	2.33 ± 0.11
^{244}Cm	17.6	1.3×10^7	1.3	2.61 ± 0.13
^{252}Cf	2.6	85	3.3×10^4	3.51 ± 0.16
^{254}Cf	long	0.17	large	—

TABLE 5-III

Neutrons from ^{252}Cf spontaneous fission

Energy (MeV)	Abundance ($n\ s^{-1}\ mg^{-1}$)
10.00 —14.92	8.28×10^6
6.70 —10.00	6.42×10^7
5.49 — 6.70	8.19×10^7
4.49 — 5.49	1.26×10^8
3.68 — 4.49	1.68×10^8
3.01 — 3.68	2.00×10^8
2.02 — 3.01	4.36×10^8
0.91 — 2.02	6.98×10^8
0.41 — 0.91	3.36×10^8
0.11 — 0.41	1.56×10^8
0.015— 0.11	2.74×10^7
0.0 — 0.015	0
Total	2.3×10^9

sulation of sources in the range of 100 μg costs in the region of $ 1000 and a further investment of approximately $ 1000 is necessary to provide an adequate shipping or storage container. The cost of larger shields is of course much higher. When high source strength is needed without adding to the weight of the isotope, a uranium blanket (Californium-252 Information Center, 1974) is used to boost the neutron output. Boosting factors approaching a factor of 35 have been obtained in such a manner. Concise information regarding the purchase, encapsulation and research done with this source are found in regular ^{252}Cf bulletins issued by the ^{252}Cf Information Center of the United States Energy Research and Development Administration (ERDA) in Aiken, South Carolina. Except for its half-life to a certain extent, this source possesses ideal characteristics for moderately long-term use in the field for mineral exploration.

TABLE 5-IV

Gamma rays from ^{252}Cf alpha decay process

Energy (MeV)	Abundance (photons $s^{-1}\ mg^{-1}$)
0.043	2.8×10^6
0.100	2.0×10^6
0.156	4.0×10^5

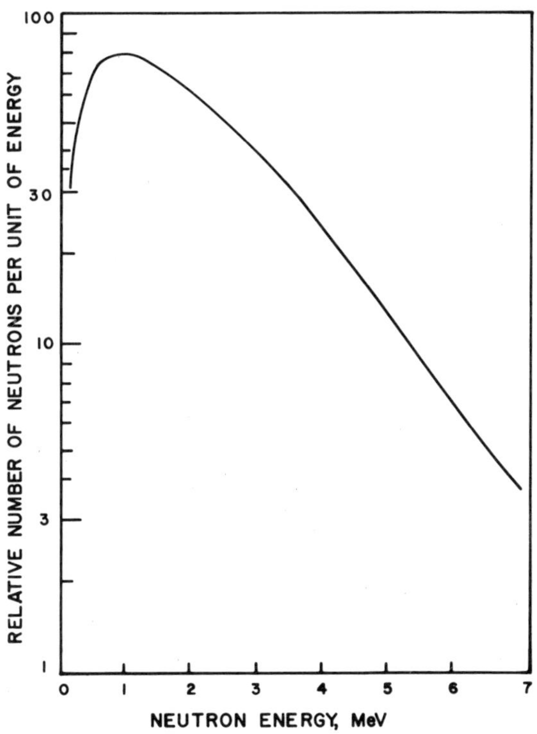

Fig. 5-19. Fission neutron spectrum of ^{252}Cf source.

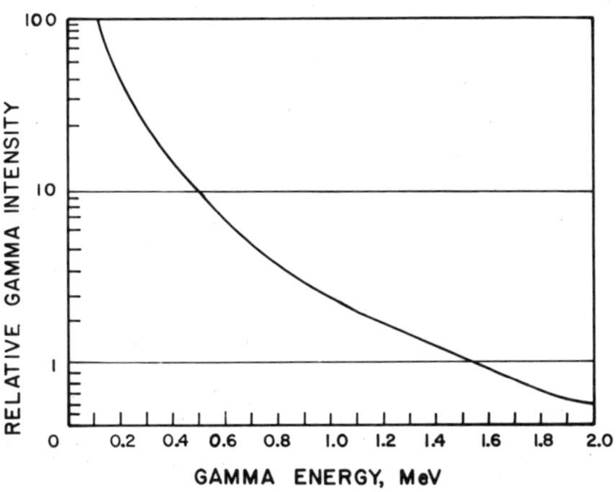

Fig. 5-20. Gamma spectrum of ^{252}Cf source.

TABLE 5-V

Gamma rays from ^{252}Cf spontaneous fission

Energy (MeV)	Prompt gamma rays (photons s^{-1} mg^{-1})	Gamma rays from equilibrium fission products (photons s^{-1} mg^{-1})	Total (photons s^{-1} mg^{-1})
0 —0.5	3.3×10^9	1.3×10^9	4.6×10^9
0.5—1.0	1.7×10^9	4.0×10^9	5.7×10^9
1.0—1.5	7.7×10^8	9.1×10^8	1.7×10^9
1.5—2.0	4.2×10^8	3.5×10^8	7.7×10^8
2.0—2.5	2.2×10^8		2.2×10^8
2.5—3.0	1.1×10^8		1.1×10^8
3.0—3.5	5.6×10^7		5.6×10^7
3.5—4.0	3.0×10^7		3.0×10^7
4.0—4.5	1.7×10^7		1.7×10^7
4.5—5.0	8.2×10^6		8.2×10^6
5.0—5.5	4.9×10^6		4.9×10^6
5.5—6.0	1.8×10^6		1.8×10^6
6.0—6.5	1.0×10^6		1.0×10^6
Total	6.6×10^9	6.6×10^9	1.3×10^{10}

Accelerator neutron sources

Neutron generators

Miniature neutron generators have been widely used in activation analysis of minerals and for exploration. The commercial instruments are scaled-down charged-particle accelerators producing moderate fluxes of 14- and 3-MeV neutrons by exoergic (d, n) reactions such as ^{3}H(d, n) ^{4}He and ^{2}H(d, n) ^{3}He. Principally a neutron generator is composed of an ion source, an accelerating tube or chamber and a target. Ions generated in the ion source are accelerated through a potential difference of 60—150 kV in the accelerator tube which is maintained at pressures of 10^{-4} to 10^{-6} mm Hg. The accelerated ions then strike the target to produce high-energy neutrons which are approximately isotropic in emission. These neutrons can be moderated by polyethylene or paraffin wax to generate thermal neutrons which can be used for activation purposes. Certain minerals are more amenable to analysis via their threshold reactions. In such cases, the moderating medium is eliminated from the measurement device. A recent treatise on the use of neutron generators for activation analysis (Nargolwalla and Przybylowicz, 1973) describes the application and mechanical aspects of this type of accelerator. The section on sealed-tube generators applicable to borehole analysis is also given. The reader is advised to refer to this text for discussions pertinent to the operation and maintenance of the accelerator. The application of this type of neutron source has the following advantages: (1) the neutron beam can be turned off by simply de-exciting the high voltage, and (2) threshold reactions often provide a unique solution to certain nuclear interference problems, not possible with thermal neutron activation. On the other hand the instrument also presents some problems in terms of accurate flux monitoring of the primary neutron emission from the target and the mechanical difficulties associated with the generation of high voltages of that magnitude inside small devices lowered into boreholes. Some applications that have found success using this type of neutron source are discussed in the section "The radioactivation methods for mineral exploration".

Other accelerator neutron sources

Limited use has also been made of other accelerator type neutron sources. In a few cases attempts have been made to use a Van de Graaff generator in the borehole environment. This method requires an extremely high degree of instrument sophistication and does not appear to have reached a state when it can be conveniently used for mineral exploration in downhole devices.

Gamma-ray sources

These sources are principally used for gamma-gamma devices for mineral characterization. Four useful sources employed in various devices are ^{137}Cs, ^{60}Co, ^{226}Ra and ^{75}Se. For selective gamma logging the sources described earlier are recommended. Gamma sources are cheap and easily procured. The experimenter has a great deal of flexibility in his choice of the specific activity and encapsulation. Since these sources do not activate the sonde or the irradiated matrix, the analysis is more convenient and straightforward.

Some properties of such sources are listed in Table 5-VI.

TABLE 5-VI

Nuclear properties of gamma-ray sources

Source	Mean gamma energy (keV)	Half-life	Mass absorption coefficient (cm^2 g^{-1})	
			$CaCO_3$	ZnS
^{75}Se	224	120 days	0.1223	0.1353
^{137}Cs	662	30 yr	0.0777	0.0741
^{60}Co	1250	5.3 yr	0.0577	0.0546
^{226}Ra	186	1602 yr	—	—

RADIOACTIVATION METHODS FOR MINERAL EXPLORATION

Introduction

The application of radioactivation methods for the analysis of mineral samples has been well documented in literature. Single- and multi-element studies are discussed in several hundred publications. The general activation analysis bibliography compiled by Lutz et al. (1971) issued by the National Bureau of Standards in 1971 lists over 400 references related to the analysis of mineral samples. However, by far the majority of this research has been carried out in highly controlled laboratory environments. Despite the fact that many of these studies purport direct applications to the mineral industry, very few have been tested under typical field conditions. Furthermore, of the limited number of such tests, only a handful are actually being utilized for mineral exploration and mine development. Reasons for the rather meagre success index are many. Some of the principal causes for the difficulties experienced by researchers in attempting to "sell" their individual developments to the mining industry are:

(1) The general reluctance on the part of prospective users to displace or complement existing chemical and non-nuclear instrumental methods in current use for the characterization of minerals. The chemical analyses of drill-core, chips, cuttings and other types of recovered geological samples still form the basis of orebody evaluation and mineral exploration.

(2) Insufficient examination on the part of the researcher of the specific needs of a particular exploration or prospecting technique. In all fairness, it might be emphasized that these requirements change radically for a given type of mineral deposit operated by two different mine developers. For example, in drill-hole logging for nickel laterites and porphyry copper, one mining organization may only be interested in the principal elements, i.e. nickel and copper; on the other hand, another company would require compositional knowledge of subsidiary elements, such as iron and silicon, since the particular mineral process it uses is extremely sensitive to these matrix elements. Only under these conditions would it consider a new method as a viable alternative to the current chemical and instrumental methods employed.

(3) An investigation into the "field" analytical speed of a developed radioactivation method in the laboratory is often neglected. In mining circles, this factor is of utmost importance, and almost always is directly proportional to the ultimate cost of exploration.

(4) The researcher often overlooks the fact that instrumentation used to analyze samples in the laboratory may not function in the hostile environment so synonymous of field conditions. Enough forethought in this area is often not given, with the result that both electronic and mechanical hardware fail miserably during the early stages of perhaps a hard-to-organize and costly field test.

(5) The high degree of sophistication built into radioactivation methods often acts as a deterrent for its acceptance by field geologists acclimatized to the day-to-day routine of core extraction, bag sampling and mine laboratory analyses.

(6) The researcher often develops a technique under the best possible conditions which do not allow for a sufficient degree of flexibility to research the effects of various physical and mechanical constraints usually imposed by field conditions.

It is the general experience of the few researchers who have found success that, before a research study is initiated, a full understanding of a single application be thoroughly explored, to satisfy the needs of a single user; rather than develop a system which claims to have a general application in the mining industry. Any additional benefits derived from a successful field technique can always be applied to strengthen the case in favor of similar requirements expressed by another mine developer of a similar ore deposit.

In this section, an attempt has been made to select radioactivation methods with a potential for their application to the mineral exploration industry. As such, only those applications having a direct bearing on exploration and mine development, have been considered. However, mention of techniques which are considered to have potential for future applications have also been discussed. A short treatment of the significance of assay data to the evaluation of ore deposits is given. The intention in this chapter is not to list applications on an element-to-element basis, but rather to emphasize on radioactivation techniques that have broken ground with demonstrations of successful field case histories. In this respect, the authors have exercised their own judgment in forming selection criteria for the inclusion of successful studies. As far as possible, nuclear data information, such as sensitivity, cross-section and the like, have been excluded, and the reader is referred to pertinent references for such information. Finally, on-going research into radioactivation methods pertinent to geological exploration is briefly described. In this regard, the authors have offered some extrapolations into the future for these techniques and their importance to the mining industry from economic and analytical viewpoints.

Significance of borehole analysis for deposit evaluation

In the mineral evaluation of ore deposits, it is necessary to know the ore grade of the mineralized layers located between any two sampling points from the wide-spaced drilling and sampling in the exploration stage of mining. The delineation of an area of influence in an assay is thus important for grade estimation in underground mine sections, and for reporting daily grades and tonnages for production scheduling.

In the statistical analysis applicable to mine sampling, exploration, grading of deposits or ore occurrences, special techniques are used in the data analysis. Two such techniques, i.e. the mean-square-successive-difference test and linear correlation, are applied to designate an area of influence to a particular assay and continuity of assays obtained, and mineral deposit samplings.

Hazen (1967) applied the mean-square-successive-difference and linear correlation techniques to investigate the area of influence and the continuity of assays. The linear correlation technique is used to relate a series of measurements of two variables, e.g. sequence of assays in one drill hole to that in an adjacent drill hole in order to establish ore continuity, ore grade between holes and two samples in layered deposits. The mean-square-successive-difference technique is used to provide an indication of the difference between successive assays of successive samples and is used as a standard test procedure for both long- and short-range trends or fluctuations. A brief description of the two statistical procedures follows.

Linear correlation

In this technique, the linear correlation coefficient, r, is related to N, the number of items in each set of data, i.e. X values being individual items in one set of data and Y values being individual items in the second set of data, as follows:

$$r = \frac{N\Sigma XY - (\Sigma X)(\Sigma Y)}{\sqrt{[N(\Sigma X^2) - (\Sigma X)^2][N(\Sigma Y^2) - (\Sigma Y)^2]}} \tag{34}$$

The linear correlation, r, may be expressed in terms of the variance, based on the assumption that the correlation between Y and X values is linear and that variation in X is related to variation in Y. The amount of variation is expressed by r^2.

The residual variation, $(1 - r^2)$, is often considered as a random element in the two data sets.

Mean-square-successive difference

This difference, δ^2, which relates the ith assay X_i, the successive $(i + 1)$th assay X_{i+1}, and the number of assay items, N, is obtained by taking the mean of the square of the $(N - 1)$ difference between two successive assays, or:

$$\delta^2 = \frac{\Sigma(X_{i+1} - X_i)^2}{N - 1} \tag{35}$$

The variance of the assays is dependent on the ith assay X_i and the mean of all assays $\overline{X}$ as follows:

$$s^2 = \frac{\Sigma(X_i - \overline{X})^2}{N - 1} \tag{36}$$

The comparison between δ^2 and s^2 is that the value of δ^2 is based on the difference between two successive assays, whereas the value of s^2 is dependent on the deviates of each assay from the mean of all assays. When $N > 25$, the formula for the mean square succession difference, Z, is used (Bennett and Franklin, 1954):

$$Z = \frac{\epsilon}{\sigma_\epsilon} \qquad (37)$$

where:

$$\epsilon = 1 - \frac{\delta}{2\,s^2}; \qquad \sigma_\epsilon = \sqrt{\frac{N-2}{N^2-1}} \qquad (38)$$

For long-period fluctuation in the assay data, ϵ will be positive when the ratio $\delta^2/s^2 = \eta$ is less than 2. However, for short-period fluctuation in the assay data, ϵ will be negative for values greater than 2. These positive and negative values of ϵ are useful in the mean-square-successive-difference test in determining the independence and non-randomness between successsive assays, and also to differentiate between the types of fluctuations in the assay data as well as the prediction of trend in the assay data.

A process, referred to as Kriging, of finding optimal weighting factor for sample grades used in the estimation of blocks of ore has been recently described by Dowd (1975). This procedure uses a cathode-ray display screen in the conversational mode with the operator. The method of Kriging in a two-dimensional system and the use of the variogram, a major tool of geostatistics, is applied to basic mining problems. The variogram is a function which represents the variance of samples with respect to the differences between them. Several examples of this technique in which the operator is intimately involved in the estimation process via the CRT display are given. The use of such computer techniques are necessary to bring operators in close touch with the problem, rather than depend on the service of expert computer specialists who are seldom aware of the many variables affecting the estimation of ore grades in a complex geology.

Application of radioactivation methods

Although the petroleum industry has been using nuclear techniques for well logging of oil-bearing formations since the 1930's, the mineral industry, on the other hand, has been rather slow to take advantage of such techniques. Consequently applications and appropriate documentation of field case histories have been meagre to date. In the literature, it is difficult to search for such applications for specific elements since the methods themselves are generally classified according to the nuclear principles used. A recent review by Scott and Tibbetts (1974) summarizes nuclear methods for mineral deposit evaluation. The radioactivation methods as applied to the determina-

TABLE 5-VII

Radioactivation methods in well logging

Logging technique	Metals														
	general	Al	Sb	Be	B	Cu	Fe	Pb-Zn	Mn	Hg	Ni	Ag-Au	Sn	W	U
Natural gamma ray	×	×					×								×
Gamma-ray spectral															×
Gamma-gamma ray						×	×	×		×	×		×		×
Selective gamma-gamma ray						×	×	×	×	×				×	×
X-ray fluorescence			×			×		×	×	×		×	×		×
Neutron absorption		×			×				×	×					
Neutron activation		×				×			×		×	×		×	×
Photoneutron				×											
Resonance gamma absorption													×		
Neutron fission															×

	Non-metals						
	general	coal	evaporites	fluorspar	oil shale	rock salt	S
Natural gamma ray	×	×	×			×	
Gamma-ray spectral							
Gamma-gamma ray	×	×	×		×	×	×
Selective gamma-gamma ray		×					
X-ray fluorescence							
Neutron absorption	×	×	×				
Neutron activation				×			×
Photoneutron							
Resonance gamma absorption							
Neutron fission							

tion of elements by well logging techniques are abstracted from this review and tabulated (Table 5-VII).

These techniques are within the general scope of this chapter and therefore described. In order to reduce the complexity of this discussion, an attempt is made to classify applications in terms of the nuclear process described in the section "Theoretical considerations". Furthermore, since the utilization of instruments, logging devices, mobile laboratories, and other logging hardware are closely tied in with their individual applications to the determination of specific elements, such descriptions are included in the discussion of pertinent applications. Properties of special materials used for the construction of logging devices or sondes are given to aid in the design and fabrication of similar measuring tools. Such discussions are again restricted to the application in question. An attempt is made to minimize the inclusion of large tabulations of nuclear data, sensitivity and the like. As far as possible the sources of such information have been referenced and only appropriate abstractions from such sources have been tabulated. It is hoped that the approach described above will be less confusing to the reader and still maintain continuity of discussion relevant to application of radioactivation techniques to mineral exploration. It may also be recalled that the selection of techniques is based on the authors' judgement as to which technique had been adequately researched and had produced, or had the potential of producing satisfactory field case histories. Before discussing the application of individual radioactivation techniques to mineral exploration, the reader's attention is drawn to some excellent reviews by Berzin et al. (1966), Keys and Boulogne (1969), Czubek (1971), and Gjibels (1973), which summarize applications; and to sources of nuclear data compiled by Greenwood and Reed (1965), Brookhaven National Laboratory (1960—1966), Rasmussen et al. (1969), Duffey et al. (1970), and Nargolwalla and Przybylowicz (1973), applicable to radioactivation methods of elemental analysis.

Gamma-gamma techniques

As mentioned earlier, the gamma-gamma backscattering technique has been used for element determinations by three possible methods: (1) density logging, (2) selective gamma logging, and (3) spectrometric gamma-gamma logging.

Perhaps one of the best examples of the utilization of this general technique is documented by Donhoffer (1968). The evaluation in a horizontal drill hole concerns the analysis of zinc in a zinc-lead deposit. Because of the special geology in the mine operated by Bleiberger Bergwerksunion, Austria, calibration curves relating density to the zinc content were developed for an optimum source-detector geometry and an optimally selected primary gamma-ray energy source. The backscatter probe used by Donhoffer is shown in Fig. 5-21. A family of calibration curves was developed for various source-detector geometries and source types using fundamental

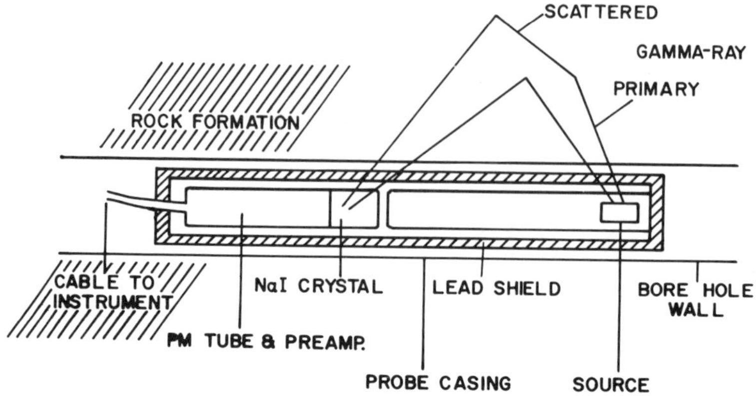

Fig. 5-21. Schematic view of backscatter probe (Donhoffer, 1968).

gamma-ray transport principles applied to a two-component system. The particular geology investigated was a good example of a two-component system. The zinc ore occurs mainly as the sulphide in a limestone matrix. Lead concentrations were generally of an order of magnitude less than that of zinc. The analysis was conducted in 48-mm-diameter boreholes. For comparison purposes fourteen holes, some of which were 50 m long, and for which core analyses were available, were selected. The total length of boreholes logged was about 940 m.

The probe (Fig. 5-21) consisted of a source holder in which three types of gamma-ray sources were used. These were ^{75}Se, ^{137}Cs and ^{60}Co. A 14-cm-long lead shield separated the source from a 0.75-inch by 0.75-inch NaI(Tl) scintillation detector. The distance between the source and detector could be varied by introducing acrylic spacers. The overall diameter of the probe was 36 mm. The probe was enclosed in a close-fitting plastic tube of 42 mm outside diameter, so that the probe was essentially centered in the borehole.

A ratemeter with a time constant of 15 seconds was employed to measure count rates up to 2000 counts per second. Incremental logging in 0.25-m steps was performed. This distance corresponded to the linear length of the sensitive volume scanned by the probe. Nominally the scans were done at a speed of 25 m per hour. The logging data were normalized to measurements made on only the rock matrix. These data were obtained from the scanning of boreholes in which only the matrix was present. This so-called zero value (I_0) was found to be essentially constant. Plotting I/I_0 against the mineral content, Donhoffer obtained results as shown in Figs. 5-22 to 5-24. Based on his studies, Donhoffer concluded that a 37-mCi ^{137}Cs gamma source separated from the detector by 30 cm not only gave almost an optimum sensitivity, but the probe was also insensitive to other interferences from anomalies. Several conclusions are drawn from the results: (1) if zinc was the only

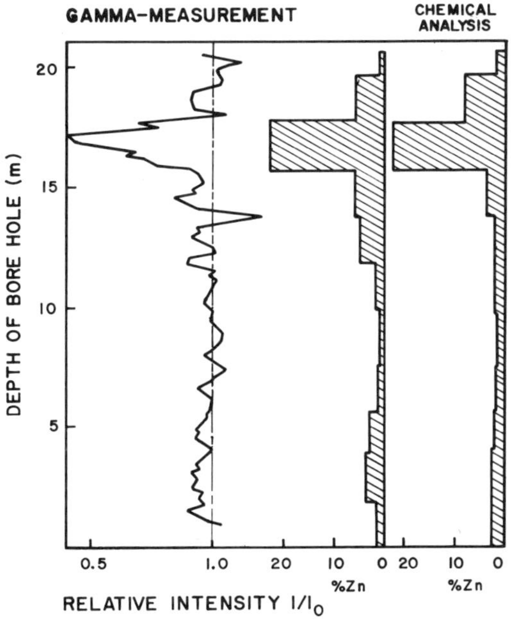

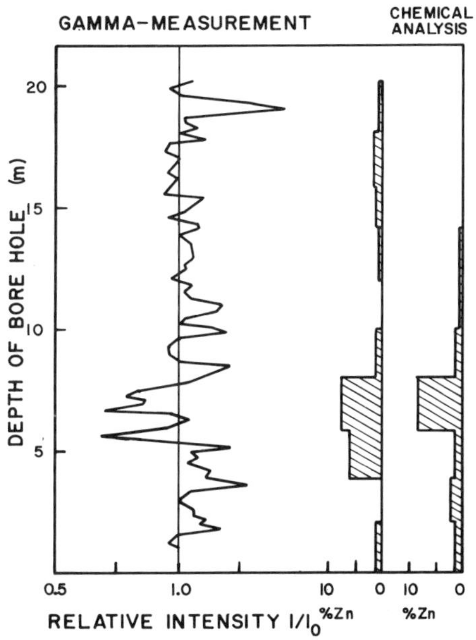

Fig. 5-22. Borehole profile, showing high ore concentration at the end (Donhoffer, 1968).

Fig. 5-23. Borehole profile with strong anomalies (Donhoffer, 1968).

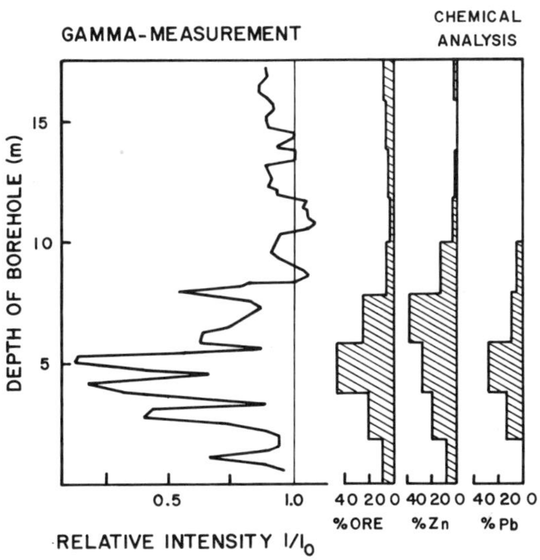

Fig. 5-24. Borehole profile with high-grade zinc- and lead-bearing zone (Donhoffer, 1968).

element present in the matrix, ore concentration in the range of 1—10% Zn could be analyzed to an accuracy within ±1% Zn; (2) if lead is present some knowledge of Pb/Zn ratios is necessary; (3) the method has the capability of analyzing a much larger ore sample than that used for core analysis; and (4) the technique offers distinct advantages in terms of the speed of analysis and the almost instantaneous data retrieval.

In assessing the merits of Donhoffer's study, it can be said the presence of lead can have a very serious effect on zinc determinations. The sensitivity for lead determinations is of the order of 3—5 times that for zinc. The possibility of determining both lead and zinc also exists. In this case two sources, ^{137}Cs and ^{75}Se, are used. The primary source energies essentially straddles the K-edge of lead. By using a very short source-detector distance the response from each element can be measured. However, such experiments are still in the laboratory stage.

The use of density gamma-gamma and selective gamma-gamma methods have been made for the determination of iron. Czubek (1971), reporting on the success of several investigations in this area, describes a method employing the two-source principle. The logging sonde contains a ^{75}Se and a ^{137}Cs source. For qualitative identification of ore zones, the total scattered gamma-ray intensity is measured. Ore zones containing iron are characterized by minimum intensity as measured by the detector system. Since a total intensity measurement is not adequate for an iron correlation, two approaches are taken: (1) developing a relationship between the metal concentration and the average atomic number of the matrix, Z_{eq} (for iron there appears to be a good correlation), and (2) developing a correlation between the in-place density and the iron content. Results reported by Czubek for the latter technique are shown in Fig. 5-25. However, Czubek points out that if no correlation between the metal content and the in-place density exists, a relative spectral intensity measurement method can be used. In this case the ratio of soft to the hard scattered components is correlated with the element concentration. An example of this application for the determination of iron is shown in Fig. 5-26. It should be recognized that logging instrumentation required for this type of operation is considerably more sophisticated and in the development stage.

Recalling Donhoffer's study (1968) it was emphasized that the presence of lead in the deposit would influence determinations for zinc, and the possibility of using the two-source principle for determining a correction for the Pb/Zn ratios was considered. Czubek's report (1971) describing the ^{75}Se-^{133}Ba system is encouraging. By choosing an optimum source-detector distance expressed in g cm^{-2}, the equation:

$$r^2 R_{\gamma\gamma} = f(r\rho) \tag{39}$$

can be used to design a probe which is relatively insensitive to changes in the in-place density of the rock matrix. In the above expression, r, is the source-

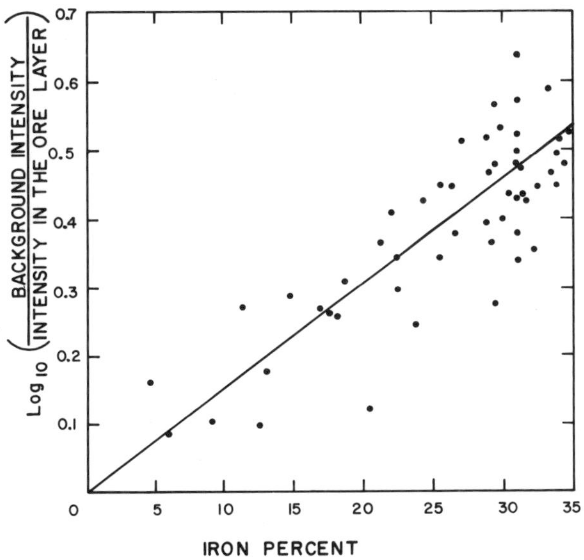

Fig. 5-25. Correlation between the logarithm of the amplitude of the selective gamma-gamma anomaly and iron content (Czubek, 1971).

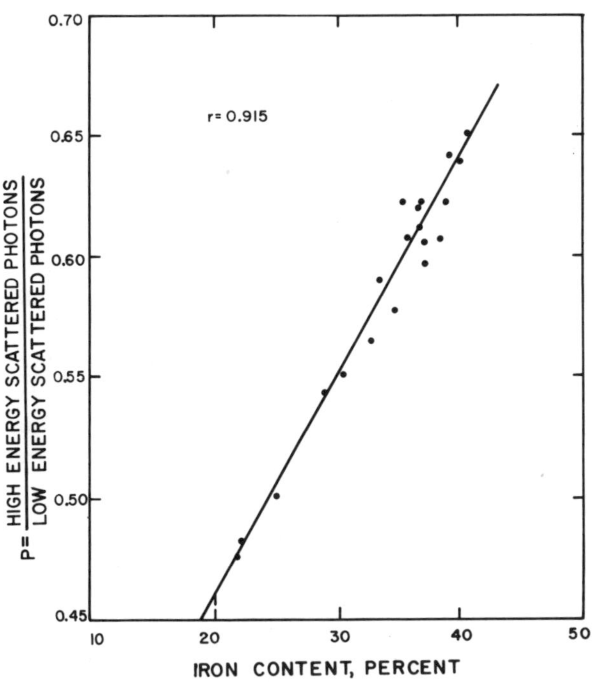

Fig. 5-26. Correlation between the P value and iron content (Czubek, 1971).

detector distance, $R_{\gamma\gamma}$, the measured intensity of scattered photons, and ρ is the rock density. The method used to obtain Fig. 5-25 can now be used to interpret the results. Such an approach is particularly successful for the determination of lead in the presence of zinc and iron. However, for the technique to be completely independent from density effects, supplemental gamma-gamma density logs are necessary. The uncertainty in lead determinations using the selective gamma-gamma method is of the order of 14% with a detection limit of about 0.2—0.3% Pb. In the Soviet Union and Czechoslovakia, lead determinations are routinely carried out by this method and have afforded a 10% savings in the total cost of production of the mine. Simon (1969) has described a system of Czechoslovakian manufacture in which a probe 34 mm in diameter is used. The primary gamma source is ^{60}Co. The detector system uses a Geiger-Müller tube. This battery-operated device is capable of making point-to-point measurements in 25-cm steps. A 30-m borehole can be logged in six hours. Quadruplicate measurements each of three-minute duration are taken at each meter. This system has undergone exhaustive tests in the Czechoslovakian tin deposits. From 1.5 to 3 mCi of ^{60}Co and from 1.5 to 2 mCi of ^{75}Se are used as excitation gamma-ray sources.

In order to reduce the systematic errors in the gamma-gamma type measurements, attempts have been made to reduce the effect of a changing borehole diameter, uneveness of the borehole walls and the effect of drilling mud on the measurements. Special efforts have been made to properly collimate both the source and the detector, to reduce some of the above sources of error. Today the state-of-the-art offers equipment which is stable even at temperatures up to 150°C. Blyumentsev and Feldman (1962) have developed an instrument which adequately corrects for the effect of drilling mud. Their system is essentially a two-detector system, each detector is of a different size and is situated at different and adjustable distances from the gamma-ray source. The larger detector response is converted into rock density values. By analog computing devices (constructed from the small detector measurements), a correction curve is generated. This curve is added to, or subtracted from (in the case of heavy solutions), the density curve. Such equipment has been known to operate under a pressure of 1400 kg cm^{-2} and at temperatures up to 200°C and provide logging speeds of 550 m per hour. However, the sonde requires a minimum borehole diameter of at least 5 inches.

Experimental conditions pertinent to selective gamma-gamma logging include compensatory parameters, to reduce the effect of changing borehole diameter. Experimentally it has been established that for a 32-mm diameter sonde employing two sources, one of which is ^{75}Se, no effect is observed for a borehole diameter change between 40—230 mm when the optimum distance of ^{75}Se sources from the detector are 190 mm and 19 mm, for an activity ratio of 25 : 1. These parameters are so chosen that as the borehole diameter varies the reduction in scattered radiation from one source is compensated by an increase in intensity from the other. Furthermore, it has been established that the double probe is essentially independent of density within a substantial range of 1.5—3.0 g cm^{-3}. Although the "double probe" concept provides comparable sensitivities to those obtained from normal probes, the rock penetration is approximately one half when compared to that of the normal probe.

In spectrometric gamma-gamma logging, the spectral analysis of scattered photons makes it possible to determine the discontinuity of the spectrum due to the K-absorption edge. Some advantages realized by this approach include high sensitivity, differentiation of Compton scattering events and less susceptibility to the effect of cavities than methods in which only integral intensity measurements are made. An important consideration in coal mining is the determination of the ash content. In the 0—30% ash content range, the accuracy for its determination, using spectrometric gamma-gamma logging, is of the order of 1—15%. In this specific application carried out by Utkin (1965), spectral ratios of counts recorded in a channel set at 60 keV and the other at 140 keV are computed; the band width of both channels is about 25 keV. The instrument included a source (^{204}Tl, 40 mCi) separated from a detector by only 10 cm. Spectrometric instrumentation that can be used in boreholes up to 1500 m in depth, has been developed.

Gamma-ray resonance scattering

Sowerby and Ellis (1973), in an elegant demonstration of the gamma-ray resonance scatter phenomenon, showed a promising application of this technique to plant-stream monitoring of certain base metals in the mineral processing industry. Another possible application is the borehole analysis for the elements copper and nickel. Inferences drawn from laboratory studies carried out in simulated media indicated that gaseous sources of $^{65}ZnI_2$ and $^{60}CoBr_2$ could be used for the determination of copper and nickel, respectively. Borehole diameters from 7.5—20 cm could be used without inducing significant excess errors. The sensitivity of the method for copper analysis was 4 counts per second per 1% Cu in the photopeak for a 10-Ci $^{65}ZnI_2$ source and a NaI(Tl) detector. At this sensitivity the background under the photopeak was also 4 counts per second. Similar data were obtained for the element nickel. It was contended from their results that logging speeds of about 0.5—1.0 m per minute were possible in boreholes $>$14 cm in diameter. Under these conditions 0.5% Cu or Ni could be determined with a relative accuracy of 20%. Advantages of this method over the diamond-core drilling process included a 50% cost saving and analysis of samples twenty times larger than equivalent core samples.

In the comparison of this method with other techniques, Sowerby (1973) is perhaps unduly harsh, since the gamma resonance scatter technique does not appear to have gone past simulated borehole testing. Furthermore, several methods criticized have already proven themselves in production logging under severe environmental conditions. Moreover, the object of using a high activity of gaseous sources for borehole work may raise difficult questions in terms of health physics precautions and radioactivity licensing administrations in various countries. However, the intrinsic advantages of this technique under well-controlled conditions, such as plant-stream monitoring in the process industry are significant. The technique is probably the most specific for copper and nickel. The application is unique, and with time will find a significant place in the analytical arsenal.

Photonuclear methods

As emphasized in the section "Theoretical considerations", photoneutron logging is limited to the determination of beryllium. Currently this technique is used to re-evaluate known beryllium deposits, to extend the thickness of these deposits and re-determine the depths. Prior to logging, standardization of beryllometers is necessary. Normally this is performed by calibrating the instrument using an experimental model made up of a mixture of beryllium oxide and paraffin. In the field, high-speed scans at rates up to 150 m per hour are carried out. Interesting ore regions isolated are carefully examined at a much reduced rate of ca. 20—40 m per hour. Berzin et al. (1966) define a relationship in saturated strata between the area S of the anomaly, the true thickness of the ore body H, and the beryllium content g by the equation:

$$S = KHg \tag{40}$$

where K is a proportionality coefficient whose value depends on the parameters of the borehole sonde, e.g. source-detector separation, detector geometry and efficiency, source strength, drilling conditions, density and composition of the rock, etc. Should the orebody contain high neutron-absorbing elements such as B, Li, etc., the gamma-gamma or neutron-neutron logging is used to supplement the gamma-neutron beryllium log. Except for the environmental factors, the technique is specific for beryllium. At the threshold energies used, beryllium is the only element which will undergo the (γ, n) reaction. The present beryllometers, as compared to the 1950 models, have not really changed too much. The sensitivity of the method is of the order of 0.001% BeO when an irradiation source strength of 10—20 mCi of ^{124}Sb and a ^{3}He neutron detector are used. Beryllometers incorporating source strengths up to 50 mCi of ^{124}Sb have been used to determine beryllium with purported accuracies of 0.003% with a logging speed of about 20—30 m per h. Much of the development of beryllometers is done by Soviet researchers (Saltsevich et al., 1968a, b) whose work is highly recommended for interested log analysts.

Neutron methods

Of all the nuclear techniques applied to in-situ elemental determination in boreholes, the neutron activation and the neutron-capture-gamma methods are considered most promising. Basically, these techniques involve the irradiation of the borehole walls and surrounding matrix with neutrons. Upon interaction, several types of nuclear reactions can be induced. The most frequently used in neutron activation analysis is the (n, γ) reaction. Measurements of characteristic gamma radiation resulting from this reaction can be made simultaneous (i.e. prompt gamma emissions from the de-excitation of the compound nucleus) or subsequent (i.e. gamma emissions from the product nuclide decaying with a characteristic half-life). By and large, applied research in neutron-gamma methods has been limited to the measurements of gamma emissions from decaying product nuclide (i.e. delayed gamma-ray measurements). Consequently, the logging process can be severely hampered by the existence of nuclear interferences, environmental background difficulties, and relatively slow and uneconomical logging speeds. On the other hand, the use of prompt gamma emissions for analytical purposes has been neglected. Recently, however, the work of Senftle et al. (1971b) generated considerable interest in the application of this technique to the borehole logging of base metals. In this semi-quantitative study, Senftle et al. determined the element nickel in ore samples by a measurement of the single and double escape fractions of the primary (8.53 MeV) prompt emission from thermal neutron capture in nickel nuclei. A high-resolution solid-state detector was used in their laboratory study. Results obtained in a test borehole indicated the feasibility of the technique. In comparing the above technique with those in which delayed activation products have been measured by Eisler et al. (1971), Hoyer and Locke (1972), Landström et al. (1972), and Moxham et al. (1972), some general conclusions can be drawn (Table 5-VIII).

TABLE 5-VIII

Comparison of in-situ gamma measurement techniques

Analytical property	Analysis technique, dry matrix (density = 2 g cm^{-3})	
	prompt gamma, $E_\gamma = 7$ MeV	delayed gamma, $E_\gamma = 2$ MeV
Useful sample radius	40—50 cm	4—6 cm
Effective sample volume	$\sim 10^6$ cm^3	$\sim 10^3$ cm^3
Borehole effects	minimal	significant
Inhomogeneity effects	minimal	significant
Density effect (1.5—2.0 g cm^{-3})	minimal	significant
Moisture effect	significant	significant
Logging speed	fast, 50—100 ft/h	slow, 5—10 ft/h
Environmental radioactivity interference	not pertinent	severe

Neutron-gamma methods used in mineral exploration have depended upon isotopic (e.g. ^{241}Am-Be), fission (e.g. ^{252}Cf) and accelerator (e.g. sealed-tube 14-MeV neutron generators) neutron sources. The detectors used are essentially of two types, i.e. sodium iodide scintillator and the lithium-drifted germanium detector. Properties and operation of these detectors, in conjunction with nuclear pulse-handling systems can be found in a text by Krugers (1973). To date there are several exploration case histories developed from the use of the sodium iodide detector. An adequate production logging case history (Fanger et al., 1973), based on the use of the Ge(Li) detector, has been reported. Recent developments concerning long-term field use of the Ge(Li) detector described by Lauber and Landström (1972), Tanner et al. (1972), and Fanger et al. (1973), are being actively pursued and it is hoped that in the very near future this type of detector with acceptable efficiency will be available for operation on a continuous day-to-day basis in production logging. The applications of neutrons as a base for mineral explorations are classified according to the nuclear reactions being utilized for elemental analysis.

Inelastic neutron scattering — prompt gamma-ray measurement. This method possesses serious limitations for its direct application to borehole exploration. Studies carried out for the development of this mode of analysis are essentially of an exploratory nature. To construct a viable borehole sonde to include a high-energy pulsed neutron source presents considerable fabrication and analytical problems. Normalization and calibration of such devices in terms of the incident neutron flux delivered per burst to the matrix, give data that is at best within ±20% in accuracy. This discrepancy in the principal controlling activation parameter thus affects the final result accordingly. Caldwell et al. (1960) measured the inelastic gamma-ray spectra of nine elements representing the common earth. Except for some of the light elements (i.e. C, O, Mg, Si and P), the majority of gamma rays resolved were below 2.5 MeV. The neutron source used consisted of a Van de Graaff accelerator in which 325-keV deuterons were allowed to impinge upon a thick zirconium-tritium target. The gamma-ray measurements were made with a 2-inch diameter by 2-inch long NaI(Tl) crystal. Stinner (1961) conducted an extensive feasibility study to assess the potential of this technique for lunar exploration. A high-energy pulsed neutron source and a cylindrical 1.75-inch diameter by 2-inch long NaI(Tl) scintillator were used to generate and measure the inelastic gamma-ray spectra, respectively.

A summary of Stinner's study in which the minimum detectable abundance of nine common earth elements is tabulated in Table 5-IX. A time diagram illustrating the basis of the inelastic scattering technique is shown in Fig. 5-27. In this study, no mention of instrumental difficulties or probable errors to be expected has been made. A more elaborate study made by Waggoner and Knox (1968) is worthy of notice. In this research, a great deal

TABLE 5-IX

Analysis of common lunar elements by nuclear inelastic scattering

Possible elements found on the moon surface	Principal isotope	Isotopic abundance (%)	Prominent energy level (MeV)	Minimum detectable abundance (%)
Iron	^{56}Fe	91.5	0.84	0.1
Silicon	^{28}Si	92.3	1.78	0.5
Magnesium	^{24}Mg	78.6	1.37	0.3
Aluminum	^{27}Al	100	2.20	2.0
Oxygen	^{16}O	99.8	6.14 (triple peak)	3.7
Potassium	^{39}K	93	2.15	1.0
Calcium	^{40}Ca	97	0.77	1.2
Sodium	^{23}Na	100	0.63	1.3
Nickel	^{58}Ni	67.8	1.45	1.0

of effort was expended in developing a suitable fast neutron generator which would potentially function on the lunar surface. The total weight of the instrument was just 14 pounds including a pulsed source of 14-MeV neutrons, a 512-channel gamma-ray spectrometer and associated electronics. Special attention has been given to detector shielding requirements. Readers interested in this mode of analysis will find this report very useful.

Fast neutron-induced reactions. To fully appreciate the significance of principal fast neutron interactions, i.e. (n, p), (n, α) and (n, 2n), as applied to

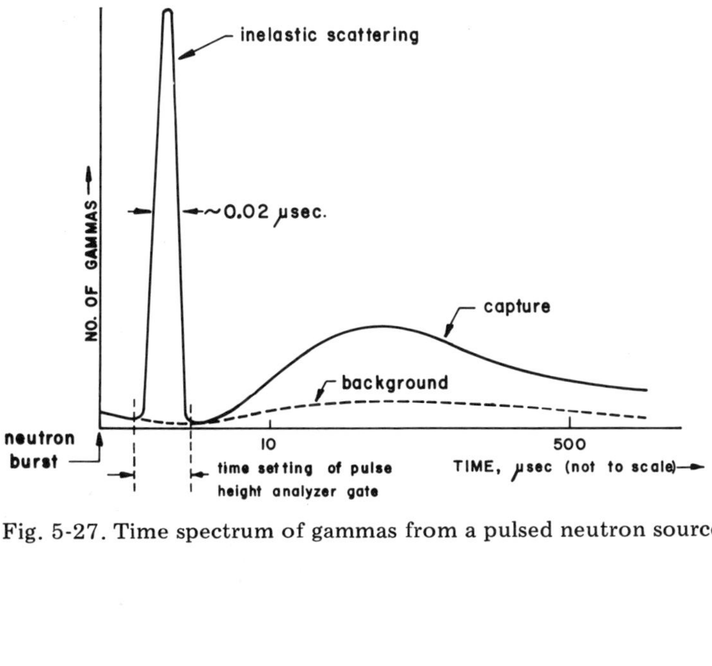

Fig. 5-27. Time spectrum of gammas from a pulsed neutron source (Stinner, 1961).

mineral exploration, one must compare their merits with the operational parameters of both standard chemical and non-nuclear instrumental techniques, and to similar considerations pertinent to other modes of nuclear analysis. Activation analysis with fast neutrons, continuous or pulsed, is principally conducted with portable sealed-tube accelerators described extensively by Nargolwalla and Przybylowicz (1973). Due to the high incident neutron energy, the most probable nuclear reactions are those in which particles are emitted. The basic method is highly selective, particularly for the light elements, and the neutron source, through a choice of different target materials, can provide copious fluxes of intermediate-energy (~3 MeV) or high-energy (~14 MeV) neutrons. With the rapid advancement in electronic technology, it is possible to construct measurement devices that can be introduced into boreholes of only 3 inches in diameter.

In its application to mineral analysis, Senftle and Hoyte (1966) have discussed the use of a fast neutron accelerator source for elemental determinations. Their research shows that when exploring for elements of economic interest, the neutron-producing reaction $^{2}H(d,n)^{3}He$ (3-MeV neutrons) is more practical than 14-MeV neutrons produced from the $^{3}H(d,n)^{4}He$ reaction. From their feasibility study, analytical conclusions regarding the use of 14-MeV neutrons for mineral explorations are shown in Table 5-X. Considerable interest in the use of accelerator neutrons for prospecting has been sponsored by Australian research organizations. Wylie and Eisler (1969) used a sealed-tube accelerator manufactured by Philips, and conducted carefully controlled laboratory studies. From the results obtained, requirements for a viable technique applicable to borehole exploration were expressed. These requirements included a desired rapid scanning logging speed up to 250 m per hour, followed by a more detailed examination of interesting sections of the borehole. A suggested detector system favoured the selection of the NaI(Tl) scintillator over the high-resolution Ge(Li) detector. The importance of shielding the detector with a blanket of ^{10}B or ^{6}Li was emphasized to reduce the measurement of gamma photons created by neutrons scattered into the detector itself. This work was subsequently followed up by a study conducted by Eisler et al. (1971) concerning the analysis of copper. In this study, experimental models which simulate boreholes were used. The measurement of the strong annihilation peak due to positron decay of the activated species produced from $^{63}Cu(n,2n)^{62}Cu$ ($T_{1/2}$ = 9.8 min) and $^{63}Cu(n,\gamma)^{64}Cu$ ($T_{1/2}$ = 12.8 h), respectively, were used for the analysis of copper. According to Dibbs (1970), the elements which interfere in this method are tabulated in Table 5-XI. Eisler et al. (1971) used a 2-inch diameter by 2-inch long NaI(Tl) scintillator with a resolution of 8% for the 0.662-MeV gamma emission from ^{137}Cs. This spectral quality was maintained at the end of a 35-m logging cable. The neutron tube was operated at a continuous output of 10^{8} n s^{-1}. Spectra were recorded by a 400-channel pulse-height analyzer, and digital gain stabilization was used to maintain spectral integrity

TABLE 5-X

The counting rates * for some of the nuclides ** most easily activated by 14-MeV neutrons (10^9 n cm^{-2} s^{-1}) after 2-minute irradiation and a 5-second delay time (Senftle and Hoyte, 1966)

Parent nuclide	Reaction	Daughter nuclide	Half-life	Counting rate (cps/g) after 2-minute irradiation	Counting rate (cps/g) after followed by 5-second delay time	Major γ radiation (MeV)
^{23}Na	n, α	^{20}F	11.2 s	5.8×10^5	5.1×10^5	1.63
	n, p	^{23}Ne	40.2 s	8.0×10^4	7.3×10^4	0.44
^{19}F	n, p	^{19}O	29.4 s	4.2×10^5	3.7×10^5	1.37, 0.20
^{16}O	n, p	^{16}N	7.35 s	3.5×10^5	2.2×10^5	7.1, 6.1
^{28}Si	n, p	^{28}Al	2.27 min	2.0×10^5	1.9×10^5	1.78
^{29}Si	n, p	^{29}Al	6.56 min	5.0×10^4	4.9×10^4	1.28, 2.42
^{11}B	n, α	^{8}Li	0.83 s	1.5×10^5	2.3×10^3	4.43
^{31}P	n, α	^{28}Al	2.27 min	1.4×10^5	1.4×10^5	1.78
^{109}Ag	n, 2n	^{108}Ag	2.3 min	1.0×10^5	1.0×10^5	0.44
^{79}Br	n, 2n	^{78}Br	6.4 min	9.5×10^4	9.4×10^4	0.11, 0.049
^{25}Mg	n, p	^{25}Na	60 s	8.5×10^4	8.0×10^4	1.37, 2.75
^{37}Cl	n, α	^{34}P	12.4 s	7.8×10^4	5.9×10^4	2.1
^{63}Cu	n, 2n	^{62}Cu	9.8 min	4.5×10^4	4.5×10^4	0.66 (weak)
^{103}Rh	n, α	^{100}Tc	15.7 s	3.6×10^4	2.9×10^4	0.54, 0.60
^{27}Al	n, p	^{27}Mg	9.5 min	2.4×10^4	2.4×10^4	0.84, 1.02
^{26}Mg	n, α	^{23}Ne	40.2 s	2.0×10^4	1.8×10^4	0.436
^{34}S	n, p	^{34}P	12.4 s	6.5×10^3	4.9×10^3	2.1
^{54}Fe	n, 2n	^{53}Fe	8.9 min	1.2×10^3	1.2×10^3	0.38

* Based on 10% counting efficiency, flux of 10^9 n cm^{-2} s^{-1}.
** Calculations based on amount of parent nuclide in 1 g of element having normal nuclidic abundance.

Approximately 5 seconds were needed to dump the analyzer memory, thus releasing the memory for the collection of subsequent spectra. The measurement of the delayed gammas was done by first irradiating the matrix for about 20 min and followed by the rapid insertion of the detector to the location of the irradiated area. An evaluation of this work indicated that the sensitivity for copper determinations was increased by allowing interfering elements such as ^{28}Al from the ^{28}Si(n, p)^{28}Al reaction to decay out. The analysis of the 0.51-MeV peak was done by integrating it between two selected channels. The neutron flux normalization technique used involved the summation of counts under the ^{24}Na (2.75-MeV) photopeak. The isotope ^{24}Na is produced from ^{23}Na(n, γ)^{24}Na, ^{27}Al(n, α)^{24}Na, and ^{24}Mg(n, p)^{24}Na. It was assumed that the concentrations of these elements in the simulated ore were constant. In Table 5-XII are given the results of this study. In view of the

TABLE 5-XI

List of positron emitters produced by fast-neutron (n, 2n) activation that could cause interference in copper determination (Dibbs, 1970)

Target nuclide	Isotopic abundance (%)	Product isotope	Half-life	Positron emission (%)	Other gamma energies (MeV)
^{31}P	100	^{30}P	2.53 min	100	—
^{39}K	98.08	^{38}K	7.7 min	100	2.16 (100%)
^{54}Fe	5.82	^{53}Fe	9.0 min	98	0.38 (32%)
^{14}N	99.63	^{13}N	10 min	100	—
^{121}Sb	57.25	^{120}Sb	15.7 min	43	1.17 (1.3%)
^{35}Cl	75.53	^{34m}Cl	32.4 min	50	1.17 (12%), 2.12 (38%), 3.3 (12%)
^{64}Zn	48.89	^{63}Zn	38.1 min	93	0.67 (8%), 0.962 (6%), 1.42 (0.9%)

large number of potential sources of error, it is difficult to assess the accuracy of this method for copper analysis.

Much of the development in fast activation logging has emanated from research groups in industry. Sophisticated logging sondes, and associated pulse sorting and data processing devices, have resulted. Several companies, predominantly in the United States, offer contract services in many areas of nuclear logging. Most of these services have resulted from many years of research in the field, and the final products bear the evidence of concise thinking and innovative instrument design. Even though the accelerator neutron source was developed by a few major electronic companies, the source which could be profitably used was developed by industries involved in providing contract services. Wichmann (1972) describes several applications in which fast neutron interactions were used to produce adequate case histories. The elements amenable to fast activation analysis are oxygen, silicon and aluminum. In Table 5-XIII are given pertinent activation data.

TABLE 5-XII

Copper content of samples (Eisler et al., 1971)

Estimated least-squares fitted Cu value (%)	Cu value determined by chemical analysis (%)
0.52	0.53
1.11	1.04
1.86	1.98
3.79	3.76

TABLE 5-XIII

Nuclear data for fast activation analysis with 14-MeV neutrons (Wichmann, 1972)

Element	Reaction	*E* (MeV)	Half-life
Oxygen	$^{16}O(n, p)^{16}N$	6.1	7.2 s
Silicon	$^{28}Si(n, p)^{28}Al$	1.78	2.3 min
Aluminum	$^{27}Al(n, p)^{27}Mg$	0.84, 1.01	9.5 min

The activation buildup of the irradiated products as a function of activation time is shown in Fig. 5-28. The logging device described by Wichmann (1972) is shown in Fig. 5-29. A standard 12-foot source-detector spacing is used to

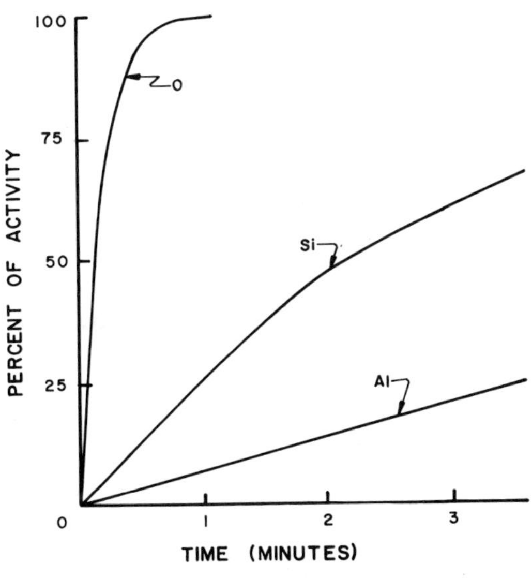

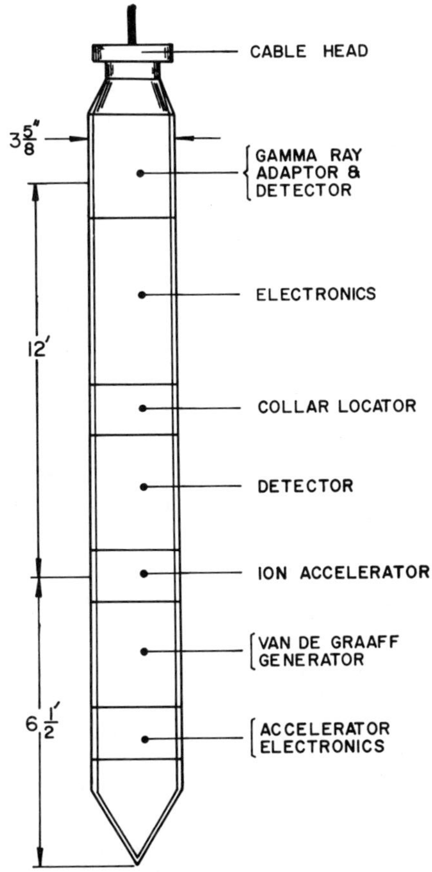

Fig. 5-28. Activation buildup of oxygen, silicon and aluminum (Wichmann, 1972).

Fig. 5-29. Instrument configuration for activation logging (Wichmann, 1972).

log the above elements. This distance is optimally suited for silicon, and in view of the half-life of ^{16}N, oxygen logs have to be performed at speeds faster than operationally derived. To make this instrument capable of logging for oxygen unambiguously, a rather high-energy gamma discrimination level has to be used, with the result that the sensitivity for oxygen drops off significantly. To determine the optimum logging speed with respect to count rate for any activation product, Wichmann gives the equation:

$$V_0 = \frac{0.693\,L}{T_{1/2}} \tag{41}$$

where V_0 is the optimum logging speed in feet per minute; L the source-detector distance in feet; and $T_{1/2}$ the half-life of the measured nuclide in minutes. The optimum logging speed with a 6-foot source-detector spacing is approximately 34 feet per minute. Wichmann drew attention to the fact that oxygen logs for oil-water differentiation were not totally successful, since oxygen is a significant part of all major reservoir rock types. Therefore, anomalies indicated by the logs are often difficult to interpret. Silicon logging, on the other hand, is commercially used. These logs are obtained unambiguously by allowing the ^{16}N activity induced from fast activation of oxygen nuclei to decay out. Interference from the 9.5-min activity resulting from ^{27}Al activation is apparently absorbed in the error of analysis. The optimum conditions for silicon logging include a 12-foot source-detector spacing and a logging speed of 3.6 feet per minute; however, such speeds are considered uneconomical. Acceptable logs can be obtained at speeds up to 12 feet per minute. Logging for aluminum does not appear to be too satisfactory due to the very low optimum speed of ~1 foot per minute, and a background subtraction is necessary after irradiating a given section of the borehole for several minutes. This background is created by the decay of other matrix elements. For all of the three cases described above, it must be emphasized that these logs are essentially "element identification" types, and do not give quantitative results. However, these developments do demonstrate the potential of fast neutron accelerator sources for the determination of light elements. Furthermore, these "identification" logs provide significant information regarding instrumentation and hardware applicable for downhole operation, and form the basis for quantitative determinations in the future.

The possibility of applying a fast neutron accelerator source for silver exploration was investigated by Hoyte et al. (1967). A truck-mounted positive ion accelerator system was used. The applicability of both 14- and 3-MeV neutrons to this problem was assessed in the field. The high-energy neutrons were moderated to thermal energies and the reaction $^{109}Ag(n, \gamma)^{110}Ag$ was utilized for silver determinations. Even at depths down to 26 cm, pulse-height spectra of ^{110}Ag obtained clearly resolved the 0.66-MeV gamma emission. The equipment consisted of a neutron generator and a scintillation detector mounted rigidly on either side of a moderating shield. A sensitivity

Fig. 5-30. Experimental silver detector showing 3-MeV neutron generator on the tailgate assembly which is lowered and in position for irradiation (Hoyte et al., 1967).

for silver down to 1 ounce per ton was predicted. The experimental system including the 3-MeV irradiation system is shown in Fig. 5-30.

Field application of an accelerator source of neutrons to copper analysis was investigated by Hoyer and Lock (1972). Proceeding from laboratory studies, the developed technique was tested in two boreholes. Two types of neutron sources were used. A ^{239}Pu-Be low-energy isotopic source of about 10^7 n s^{-1} output or a (d, T) accelerator source producing 10^8 n s^{-1} of 14-MeV neutrons was incorporated in the field system shown schematically in Fig. 5-31. The purpose of using two types of sources was to distinguish between the ^{63}Cu(n, 2n)^{62}Cu reaction induced by 14-MeV neutrons and the ^{65}Cu(n, γ)^{66}Cu reaction which is predominantly propagated by thermal neutron capture processes. The production of ^{64}Cu can of course occur by fast neutrons, e.g. ^{65}Cu(n, 2n)^{64}Cu, and also by thermal activation, e.g. ^{63}Cu(n, γ)^{64}Cu. However, the nuclear activation parameters did not permit a significant contribution from ^{64}Cu.

The detector in this field system was a 2 inches in diameter by 2 inches long NaI(Tl) scintillator coupled to a 512-channel pulse-height analyzer. The field operation involved a survey of the borehole followed by stationary activation analysis. The results obtained were generally disappointing, in that the technique was not consistent in locating zones of mineralization. Background problems resulting from the activation of ^{28}Si and ^{27}Al were experi-

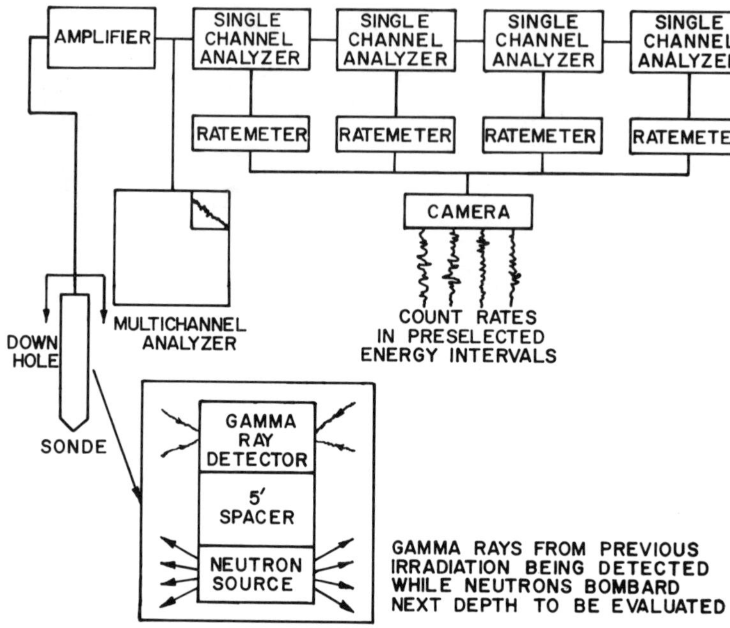

Fig. 5-31. Schematic of logging system (Hoyer and Lock, 1972).

enced. The presence of natural radiation presented additional difficulties. In the stationary runs the neutron source was positioned opposite a zone of interest and irradiated for a 5- or 10-min time period. The detector was then moved opposite to that same zone and measurements made in sequence. After waiting for a period of 2 min gamma-ray spectra were collected for 2 and 3 min each. After a further delay of 2 min, measurements for two additional 5-min time periods were made. Despite the good effort put out in this test, the conclusions were discouraging in terms of the application of delayed gamma counting for copper determination in a borehole. Perhaps an important fact to be considered here is that despite the optimization of instrumentation, the practical logging speed that could be expected, is about 2 m per hour. Such speeds would require considerable economic justification for the acceptance of this particular technique by industry.

In the present day and age, the availability of sufficient energy is of prime concern to most countries. The difficult oil situation in particular has led to a burst of concentrated effort on the part of many governments to find ore reserves for the element uranium. The recent activity in airborne surveys will very likely be followed up by detailed ground evaluations and drill-hole logging. Although the applications of pulsed neutron logging techniques as an economic method for quantitative elemental analysis are limited, the element uranium is one which can be analyzed by such methods. An excellent investigation reported by Czubek (1972) discusses the theoretical concepts and practical implications of this technique as applied to borehole logging for uranium. The method employed the measurement of the time distribution of neutrons after a burst of fast neutrons. Czubek developed a time- and energy-dependent theory, but independent from space, for the four distributions resulting from the prompt thermal fission of ^{235}U and delayed fission of uranium. Theoretical derivations are included in this study. From the experimental verification of the theory the following conclusions were deduced: (1) the application of prompt fission neutrons in the epithermal region requires a short- and high-frequency neutron burst, and (2) a long low-frequency neutron burst is necessary if a measurement of delayed fast or thermal fission neutrons is to be made. This technique, at least in the laboratory, appears to have good potential for the borehole analysis of uranium. Its success in the field, however, is yet to be demonstrated.

Thermal neutron absorption. Recently, the spontaneous fission neutron source ^{252}Cf was made available for sale by the U.S. Atomic Energy Commission (now ERDA). This development has resulted in increasing attention to the use of this source in logging sondes. Reference is made to several investigations carried out by Philbin et al. (1971), Senftle et al. (1971b), Lauber and Landström (1972), and Tanner et al. (1972), which discuss the properties of materials with a view to their selection for sonde construction. Pertinent details abstracted from these investigations are included in the discussion of methods to follow.

For the determination of copper by thermal neutron activation, Landström et al. (1972) conducted field tests similar to those carried out by Hoyer and Lock (1972). Tests were performed in both 46-mm diameter diamond-core drilled holes and 250-mm diameter production holes. Aside from the fact that Landström et al. (1972) used a 10-Ci ^{210}Po-Be neutron source exclusively, and that their logging sonde included a device consisting of an inflatable rubber cylinder to preclude the measurement of radon, both investigations have much in common. A schematic of the sonde is shown in Fig. 5-32. The important conclusions drawn from this test were that a high-resolution Ge(Li) detector be used in future work in conjunction with a neutron source about five times greater in source strength. However, for the boreholes investigated the log (Fig. 5-33) shows evidence of satisfactory calibration between ^{64}Cu (0.51 MeV) counts and the core chemical analysis. The regions of mineralization are clearly defined. However, as emphasized earlier, the utilization of this technique for low-abundant (0.2—1.5%) copper ore does not project a good economic viability, even with the modifications suggested in this test. This statement is clarified by examining the outcome of another test conducted by Lauber and Landström (1972) in which a Ge(Li) detector system was employed. The fabricated sonde was 8.9 cm in diameter and 160 cm long. A 22-cm^3 coaxial Ge(Li) detector, cooled by liquid nitro-

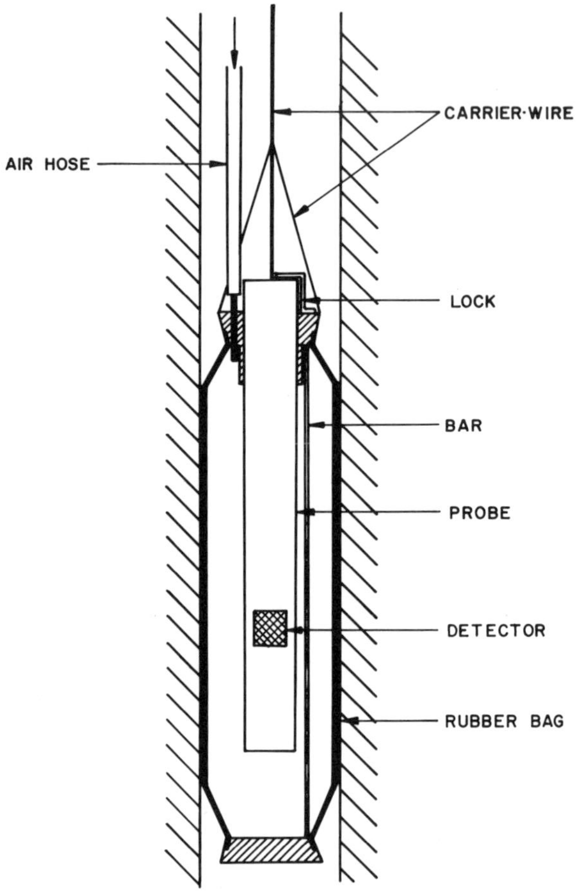

Fig. 5-32. Logging sonde showing arrangement with rubber device (Landström et al., 1972).

gen, had a hold time of about 10 h. The energy resolution of the detector was 2.3 keV for 122-keV gamma rays and 2.6 keV for 1.33-MeV photons. A schematic of this probe is shown in Fig. 5-34. Field tests with the system demonstrated the good performance of the detector system in a borehole environment. As to whether a Ge(Li) detector system can stand the rigorous of day-to-day production logging over several months in the field without expensive down times is an important factor in the economics of neutron activation logging.

The application of both delayed and prompt gamma-ray measurements to borehole analysis for elements of interest has been of special interest to the research team of the U.S. Geological Survey. These workers have placed a great emphasis on the use of the Ge(Li) detection system and have pioneered

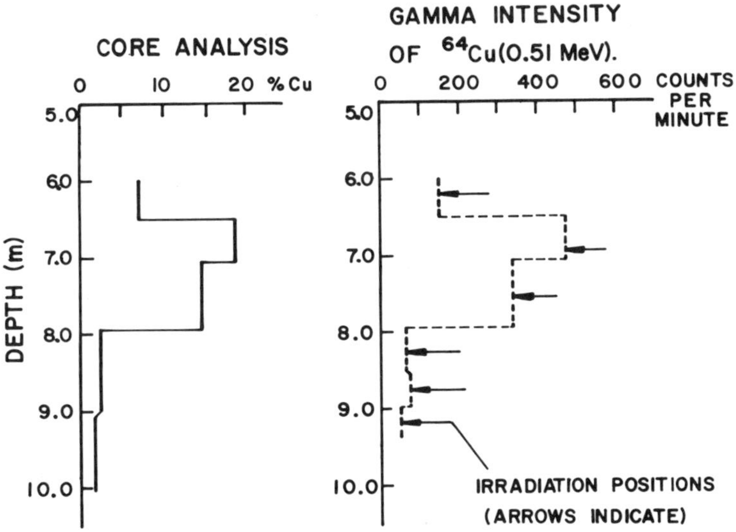

Fig. 5-33. Comparison between core analyses and borehole measurements of induced activity in copper (0.51 MeV from ^{64}Cu) after irradiation in 250-mm borehole (Landström et al., 1972).

much of the hardware development for its down hole application. Tanner et al. (1972) have described a borehole sonde incorporating such a system of detection. Applying the principle of a melting cryogen like propane, hold times as long as 10 h have been achieved. The choice of the cryogen was made from commercially available fluorocarbon refrigerants. The ultimate selection was restricted to the use of instrument-grade propane (>99.5%). The melting point of solid propane (−187.69°C) is the closest to that of liquid nitrogen (−195.8°C) commonly used for Ge(Li) detectors in the laboratory. The borehole sonde was 7.3 cm in diameter and about 2 m long. The modular construction of the device permits easy access to the important components such as the source holder for the ^{252}Cf neutron source, shield, cryostat holding the Ge(Li) detector and electronics. During storage and refreezing periods, the cryostat section was detached from the probe, held in a vertical position and cooled down by a 14-liter liquid nitrogen reservoir in the conventional manner. The total weight of the sonde was about 16 kg (35 pounds), excluding the shield and the source-holder filling material. The source holder itself was made from epoxy-filled fibreglass to provide mechanical strength and minimum swelling due to water absorption. The neutron source was mounted in a short acrylic rod and inserted in the annular source holder. The shadow shield was a tungsten rod which was mounted coaxially with the source. By introducing plastic spacers, the source-detector distance could be varied between 33 and 120 cm. For field

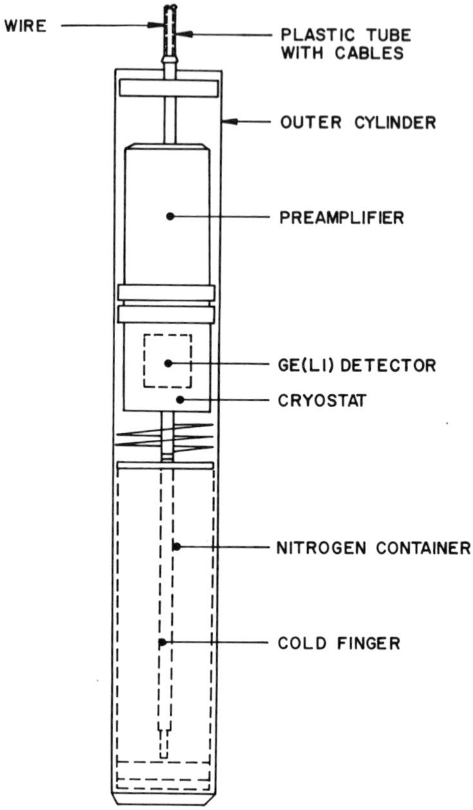

Fig. 5-34. Ge(Li) borehole probe (Lauber and Landström, 1972).

application a 300 m long cable was attached to the probe.

A portion of a prompt gamma-ray spectrum measured by this sonde is shown in Fig. 5-35. The development of this probe was based on findings by Senftle et al. (1971b). This research laid the basic foundation for the successful application of this technique to nickel exploration on a production basis. Interesting laboratory experiments conducted by Senftle et al. (1971b) are found in the report.

Moxham et al. (1972) pursued the early development carried out by Senftle et al. (1971b) described above and conducted field tests in copper- and nickel-bearing gabbro. The elements Al, Mn, Na, Mg, Cu and V were readily detected in the delayed gamma spectra. The elements H, Fe, Si and Ni were identified from the prompt gamma response in the higher energy regions of the spectra. Table 5-XIV, abstracted from Moxham et al. (1972), contains nuclear parameters pertinent to this test. The delayed and prompt gamma-ray spectral analysis indicated a better sensitivity for copper with the

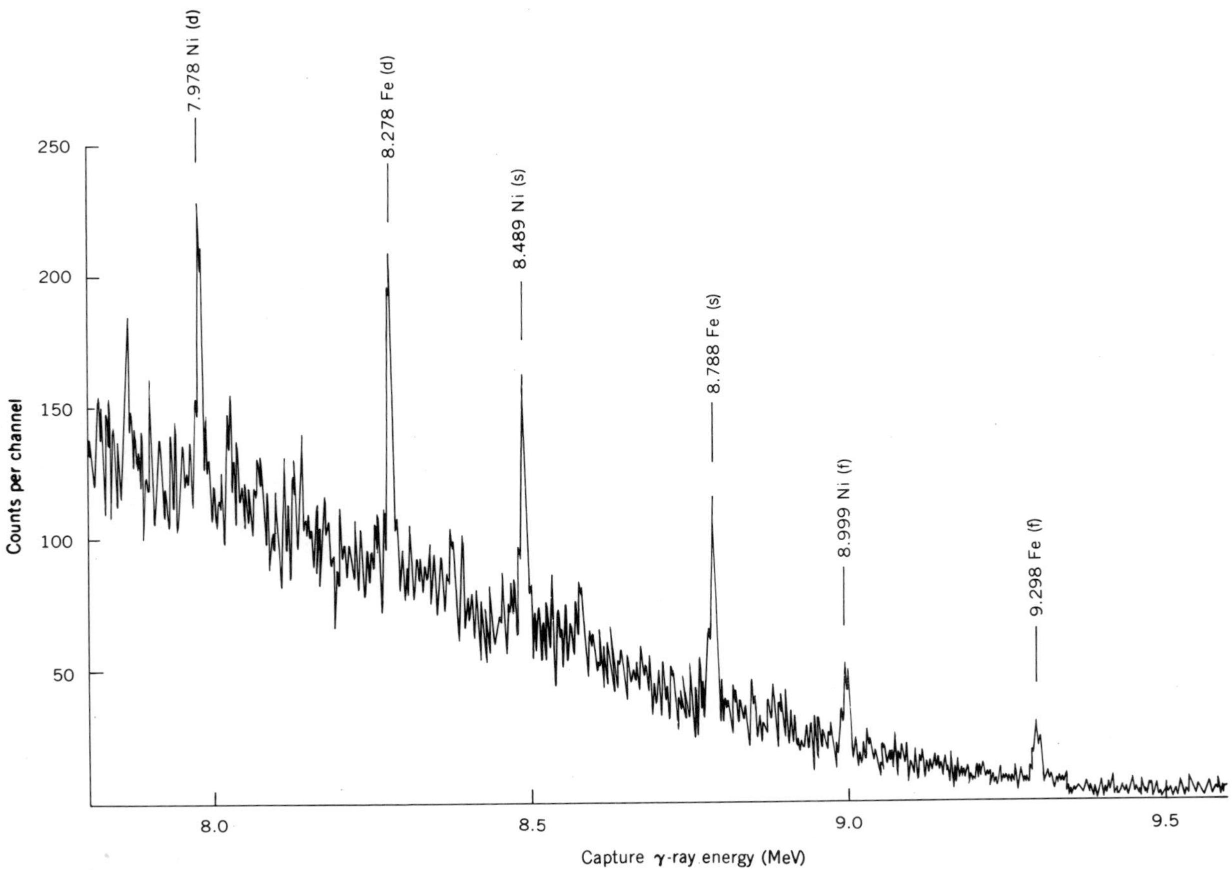

Fig. 5-35. Part of the spectrum of radiative-capture gamma rays detected by borehole sonde from ore containing about 1% Ni during $4\frac{1}{2}$ h irradiation by a 0.9μg ^{252}Cf source in a simulated borehole. (f), (s), and (d) denote full-energy, single-escape, and double-escape gamma-ray photopeaks (Tanner et al., 1972).

TABLE 5-XIV

Gamma-ray emission from copper and nickel isotopes (Moxham et al., 1972)

Target	Isotopic abundance (%)	Cross-section (barns)	Product	Gamma-ray emission				Half-life
				full-energy peak (keV)	intensity (%)	single escape (keV)	double escape (keV)	
				Neutron-capture gamma rays				
^{63}Cu	69.17	4.4	^{64}Cu	7916	27.3	7405	6894	
				7637	14.9	7126	6615	
				7307	9.3	6796	6285	
				7176	2.7	6665	6154	
^{58}Ni	67.88	4.4	^{59}Ni	8999	41.6	8488	7977	
				8533	18.74	8022	7511	
				6837	11.91	6326	5815	
				Delayed gamma rays				
^{63}Cu	69.17	4.4	^{64}Cu	511	19			12.8 h
				1343	0.5			
^{65}Cu	30.83	2.17	^{66}Cu	1039	9.0			5.1 min
				828	0.25			
^{64}Ni	1.08	1.52	^{65}Ni	1480	25.0			2.56 h

delayed technique. Experiments carried out in the laboratory showed that for delayed gamma measurement using the 1.78-MeV ^{28}Al photopeak as an index, an infinite sample was achieved when the vertical thickness of the rock layer was ~20 cm and a horizontal thickness from the borehole walls was ~10 cm. However, it should be borne in mind that in the case of copper, the optimum energy which was being measured was less than 1 MeV. Gamma-ray transport for such energies would indicate a significant decrease in the size of the infinite sample although the sensitivity would not be unduly affected. Some critical information gathered from this report indicated that the maximum probe response was obtained when the ratio $D_h/D_d \approx 2.5$, where D_h was the borehole diameter, D_d was the detector diameter, and the annulus was considered to be full of water. When this ratio approached ~3.2, the centering of the sonde in the borehole was found to be unnecessary.

The effect of water in the matrix increased the peak intensity of ^{66}Cu (1.039 MeV). This increase was further augmented when the borehole was also filled with water. It was suggested therefore that some means for correcting for the presence of water, e.g. a measurement of the prompt 2.23-

MeV emission from hydrogen, would be necessary for quantitative determinations for copper.

In the field a source-detector distance of 73.6 cm was used. The source strength of ^{252}Cf was ~85 μg (i.e. 2×10^8 n s^{-1}). In the borehole (diameter ~12.7 cm) the matrix was irradiated for about 17 min and followed up with a 17-min count. Estimates of the copper content at the 0.91% Cu level were made. Because of the poor response from copper in the prompt gamma-ray energy region, the use of the NaI(Tl) detector was considered to be futile. However, it should be realized that such deductions are by no means general and perhaps apply only to the detector system used and the data analysis methods employed. The commonly applied method of peak integration in certain cases places restrictions which automatically preclude the analysis of elements whose peaks are not visible, even though the gamma-ray contribution from such an element may be substantial for its determination. Moxham et al. (1972) indicated that a sensitivity of 0.5% Cu or Ni could be obtained using the delayed and prompt gamma-ray measurement techniques, respectively.

The natural outgrowth of the above test was an attempt to compare the two techniques. Senftle et al. (1972) produced an extensive table in an effort to derive comparison between the two techniques for a large number of elements. This table is helpful in preliminary assessments of feasibility. However, caution must be exercised in such assessments. Furthermore, care should be taken in examining all assumptions made in arriving at the tabulation.

At this stage it is necessary to pause and reflect on the applications described thus far. Several observations can be made, and pertinent questions raised. It is obvious that extremely few developments have advanced to the point of active consideration by the mineral exploration industry. It is true, however, that some studies and field case histories (as limited as they have been) offer considerable potential. Nevertheless, the acceptance of such techniques by industry requires a greater degree of field testing particularly on a production logging basis. Most important, however, it is necessary to demonstrate that the nuclear technique proposed not only satisfies the analytical parameters, but is also economical when compared to the core extraction—chemical analysis procedures commonly used for orebody evaluation. Industry cannot be expected to grasp new sophisticated techniques without adequate and extensive demonstrations, both in terms of its analytical capability and mechanical ruggedness. The borehole environment in the best field conditions is a far cry from the air-conditioned laboratory in which "ideal" samples are analyzed. Gross heterogeneity, large density variations, irregular borehole conditions, and the presence of moisture, are part and parcel of almost all geological strata. If a nuclear method based on radioactivation by neutrons is to function to the satisfaction of exploration geologists, it is imperative that thoughts be given to the sources of analytical error created by the changing hostile borehole environment. Essentially the above picture, as pessimistic as it appears, faced a Canadian research team in 1970. This team was charged with the responsibility of developing and fabricating a viable borehole analysis system for the analysis of Cu in porphyry copper, and Ni, Fe and Si in a lateritic matrix. Many discussions carried out with exploration experts soon made it apparent that laboratory studies, if any, would have to be conducted under very closely simulated field conditions. From the literature abounding at that time, the strong concensus of opinion was that Ge(Li) detectors were "in" and NaI(Tl) scintillators were "out". It was also obvious at that time that Ge(Li) detectors could ultimately present a very serious set back; because it was assessed that a detection system based on this principle would not be suitable or acceptable by the mining industry, and despite the pioneering efforts described by Lauber and Landström (1972) and Tanner et al. (1972), it would not be on the horizon in a state of development sufficient for its routine and long-term use by mineral explorationists. Furthermore, a system developed around the Ge(Li) detector with peak amplitude-type analysis would not provide adequate sensitivities, to determine low-grade cut-off levels. To add to the difficulty, the delayed gamma measurement technique does not provide results reflecting the response from a large enough sample volume. The end result from these observations at least for the base metals, Fe, Ni and Cu, was that if the mining community was to be fully convinced, any techniques developed must satisfy the following criteria.

(1) Unambiguous analysis in the actual environment.

(2) The overall experimental error from statistical and systematic considerations be $<\pm 15\%$ relative at the 0.5% Cu level with a detection limit approaching 0.1% Cu; and $<\pm 5\%$, $<\pm 10\%$, and $<\pm 15\%$ for the elements Ni, Fe and Si in nickel-bearing strata as typified by the laterite medium.

(3) The analysis must be reported on a dry weight basis.

(4) The logging speed (continuous or incremental) be reasonably fast ($>$25 m per hour for continuous logging and at least 5 m per hour for fine structure incremental logging).

(5) The method should provide analytical information that represents a much larger (50—100 times) the volume of rock when compared to drill-core samples.

(6) The logging system, whether it be hand-carried or installed in a mobile laboratory, be extremely rugged and fully adaptable to general field conditions.

With these goals in mind, the Canadian team embarked on a five-year research and development study. In early 1975 they (Nargolwalla, 1975) demonstrated the application of the first field production logging system for the simultaneous analysis of Ni, Fe and SiO_2 in a nickel laterite mining property. This matrix possesses all the qualifications which make quantitative determinations subject to error. The mobile laboratory (Figs. 5-36 and 5-37), called the "Metalogger", was put through every required test, to prove that the nuclear technique was ready to replace the core extraction method with substantial long-term savings to the lateritic nickel industry.

An entirely original method of gamma-ray spectral analysis was developed for general application. This method takes into account environmental effects which perturb the sensitivity of the elements and, therefore, the final results. The effects of heterogeneity and density variations are absorbed in the overall experimental error. The method is based on the prompt gamma-ray measurement process. Design of the sonde was such as to provide analyses for these elements averaged over approximately 680 kg of ore per determination. In the laterite application, an average logging output of ~40 m per 8-h day was maintained for almost 600—700 m of production logging. Results for all three components are provided on an incremental basis for every meter of borehole, almost instantaneously. The field geologist can, therefore, assess the quality of the hole being logged, and is in a position to make decisions as to whether to continue logging a particular hole or not. The analytical criteria set forth earlier were all satisfied.

In regard to copper analysis in the porphyry matrix, the NaI(Tl) detector was used. Copper determinations as low as ~0.2% Cu were made with an overall relative experimental error of <±15% at the 0.2% Cu level. The technique is simple and reliable in extracting the contribution of the prompt emissions from thermal neutron capture processes in copper nuclei from the gross spectra. Logging speeds up to 100 feet per hour were achieved. The

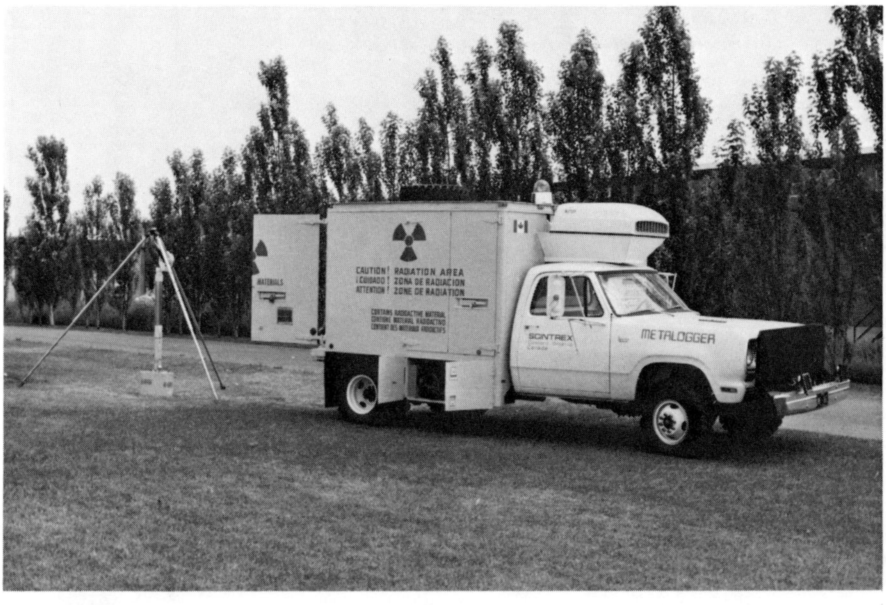

Fig. 5-36. The Scintrex Metalogger.

Fig. 5-37. Rear view of Metalogger showing sonde and tripod setup.

theoretical study described in the section "Theoretical considerations" aided in the design of the field sonde. Details of this design and the complete method of analysis are currently under patent consideration. Photographs showing the "Metalogger" in various stages of operation in the lateritic development are shown in Figs. 5-38 and 5-39. Calibration curves, reported by Nargolwalla (1975), and Nargolwalla and Robertshaw (1975), for the elements Fe, Ni and Si (as SiO_2) are shown in Figs. 5-40 to 5-42. The method indeed correlates better with the bulk sample than with the core analysis as is demonstrated in Table 5-XV. An interesting feature of the method is illustrated in Fig. 5-43. In this figure, the data were obtained by carefully logging at very small intervals in an artificial pit of 6-foot diameter containing two grades of laterite ore. The sharp physical interface is clearly identified by the measurement. A region of influence of about 70 cm may be observed.

Fig. 5-38. Lowering sonde in laterite borehole for production logging.

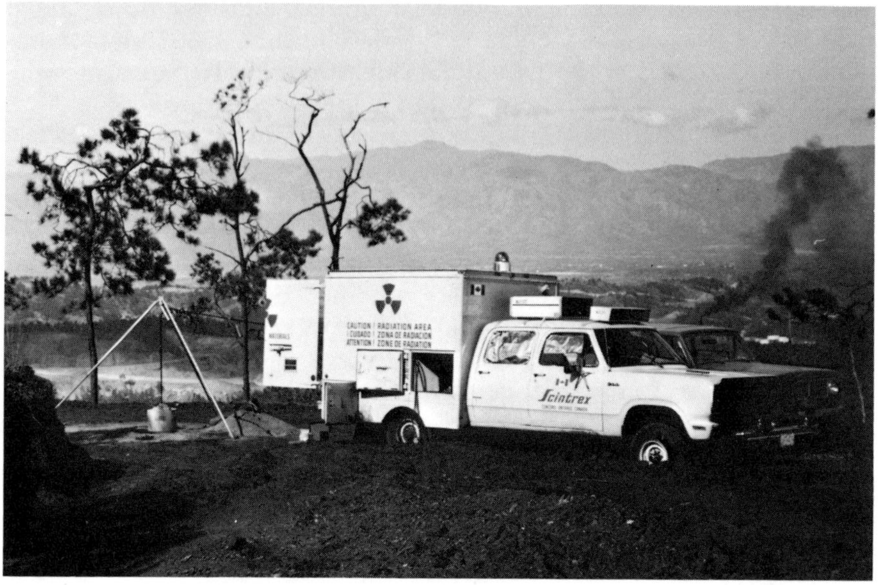

Fig. 5-39. Logging of lateritic nickel with the Metalogger.

The element iron was used as an index for this determination.

The test also included an evaluation of the reproducibility of the Metalog measurements, both in terms of counting statistics and the ability for accurate positioning of the sonde; relogging of holes to check the overall reproducibility; and the logging of water-filled boreholes. In the generation of the calibration curves, chemical analysis of tens of tons of bulk samples withdrawn from about 0.5 m radius of the borehole were chemically analyzed to assess the full capability of the method. Currently the system is building up an impressive list of field case histories which will be open to active consideration by the mineral exploration industry.

Delayed neutron measurement. During the past three decades, the search for uranium has depended heavily on the use of the Geiger monitor and the NaI(Tl) scintillator for the measurement of gamma radiation emitted from the daughter products of the uranium series. Both ^{238}U and ^{235}U are poor gamma emitters, while most of their daughter products beyond ^{222}Rn are strong emitters. Since ^{222}Rn is a gas, it can migrate through a gas permeable environment. In this migration, daughter products are continually being created. A measurement of migrated daughter products, therefore, gives a false indication as to the location of the uranium itself. Such deposits are said to be in disequilibrium. If, on the other hand, the ^{222}Rn does not escape, the deposit is considered to be in equilibrium. Consequently a measurement of the gamma activity of the radon daughters does provide a quantitative

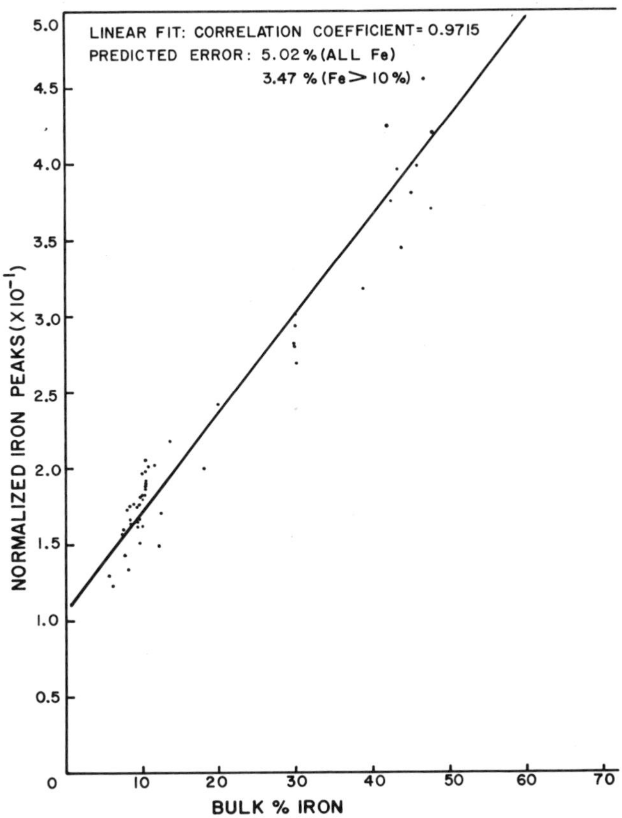

Fig. 5-40. Calibration curve for iron (Nargolwalla and Robertshaw, 1975).

determination for uranium. In view of such geological conditions, there have been continuous demands for in-situ techniques which will quantify the fissionable isotopes only. With the availability of ^{252}Cf, the possibility of in-situ determinations for uranium by the delayed neutron counting technique was thoroughly investigated.

As is well known from the nuclear fission process, both ^{238}U and ^{235}U produce six groups of delayed neutrons of comparable half-lives. Perhaps the most successful application of this technique was demonstrated by the Kerr-McGee research group (Californium-252 Information Center, 1972, No. 12). For the initial evaluation of the technique, a 1-mg ^{252}Cf neutron source ($\sim 2 \times 10^9$ n s^{-1}) was used to develop a technique with a capability of logging at a rapid rate of 120 feet per minute. A rugged field logging system was built and operated in the field for several months. During this time over 100,000 feet of both delayed neutron and natural gamma logs were obtained. Several of the boreholes investigated had been previously cored and chem-

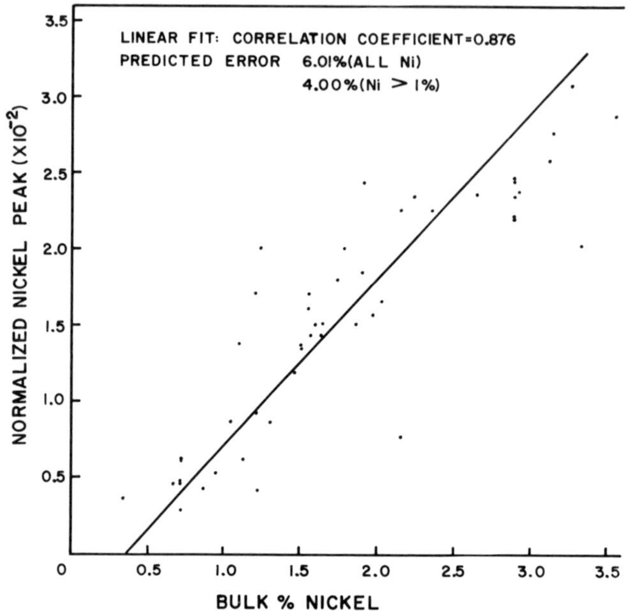

Fig. 5-41. Calibration curve for nickel (Nargolwalla and Robertshaw, 1975).

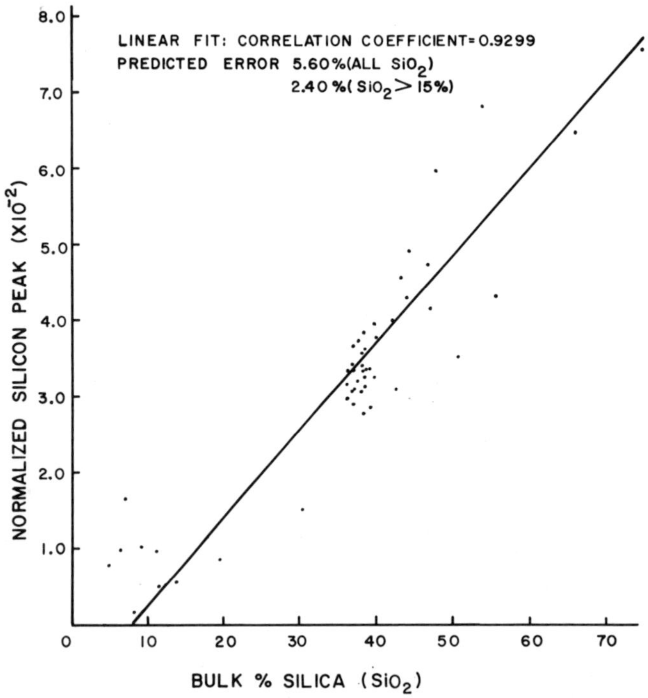

Fig. 5-42. Calibration curve for silica (Nargolwalla and Robertshaw, 1975).

TABLE 5-XV

Comparison of correlation coefficients

Sample size * (axial length, cm)	Correlation coefficient		
	element	metalog vs. core	metalog vs. bulk
100	Fe	0.912	0.962
	Ni	0.837	0.858
	SiO_2	0.812	0.903
66	Fe	0.916	0.971
	Ni	0.854	0.876
	SiO_2	0.811	0.930

* Chemical analysis of core (3-inch diameter) averaged over the axial borehole length. For bulk sample correlation, a cylindrical sample approximately 1 m in diameter and of the given axial length was withdrawn, mixed and chemically analyzed. Total weight of the sample was approximately 680 kg for the 100-cm sample.

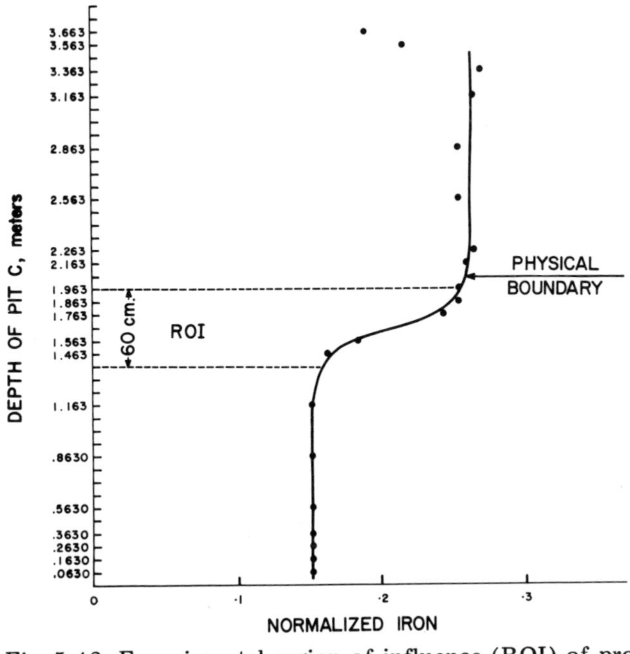

Fig. 5-43. Experimental region of influence (ROI) of prompt gammas from iron (Nargolwalla and Robertshaw, 1975).

ically analyzed. To perform this impressive task, Kerr-McGee developed a whole new system of operations and equipment so that operators could be adequately protected from the high-intensity neutron source and also meet the relevant transport licensing regulations. The development of this technique can be followed in chronological sequence by a review of Californium-252 Progress Reports (Californium-252 Information Center, 1971, No. 7; 1972, No. 10).

In the field system the ^{252}Cf source holder, confined in a large shielding flask, is lowered several feet down the borehole and clamped. The detector section containing a gamma-ray detector and a ^{3}He-filled neutron counting

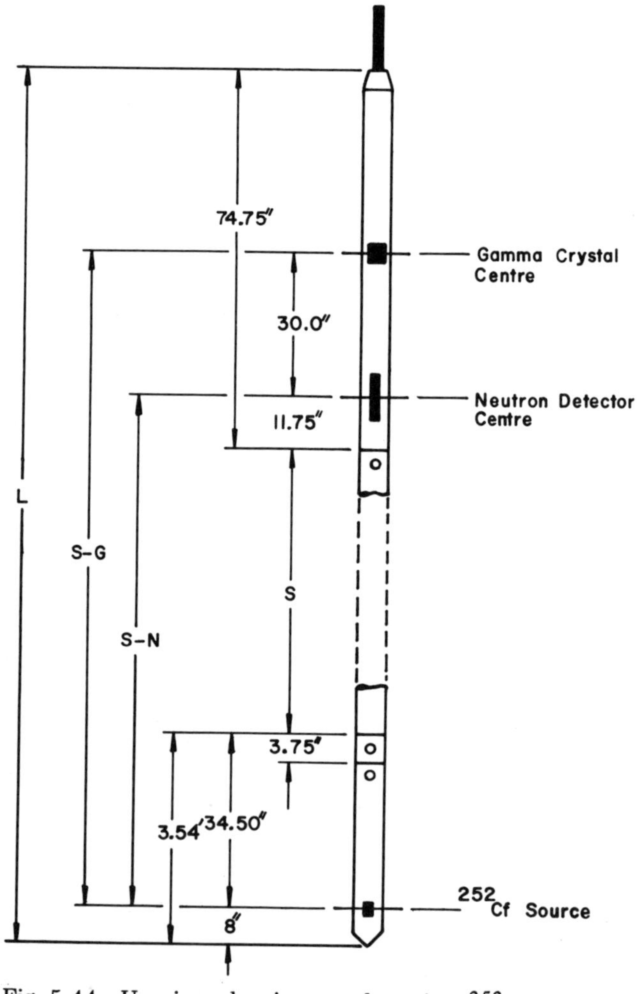

Fig. 5-44. Uranium logging sonde using ^{252}Cf (Californium-252 Information Center, 1972, No. 12).

TABLE 5-XVI

Borehole logging sonde configurations used with ^{252}Cf

Spacer length (inch)	Source-to-neutron detector		Source-to-gamma detector		Total length (inch)
	separation (inch)	time * (s)	separation (inch)	time * (s)	
95.25	141.5	23.5	171.5	28.5	212.5
56.25	102.5	17.0	132.5	22.0	173.5
39.0	85.25	14.2	115.25	19.2	156.25
30.0	76.25	12.7	106.25	17.7	147.25
25.0	71.65	11.8	101.25	16.8	142.25
0.0	46.25	7.7	76.25	12.7	117.25

* Time between irradiation and detection using logging speed of 30 feet per minute.

tube is attached to the top of the source holder. In Fig. 5-44 (Californium-252 Information Center, 1972, No. 12) is shown a schematic view of the sonde. The length of the sonde can be easily adjusted by introducing spacers between the source holder and the detector section. Various sonde configurations tested are given in Table 5-XVI. The consequence of varying these parameters is demonstrated in the decay curves (Fig. 5-45). Examina-

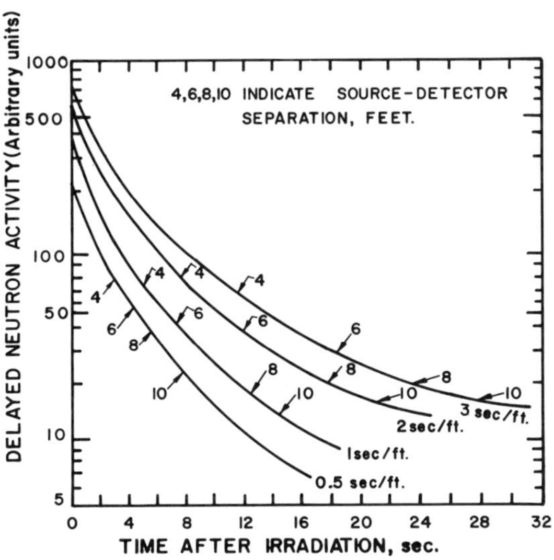

Fig. 5-45. Decay curves for the delayed neutron activity produced by ^{235}U when irradiated with ^{252}Cf; logging speeds of 20, 30, 60 and 120 feet per minute and source-detector separations of 4, 6, 8, and 10 feet (Californium-252 Information Center, 1972, No. 12).

tion of these curves indicate that the source-detector distance should be as short as possible without pickup of the scattered neutrons from the source itself. For optimum operation, a logging speed of 30 feet per minute using a source-detector distance of 76 inches is suggested.

A thorough description of other parameters which influence the delayed neutron count is given in the report (Californium-252 Information Center, 1972, No. 12) and should be studied by log analysts interested in fabricating their individual systems. A comparison of delayed neutron results with chemical analyses of the core can be gauged from Table 5-XVII. Despite the very encouraging results, the Kerr-McGee group felt the necessity of further improvements in the technique. Areas in which further investigations were needed included:

(1) Increasing the sensitivity by a factor of ten. This increase could be realized by utilizing a higher neutron source strength (2—3 mg ^{252}Cf) and enlarging the volume of the ^{3}He detector by a factor of 3—4.

(2) Investigating in greater depth the independent roles of fast, epithermal and thermal fission as a function of the depth of ore penetration by the neutron from the ^{252}Cf source.

TABLE 5-XVII

Wyoming core analyses results vs. neutron count rates (Californium-252 Information Center, 1972, No. 12)

Bore-holes	Run	Zone thick-ness (ft)	Core chemical grade (% U_3O_8)	Average n s^{-1}	n s^{-1} = 1% U_3O_8	Average n ft^{-1}	n ft^{-1} = 1% U_3O_8
1	1	25	0.088	5.71	64.9	8.5	96.6
	2	25	0.088	6.0	68.2	9.0	102.3
	1	12.5	0.082	6.0	73.2	9.5	115.8
	2	12.5	0.082	5.6	68.3	8.8	107.3
2	1	3	0.0264	2.2	83.3	3.5	132.6
	2	3	0.0264	2.8	106.1	4.2	158.0
	1	5	0.095	4.3	45.2	7.1	74.7
	2	5	0.095	6.0	63.2	9.4	98.9
	1	20	0.0725	4.4	60.7	7.3	100.7
3	1	1	0.291	15.0	51.5	25.0	85.9
	2	1	0.291	19.5	67.0	32.5	111.7
	1	2	0.064	2.75	43.0	4.8	75.0
	2	2	0.064	3.25	85.4	5.4	84.3
4	1	15	0.096	7.9	82.3	12.5	130.2
	2	15	0.096	6.4	66.7	10.1	105.2

Average conversion factors from all data: n s^{-1} = 1% eU_3O_8 = 67; n ft^{-1} = 1% eU_3O_8 = 106.

(3) Evaluating the effect of high neutron absorbers such as B, Gd, Sm, Eu and Dy.

(4) Continuing core vs. delayed neutron correlations, so that with adequate sampling statistics, a better correlation between the two methods could be established.

Further work (Californium-252 Information Center, 1973, No. 15) reported by this group incorporated some of the improvements indicated above, and was highly successful. The region of influence for this method extended about 6—12 inches from the borehole face. For a 6-inch depth of influence, the ore analyzed was about 50 times that for a corresponding length of core.

This technique still requires further investigation. The effect of the rare earth contaminants has yet to be evaluated. Furthermore, the presence of thorium could confuse the interpretation of delayed neutron counts. However, the Kerr-McGee group is to be highly commended for breaking ground in a highly complex matrix for the in-situ borehole logging for uranium. The full-scale industrial application of this technique as a routine quantitative method of ground follow-up from airborne spectrometric anomalies appears to be only a matter of time.

ONGOING AND FUTURE DEVELOPMENTS

Perhaps the most exciting developments in the application of radioactivation methods for mineral exploration are those concerned with the analysis of ocean bottoms. Research into the fabrication of analytical devices which operate at considerable depths have taken a new dimension. Although predictions of population growth may excite considerable and controversial debate, the global demand for an ever increasing requirement for mineral resources and raw materials is a matter of fact. Since the oceans cover 70% of the earth's surface, exploration of even a minimal area of the sea bed could conceivable open up mineral resources to satisfy the demand of mankind for years to come. In the long run, the technological world will be faced with the acute problem of mining either the depleted terrestrial mineral resources with very efficient methods or attempt to tap the rich marine sources, utilizing very expensive ore recovery devices. It is far too early to make such decisions at the present time. If the inherent technological difficulties pertinent to marine exploration are solved, appropriate decisions will automatically ensue. Exploratory studies made thus far indicate that mining for manganese nodules on the ocean floor at the rate of 5000—10,000 tons per day may be possible and reap reasonable benefits.

Fanger and Pepelnik (1974), in their recent review emphasizing the exploration of marine ore deposits, have classified such deposits into three main categories:

(1) "Heavy minerals" ($\rho > 2.9$ g cm^{-3}) in shelf regions. The heavy mineral

fractions in these regions are of the order of 3—6% and contain ilmenite ($FeTiO_3$), rutile (TiO_2), zircon ($ZrSiO_4$), tinstone (SnO_2), monazite [(Ce,La, Y,Th)PO_4] and traces of diamond, platinum and gold. Concentrations of Ti and Zr in the range 1—2% and 0.3—0.6%, respectively, can be expected.

(2) Hydrothermal ore slime of the Red Sea-type sediments. These sediments occur at approximately 2000 m on the ocean floor. Concentrations of 1% Cu, 5% Zn, 30% Fe, 0.1% Pb, 200 ppm Cd, Co and Mo, and 100 ppm Ag are found in such deposits.

(3) Deep-sea ores such as manganese nodules. These deposits are found at depths between 4000 and 6000 m on the ocean bed. Typical concentrations of elements in manganese nodules found on the Pacific ocean floor are 1% Ni, 0.5% Cu, 0.2% Co, 25% Mn, and 10% Fe. Orebody evaluation of the Pacific ocean bed is estimated at 10^{12} tons. An interesting observation can be made by examining the Ni and Cu concentrations. In 1970 the yearly consumption of these elements was of the order of 7×10^6 tons (Cu) and 5×10^5 tons (Ni). Conservative estimates of these elements on the sea beds are of the order of 10^9 to 10^{10} tons. This fact immediately sets up a situation that is most attractive for investigation at the expense of other types of mineral deposits.

The facts presented above have been instrumental in initiating fundamental research in this area by various groups, as presented in Californium-252 Progress reports (Californium-252 Information Center, 1970, No. 2; 1971, No. 6; 1971, No. 7; 1972, No. 12), and also described by Senftle et al. (1969), Senftle (1970), Perkins et al. (1971), Beckmann and Abdullah (1973), Wogman et al. (1973), Fanger and Pepelnik (1974), and Senftle et al. (1975). In order to organize a feasibility study, certain discrete steps must be considered. First, a general knowledge of the deposit must be acquired by rough prospecting means. These include such techniques as deep-sea television, photography, dredging and grab sampling. Secondly, some method of exploration is applied to estimate the ore reserves on a quantitative basis. In the case of manganese nodules this becomes a strictly two-dimensional case since the nodules are essentially a mono-granular layer on the sea bed. However, tonnage variations in the mineral content of nodules need to be quantitatively evaluated. This information can be combined with the nodule density, to estimate ore reserves. Some concepts exploited or considered are depicted in Figs. 5-46 to 5-49. Some of these concepts are already at the field testing stage. Fanger and Pepelnik (1974) described a pilot device soon to be tested in the Pacific. This device will concentrate on the analyses of manganese nodules. Disregarding capital costs, these workers have estimated a cost of about 85 Deutsch Marks for the analysis of about 30 kg of sample. According to them such costs are considered to be economic.

Another area where a great deal of emphasis is being applied is in the uranium exploration industry. Nuclear power reactors are springing up quite rapidly and the demand for electrical energy is acute. Recently the oil crisis

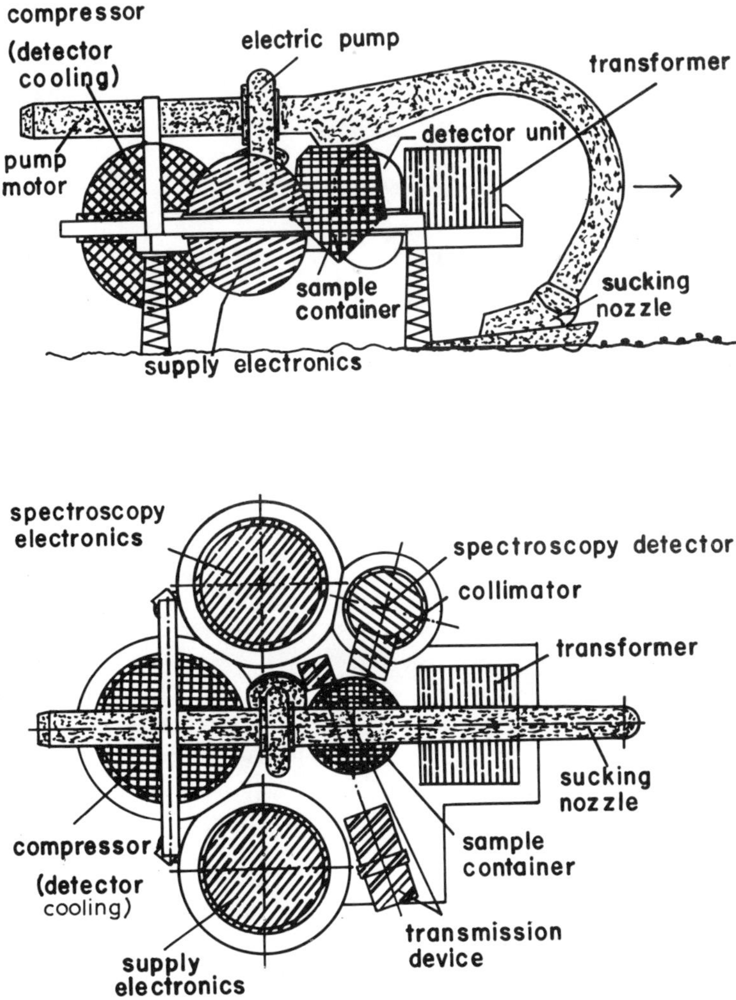

Fig. 5-46. Scheme of the technical concept for in-situ (n, γ) analysis of manganese nodules (Fanger and Pepelnik, 1974).

has spurred industry into investing funds into exploration for uranium. As described in the previous section, the delayed neutron measurement technique is being improved and will hopefully accelerate uranium exploration. It is worth mentioning, however, that another technique, i.e. nuclear track counting, could have good potential as reported by Carpenter (1970) and Carpenter (1972), for the analysis of such elements as uranium, thorium and boron. However, in the laboratory high fluxes of thermal neutrons are necessary for the feasibility of this method. Such fluxes are extremely difficult to generate in situ. Furthermore, the etching and optical counting

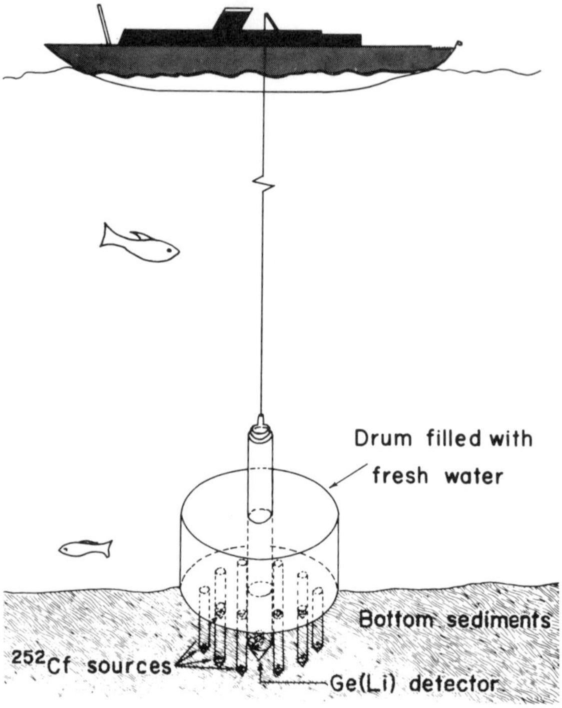

Fig. 5-47. Sketch showing a proposed improved probe using multiple ^{252}Cf sources in a more efficient array (Senftle et al., 1975).

technique for track determination could present some difficulty. The determination of uranium by the fission-track method is known to analyze uranium down to the sub-ppb levels.

Insofar as the borehole analysis of other minerals of geologic interest is concerned, the present state-of-the-art allows quantitative determination of only a few elements. There is a dire need to expand the application of radioactivation methods to other minerals, in particular, gold. Although the determination of gold in the laboratory is relatively simple, in-situ measurements which are both economical and specific for gold present unique problems in data processing. Sophisticated computer codes are necessary to extract spectral information relevant to gold. A typical gamma-ray spectrum obtained from the irradiation of a geologic matrix is generally swamped with unwanted photon responses. It is only by using certain nuclear properties of the elements, e.g. irradiation and decay times, energy discrimination etc., to the fullest advantage that one can make reasonable quantitative determinations. The problem of gold determinations in the ppm range is therefore compounded.

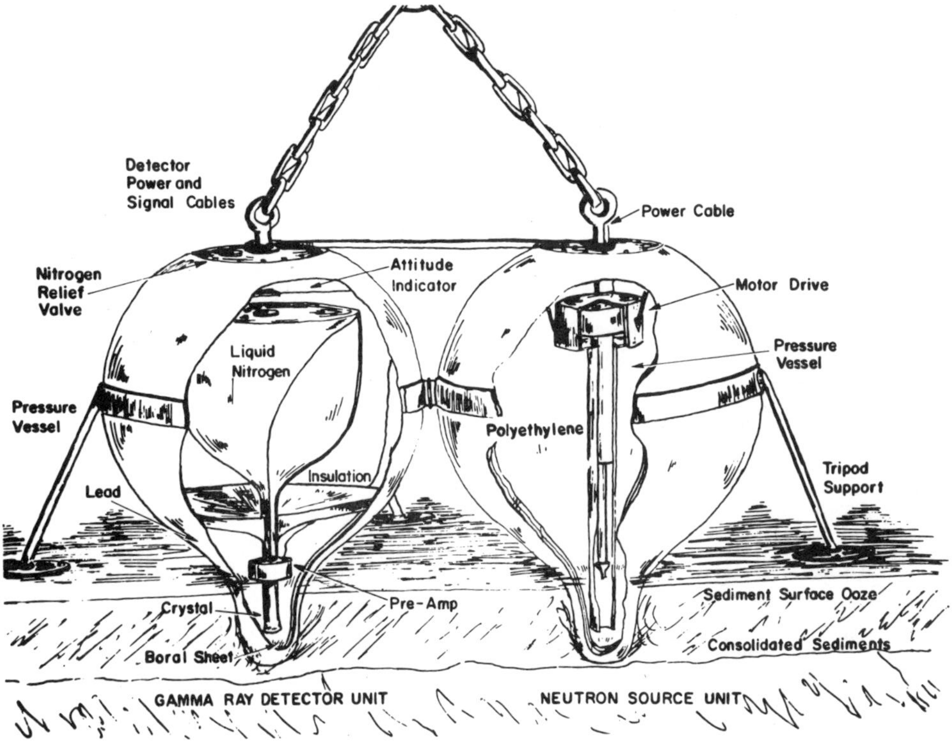

Fig. 5-48. Proposed experiment arrangement of ^{252}Cf source and Ge(Li) detector for marine use (Senftle et al., 1969).

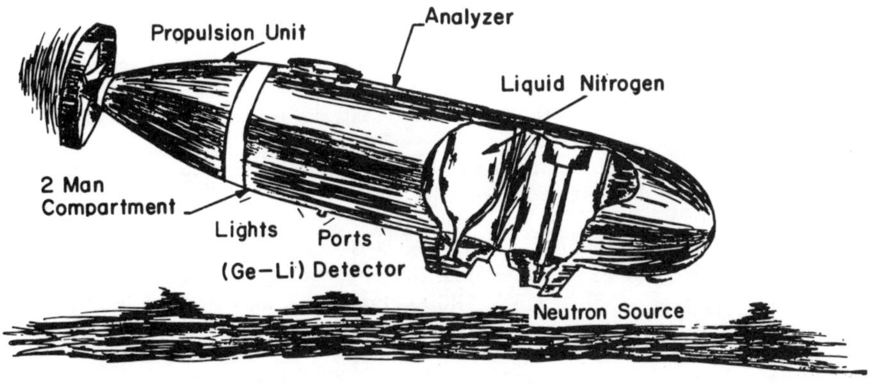

Fig. 5-49. Sketch showing similar equipment as shown in Fig. 5-48 except on a submersible, to provide greater flexibility of operation (Senftle, 1970).

The overall picture for the application of radioactivation techniques to mineral exploration is certainly very encouraging. The principal challenges are obvious. It is now incumbent upon the mineral exploration industry to invest in the development of such techniques, since the benefit gained from their successful application is invariably enjoyed by the industry itself.

REFERENCES

Allen, L.S., Tittle, C.W., Mills, W.R. and Caldwell, R.L., 1967. Dual-spaced neutron logging for porosity. *Geophysics*, 32: 60—68.

Anders, O.U., 1960. Neutron activation sensitivities. *Nucleonics*, 18: 178—179.

Anonymous, 1965. "CUNLAP". *Nucl. Sci. Eng.*, 21: 411.

Ansell, K.H. and Hall, E.G., 1971a. Recent developments in (α, n) sources. In: *Neutron Sources and Applications. Proc. Am. Nucl. Soc., Natl. Topical Meet., Augusta, Ga.*, pp. 90—99 (USAEC CONF-710402, Vol. II).

Ansell, K.H. and Hall, E.G., 1971b. High-intensity (α, n) survey. In: *Neutron Sources and Applications. Proc. Am. Nucl. Soc., Natl. Topical Meet., Augusta, Ga.*, pp. 100—111, (USAEC CONF-710402, Vol. II).

Argonne National Laboratory, 1963. *Reactor Physics Constants.* USAEC, ANL-5800, 2nd ed.

Beckmann, H. and Abdullah, H., 1973. Surveying the sea floor by nuclear logging. *Interocean Conf., Düsseldorf, 13—18 November, 1973*, Paper IO73-259/01.

Bennett, C.A. and Franklin, N.L., 1954. *Statistical Analysis in Chemistry and the Chemical Industry.* John Wiley and Sons, New York, N.Y., pp. 677—684.

Berzin, A.K., Bespalov, D.F., Zaporozhets, V.M., Kantor, S.A., Leipunskaya, D.I., Sulin, V.V., Feldman, I.I. and Shimelevich, Yu.S., 1966. Present state and use of basic nuclear geophysical methods for investigating rocks and ores. *At. Energy Rev.*, 4: 59—111.

Blyumentsev, A.M. and Feldman, I.I., 1962. Problems of the quantitative assessment of iron ore content in ores from radioactive log data. *Razved. Promoysl. Geofiz.*, 46: 102—117.

Bowen, H.J.M. and Gibbons, D., 1963. *Radioactivation Analysis.* Oxford University Press, London, 295 pp.

Brookhaven National Laboratory, 1960—1966. *Neutron Cross Section.* BNL-325, 2nd ed., Suppl. 1 (1960) and Suppl. 2; five volumes (1964—1966).

Bulashevich, Yu.P., Voskoboinkov, G.M. and Muzyukin, L.V., 1962. Nuclear geophysics in prospecting for ore and coal deposits. In: *Radioisotopes in the Physical Sciences and Industry.* IAEA, Vienna, p. 101.

Caldwell, R.L. and Mills, W.R., 1959. Activation analysis in petroleum exploration research. *Nucl. Instr. Meth.*, 5: 312.

Caldwell, R.L., Mills, W.R., Jr. and Hickman, J.P., Jr., 1960. Gamma radiation from inelastic scattering of 14-MeV neutrons by the common earth elements. *Nucl. Sci. Eng.*, 8: 173—182.

Caldwell, R.L., Baldwin, W.G., Bargainer, J.D., Berry, J.E., Salaita, G.N. and Sloan, R.W., 1963. Gamma-ray spectroscopy in well logging. *Geophysics*, 28: 617—632.

Caldwell, R.L., Allen, L.S. and Mills, W.R., 1966. Theoretical and experimental model results in neutron lifetime logging. In: *Radioisotope Instruments in Industry and Geophysics, II.* IAEA, Vienna, Paper SM-68/31, pp. 69—84.

Californium-252 Information Center, 1970. *Californium-252 Progress,* No. 2, January. Savannah River Laboratory, Aiken, S.C.

Californium-252 Information Center, 1971. *Californium-252 Progress*, No. 6, January; No. 7, April. Savannah River Laboratory, Aiken, S.C.

Californium-252 Information Center, 1972. *Californium-252 Progress*, No. 10, January; No. 12, July. Savannah River Laboratory, Aiken, S.C.

Californium-252 Information Center, 1973. *Californium-252 Progress*, No. 15, May.

Californium-252 Information Center, 1974. *Californium-252 Progress*, No. 17, May.

Carpenter, B.S., 1970. Boron and uranium determinations in simulated lunar glass by the nuclear track technique. *Trans. Am. Nucl. Soc.*, 13: 79.

Carpenter, B.S., 1972. The determination of trace concentration of boron and uranium in glass by the nuclear track technique. *Anal. Chem.*, 44: 600—603.

Chapman, E.P., Jr., 1970. Why feasibility studies for very large low-grade deposits must be accurate. *World Mining*, pp. 26—30.

Christell, R. and Ljunggren, K., 1965. Analysis of ore samples using gamma-rays emitted under irradiation with a low-level neutron source. In: *Radiochemical Method of Analysis, I.* IAEA, Vienna, pp. 263—275.

Czubek, J.A., 1971. Recent Russian and European developments in nuclear geophysics applied to mineral exploration and mining. *Log Analyst*, November/December, pp. 20—34.

Czubek, J.A., 1972. Pulsed neutron method for uranium well logging. *Geophysics*, 37: 169—173.

Dacey, P., Paine, P. and Goodman, R.A., 1949. Nuclear shielding studies, III. Shielding properties of various materials against neutrons and gamma rays. *Lab. Nucl. Sci. Eng., Mass. Inst. Technol., Tech. Rep.*, 23.

Denisik, S.A., Rezvanov, R.A. and Lukhminskii, B.E., 1963. Calculation of space-time neutron distributions by the Monte-Carlo method. *Nucl. Geophys.* (Gostoptekhizdat).

DeSoete, D., Gijbels, R. and Hoste, J., 1972. *Neutron Activation Analysis.* Wiley-Interscience, New York, N.Y., 872 pp.

Dibbs, H.P., 1970. The application of neutron activation analysis to the determination of copper in minerals. *Can. Inst. Min. Metall. Trans.*, 73: 102—108.

Donhoffer, D.K., 1968. Determination of ore concentration in exploration holes by the gamma-ray backscattering technique. *IAEA Symp. on the Use of Nuclear Techniques in Prospecting and Development of Mineral Resources, Lima, Peru*, Paper SM-112/2, pp. 23—35.

Dowd, P., 1975. Mine planning and ore reserve estimation with the aid of a digigraphic console display. *Can. Inst. Min. Metall. Bull.*, February, pp. 39—43.

Duffey, D., El-Kady, A. and Senftle, F.E., 1970. Analytical sensitivities and energies of thermal-neutron-capture gamma rays. *Nucl. Instr. Meth.*, 80: 149—171.

Eisler, P.L., Huppert, P. and Wylie, A.W., 1971. Logging of copper in simulated boreholes by gamma spectrometry. *Geoexploration*, 9: 181—194.

Fanger, H.U. and Pepelnik, R., 1974. Exploration of marine ore deposits utilizing spectroscopy of neutron-induced radiation. In: *Revue Industries Atomiques et Spatiales, 4.* Institute für Physik der GKSS, Geesthacht-Tesperhude.

Fanger, U., Lubecki, A., Pepelnik, R. and Wurz, H., 1973. Borehole analysis by means of (n, γ)- and XRF-technique demonstrated in a Scheelite deposit near Mittersill. *PACT Ber.*, 37, Gesellschaft für Kernforschung mbH, Karlsruhe.

Feld, B.T., 1953. The neutron. In: E. Segrè (Editor), *Experimental Nuclear Physics, 2.* John Wiley and Sons, New York, N.Y., Part VII.

Friedlander, G., Kennedy, J.W. and Miller, J.M., 1964. *Nuclear and Radiochemistry.* John Wiley and Sons, New York, N.Y., 2nd ed., 380 pp.

Gijbels, R., 1973. Neutron activation analysis of ores and minerals. *Miner. Sci. Eng.*, 5: 304—348.

Glasstone, S. and Edlund, M.C., 1952. *The Elements of Nuclear Reactor Theory.* Van Nostrand, Princeton, N.J., 416 pp.

Glasstone, S. and Sesonske, A., 1963. *Nuclear Reactor Engineering.* Van Nostrand, Princeton, N.J.

Glauberman, A.E. and Talianskyi, I.I., 1958. Neutron flux distributions in media separated by a cylindrical boundary. *J. Nucl. Energy, Part II,* 8: 84.

Greenwood, R.C. and Reed, J.H., 1965. *Prompt Gamma Rays from Radiative Capture of Thermal Neutrons, 1 and 2.* **IIT Research Institute, Technology Center, Chicago, Ill.,** 898 pp., USAEC, IITRI-1193-53, Vol. I and II.

Grodstein, G.W., 1957. X-ray attenuation coefficients from 10 keV to 100 MeV. *Natl. Bur. Stand. (U.S.), Circ.,* 583.

Hazen, S.W., Jr., 1967. Assigning an area of influence for an assay obtained in mine sampling. *U.S. Bur. Mines, Rep. Invest.,* 6955.

Hoyer, W.A. and Lock, G.A., 1972. Logging for copper by in-situ neutron activation analysis. *AIME Annu. Meet., San Francisco, Calif.,* Reprint No. 72-L-28.

Hoyte, A.F. Martinez, P. and Senftle, F.E., 1967. Neutron activation method for silver exploration. *Trans. Soc. Min. Eng.,* 238: 1—8.

Hughes, D.J., 1953. *Pile Neutron Research.* Addison-Welsey, Cambridge, Mass.

Hughes, D.J. and Schwartz, R.B., 1958. *Neutron Cross Sections.* Brookhaven National Laboratory, BNL-325, 2nd ed.

Keys, W.S. and Boulogne, A.R., 1969. Well logging with californium-252. *SPWLA 10th Ann. Logging Symp., Houston, Texas.*

Koch, R.C., 1960. *Activation Analysis Handbook.* Academic Press, New York, N.Y., 219 pp.

Kruger, P., 1971. *Principles of Activation Analysis.* Wiley-Interscience, New York, N.Y., 522 pp.

Krugers, J. (Editor), 1973. *Instrumentation in Applied Nuclear Chemistry.* Plenum Press, New York, N.Y., 383 pp.

Kulinkovich, A.E., 1965. On the solution of the neutron logging theory problem by the "group" method. In: F.A. Alekseev (Editor), *Soviet Advances in Nuclear Geophysics.* Consultants Bureau, New York, N.Y., p. 110.

Landström, O., Christell, R. and Koski, K., 1972. Field experiments on the application of neutron activation technique to in-situ borehole analysis. *Geoexploration,* 10: 23—39.

Lauber, A. and Landström, O., 1972. A Ge(Li) borehole probe for in-situ gamma ray spectrometry. *Geophys. Prospect.,* 20: 800—813.

Lenihan, J.M.A. and Thomson, S.J. (Editors), 1965. *Activation Analysis, Principles and Applications.* Academic Press, New York, N.Y., 211 pp.

Lutz, G.J., Boreni, R.J., Maddock, R.S. and Wing, J., 1971. Activation analysis: a bibliography through 1971. *Natl. Bur. Stand. (U.S.), Tech. Note,* 467.

Lyon, W.S., Jr. (Editor), 1964. *Guide to Activation Analysis.* Van Nostrand, Princeton, N.Y.

Moxham, R.M., Senftle, F.E. and Boynton, G.R., 1972. Borehole activation analysis by delayed and capture gamma rays using a ^{252}Cf neutron source. *Econ. Geol.,* 67: 579—591.

Nargolwalla, S.S., 1975. In-situ borehole logging of geologic materials by neutron capture-gamma-ray measurement. *Trans. Am. Nucl. Soc.,* 21: 107.

Nargolwalla, S.S. and Przybylowicz, E.P., 1973. *Activation Analysis with Neutron Generators.* Wiley-Interscience, New York, N.Y., 662 pp.

Nargolwalla, S.S. and Robertshaw, P., 1975. Metalog nuclear grade logging in lateritic nickel and porphyry copper deposits. *1975 Soc. Min. Eng. Fall Meet. Exhibit, September 10—12, 1975, Salt Lake City, Utah.*

Nargolwalla, S.S., Crambes, M.R. and DeVoe, J.R., 1968. A technique for the evaluation of systematic errors in the activation analysis for oxygen with 14-MeV neutrons. *Anal. Chem.,* 40: 666.

Nargolwalla, S.S., Niewodniczanski, J. and Suddueth, J.E., 1970. Experimental sensitivities for 3-MeV neutron activation analysis. *J. Radioanal. Chem.*, 5: 403.

Nargolwalla, S.S., Rehman, A., St. John-Smith, B., Legrady, O. and Strever, J., 1973. Neutron activation borehole logging of geologic materials. *Annual Progr. Rep. 15 December 1972 — 15 December 1973.* Scintrex Ltd., Concord, Ont.

Perkins, R.W., Haller, W.A., Reich, H.G., Rancitelli, L.A. and Wogman, N.A., 1971. Californium-252 neutron activation for terrestrial, sea floor, and possible planetary surface analysis. In: *Neutron Sources and Applications. Proc. Am. Nucl. Soc. Natl. Topical Meet., Augusta, Ga.*, USAEC Rep. CONF-710402, Vol. III.

Philbin, P., Senftle, F.E., Duffey, D., Wiggins, P. and Evans, A.G., 1971. Comparison of stainless steel and zircaloy-2 as encapsulation materials for californium-252. *Nucl. Technol.*, 12: 404—406.

Radiochemical Center, 1972. Californium-252 neutron sources. *Tech. Bull.*, 72/7, Amersham, Buckinghamshire.

Rasmussen, N.C., Hukai, Y., Inouye, T. and Orphan, V.J., 1969. Thermal neutron capture gamma-ray, spectra of the elements. *Mass. Inst. Technol. Rep.*, AFCRL-69-0071, 139 pp.

Rezvanov, R.A., Denisik, S.A. and Lebedev, V.E., 1968. Results of mathematical modeling of the problem of space-time distribution of thermal neutrons in a two-layer medium with cylindrical interface. *Prikl. Geofiz.*, 51: 124—144.

Saltsevich, V.B., Suvorov, A.D. and Krapivskiy, E.I., 1968a. Method of gamma-neutron logging execution. *Bull. Sci. Technol. Inf., Minist. Geol. USSR, Ser. Region., Prospect. Logging Geophys.*, 13: 8—12.

Saltsevich, V.B., Bakhtiarov, A.V., Krapivskiy, E.I. and Suvorov, A.D., 1968b. Qualitative interpretation of the gamma-neutron logging. *Bull. Sci. Technol. Inf., Minist. Geol. USSR, Ser. Region., Prospect. Logging Geophys.*, 13: 13—19.

Scarborough, J.B., 1966. *Numerical Mathematical Analysis.* John Hopkins Press, Baltimore, Md., 6th ed.

Scott, J.H. and Tibbetts, B.L., 1974. Well-logging techniques for mineral deposit evaluation: a review. *U.S. Bur. Mines, Circ.*, 45 pp.

Senftle, F.E., 1970. Mineral exploration by nuclear techniques. *Min. Congr. J.*, January.

Senftle, F.E. and Hoyte, A.F., 1966. Mineral exploration and soil analysis using in-situ neutron activation. *Nucl. Instr. Meth.*, 42: 93—103.

Senftle, F.E., Duffey, D. and Wiggins, P.F., 1969. Mineral exploration of the ocean floor by in-situ neutron absorption using a californium-252 (^{252}Cf) source. *Mar. Technol. Soc. J.*, 3(5): 9—16.

Senftle, F.E., Evans, A.G., Duffey, D. and Wiggins, P.F., 1971a. Construction materials for neutron capture-gamma ray measurement assembly using californium-252. *Nucl. Technol.*, 10: 204—210.

Senftle, F.E., Wiggins, P.F., Duffey, D. and Philbin, P., 1971b. Nickel exploration by neutron capture gamma rays. *Econ. Geol.*, 66: 583—590.

Senftle, F.E., Moxham, R.M. and Tanner, A.B., 1972. A comparison of radiative capture with decay gamma-ray method in borehole logging for economic minerals. *Nucl. Instr. Meth.*, 104: 485—492.

Senftle, F.E., Tanner, A.B., Philbin, P.W., Noakes, J.E., Spaulding, J.D. and Harding, J.L., 1975. In-situ capture γ-ray analyses for seabed exploration: a feasibility study. *IAEA Symp. on Nuclear Techniques in Geochemistry and Geophysics.*

Simon, L., 1969. Quantitative evaluation of ore mineralization in drill holes by gamma-gamma logging. *Can. Min. J.*. 90(7): 57—59.

Sowerby, B.D., 1973. A comparison of gamma-ray resonance scattering techniques for borehole analysis. *Nucl. Instr. Meth.*, 108: 317—326.

Sowerby, B.D. and Ellis, W.K., 1973. Industrial on-stream analysis using gamma-ray resonance scattering. *IAEA Symp. on Nuclear Techniques in the Base Metal Industries*, Paper SM-159/24, pp. 479—492.

Stinner, R.J., 1961. A feasibility study of a method to analyze the moon's surface using nuclear inelastic scattering. University of California, Lawrence Radiation Laboratory, Calif., 46 pp., UCRL 6487.

Storm, E. and Israel, H.I., 1967. Photon cross section from 0.001 to 100 MeV for elements 1 through 100. *Los Alamos Sci. Lab. Rep.*, LA-3753.

Strain, J.E. and Lyon, W.S., 1965. The use of isotopic neutron sources for chemical analysis. In: *Radiochemical Methods of Analysis, I.* IAEA, Vienna, pp. 245—258.

Tanner, A.B., Moxham, R.M., Senftle, F.E. and Baicker, J.A., 1972. A probe for neutron activation analysis in a drill hole using ^{252}Cf and a Ge(Li) detector cooled by a melting cryogen. *Nucl. Instr. Meth.*, 100: 1—7.

Taylor, D., 1964. *Neutron Irradiation and Activation Analysis.* Oxford University Press, London.

Tittle, C.W., 1961. Theory of neutron logging, I. *Geophysics*, 26: 27—39.

Tittle, C.W. and Allen, L.S., 1966. Theory of neutron logging, II. *Geophysics*, 31: 214.

Tittman, J. and Wahl, J.S., 1965. The physical foundations of formation density logging (gamma-gamma). *Geophysics*, 30: 284—294.

Utkin, V.I., 1965. Determination of the ash content of coals by selective logging. *Fiz. Zemli*, 3: 109—114.

Wachpress, E.L. and Habetler, G.J., 1960. *J. Soc. Ind. Appl. Math.*, 8: 403—424.

Waggoner, J.A. and Knox, R.J., 1968. Elemental analysis using neutron inelastic scattering. UCRL-14654 (Rev. 1). University of California, Lawrence Radiation Laboratory, California, 47 pp.

Wichmann, P.A., 1972. Neutron activation for elemental determination in boreholes. *Log Analyst*, January/February, pp. 10—18.

Wing, J. and Wahlgren, M.A., 1965. Detection sensitivities in thermal neutron activation. *USAEC Rep.*, ANL-6953.

Wing, J. and Wahlgren, M.A., 1966. Detection sensitivities in nuclear activation with an isotopic neutron source (with a collection of gamma-ray spectra). *USAEC Rep.*, ANL-7242.

Wing, J. and Wahlgren, M.A., 1967. Detection sensitivities in nuclear activation with an isotopic fast neutron source. *Anal. Chem.*, 39: 85—89.

Wogman, N.A., Rieck, H.G. and Nielson, H.L., 1973. In-situ analysis of the major and minor elements in manganese fields. *Mar. Technol. Soc. J.*, 7(6): 35—40.

Wylie, A.W. and Eisler, P.L., 1969. Prospecting with neutrons. *Philips Tech. Rev.*, 30: 97—105.

Youmans, A.H., Hopkinson, E.C. and Stewart, R.M., 1959. (D, T) neutron activation logging. *34th Annu. SPE Meet., Dallas, Texas*, SPE Paper 130-G.

Youmans, A.H. and Zimmerman, C.W., 1959. Recent advances in the use of nuclear physics in oil well surveys. *5th World Pet. Congr., Sect. X*, Paper 17.

Yule, H.P., 1965. Experimental reactor thermal-neutron activation analysis sensitivities. *Anal. Chem.*, 37: 129—132.

Chapter 6

NUCLEAR WELL LOGGING FOR PETROLEUM

DOUGLAS W. HILCHIE

INTRODUCTION

Well logging as used in the petroleum industry has several separate objectives which are combined to give the overall final result of finding hydrocarbons in the porous and permeable rocks. Wells are logged either in "open-hole" in which case the hole is filled with some drilling fluid, usually bentonite-water mixture, or are logged in "cased-hole" in which the hole has iron casing in it which may or may not have been cemented in place. Most of the logging that takes place is done in open-hole although there is considerable cased-hole logging.

At the time of logging in open-hole the porous and permeable beds that contain hydrocarbons are often invaded by the drilling fluids and the hydrocarbons to be detected are not close to the wellbore. In cased-hole situations there is usually no invasion of drilling fluids but the iron casing and the annulus that may be filled with cement interfere with the measurements, introducing unknowns.

At present time in open-hole logging nuclear tools are used almost exclusively for the determination of rock type and porosity (the capacity of the rock to contain fluids). These measurements are used in conjunction with electrical resistivity measurements to detect the presence and volume of hydrocarbons present.

In cased-hole applications nuclear tools are used to detect the rock type and the presence and volume of hydrocarbons.

Nuclear logging in this text is the well logs classified as natural and induced.

NATURAL RADIATION

Natural gamma-ray log

One of the prime nuclear measurements is the measurement of natural gamma-ray activity of the formations. The gamma rays, primarily from ^{40}K although significant contributions are made from the thorium and the uranium-radium series. The energy of the ^{40}K contribution as shown in Fig. 6-1

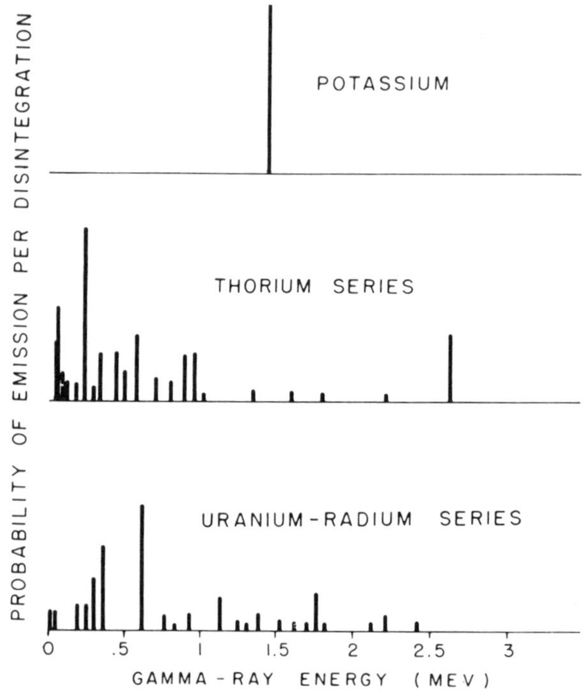

Fig. 6-1. Energy spectra of three components making up to natural gamma-ray log.

is 1.46 MeV. The radioactive materials in the formations are generally associated with the shales although other rock types are radioactive on a local basis. The gamma-ray log is considered a shale log. Using this general characteristic the formations may then be separated into shales and non-shales. The non-shales generally being the potential hydrocarbon-bearing formations.

Most gamma-ray logging devices use sodium iodide crystals (thallium-activated) and photomultipliers with the whole assembly kept in a dewar flask. In extremely hot or rugged operations Geiger-Mueller counters are generally used.

In some areas it is difficult to correlate between one formation and another or the zones of interest are radioactive and a spectral natural gamma-ray log is used. This device breaks up the gamma rays being measured into their three basic energy classifications, potassium, thorium series, and uranium-radium series. This is done on a energy-selective basis. The ^{208}Tl peak occurs at 2.62 MeV, ^{214}Bi occurs at 1.76 MeV, and the ^{40}K peak occurs at 1.46 MeV. Fig. 6-2 shows a gamma-ray log and a spectral gamma-ray log run in the same hole.

The spectral gamma-ray log is not a routinely used device as it is primarily used in areas where correlation is a problem or in areas where the formation is radioactive and better information is needed.

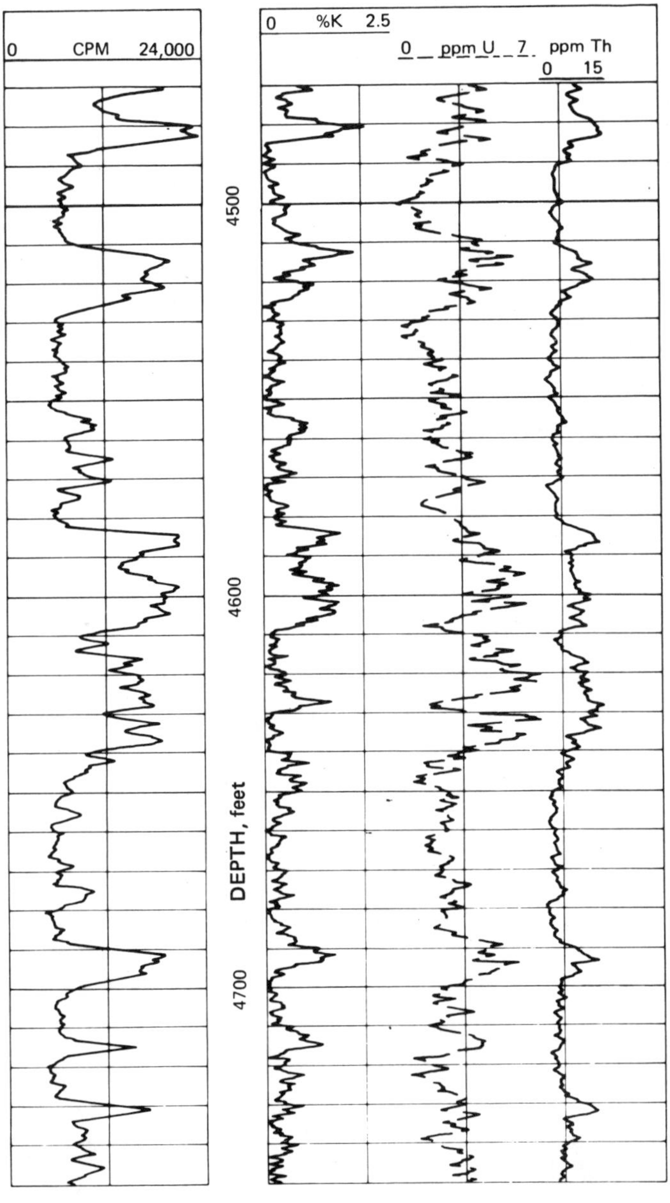

Fig. 6-2. Spectral (left) and natural (right) gamma-ray logs.

INDUCED LOGS

The induced radioactivity logs fall into those run with neutron sources and those run with gamma-ray sources.

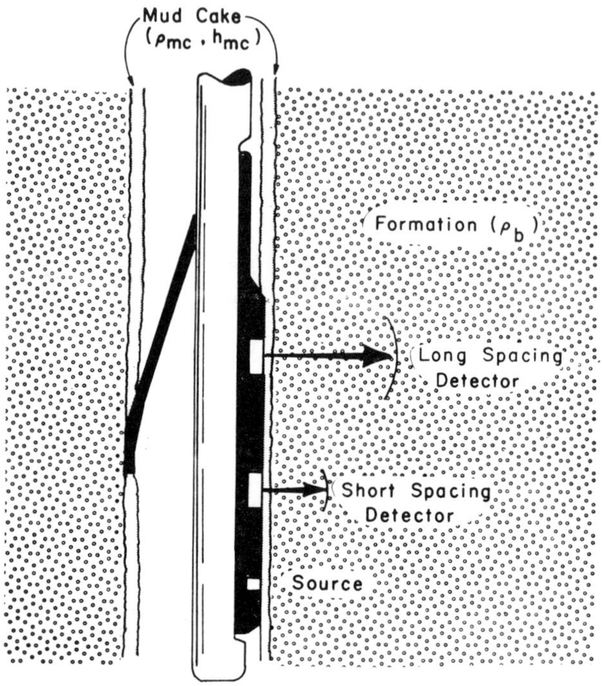

Fig. 6-3. Schematic of a compensated formation density logging tool.

Induced gamma-ray logs

The only induced gamma-ray log being commercially run is the density log. The tool consists of a ^{137}Cs source and two (commonly scintillation) detectors as shown in Fig. 6-3. The long-spaced combination measures deeper into the formation whereas the shorter space is more influenced by mud cake and other materials that exist between the pad (containing the source and detectors) and the formation. A plot of typical laboratory results showing the response of the long- and short-spaced count rates is shown in Fig. 6-4. In Fig. 6-4 the solid line is the response with no material between the pad and the formation. As the pad is held away from the rock the data points shift off the primary "spine" and follow the "rib" curves. The direction (left or right) depends on the density of the mud cake relative to the formation density. A computer is used to make the correction for the material between the pad formation (mud cake) and the log reads out in apparent density.

The process that is measured is primarily Compton scattering. The source has sufficiently low energy to effectively eliminate pair production and the detectors are set so that they do not see gamma rays with energies that significantly are affected by photoelectric effects.

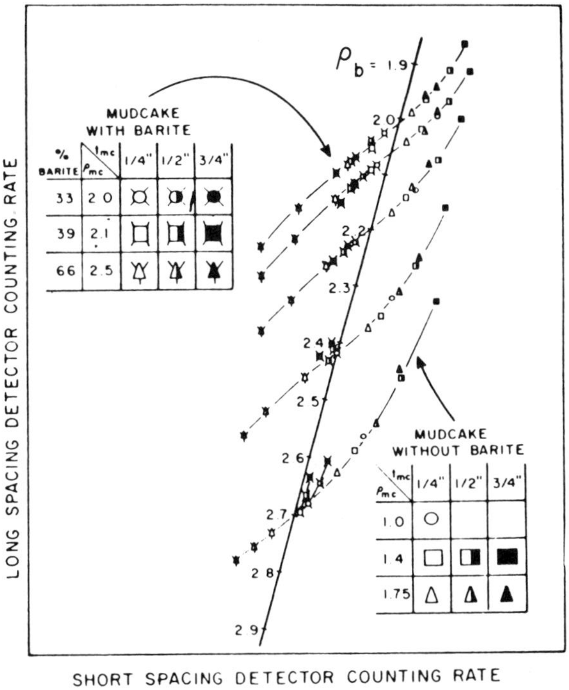

Fig. 6-4. A compensated formation density log "spine and ribs" chart showing the effect of mud cake on the short- and long-spaced counting rates.

In effect what is being measured is electron density. The relation between electron density and bulk density is apparent in equation (1):

$$\rho_e = 2 \sum_i \frac{Z_i}{A} \rho_b \tag{1}$$

where ρ_e = electron density, Z = the atomic number, A = the atomic weight, and ρ_b = the bulk density.

For most reservoir type rocks, for example sandstone, limestone, and dolomite, Z/A is equal to 0.5 and the relationship between bulk density and electron density is apparently the same. Unfortunately, the Z/A is not equal to 0.5 for water but by calibrating the density log in limestone filled with water this effect is eliminated so that the relationship between electron density and bulk density is further modified by equation (2):

$$\rho_b = 1.07\,\rho_e - 0.188 \tag{2}$$

The density log readings are converted to porosity, ϕ, using equation (3):

$$\phi = \frac{\rho_b - \rho_{ma}}{\rho_f - \rho_{ma}} \tag{3}$$

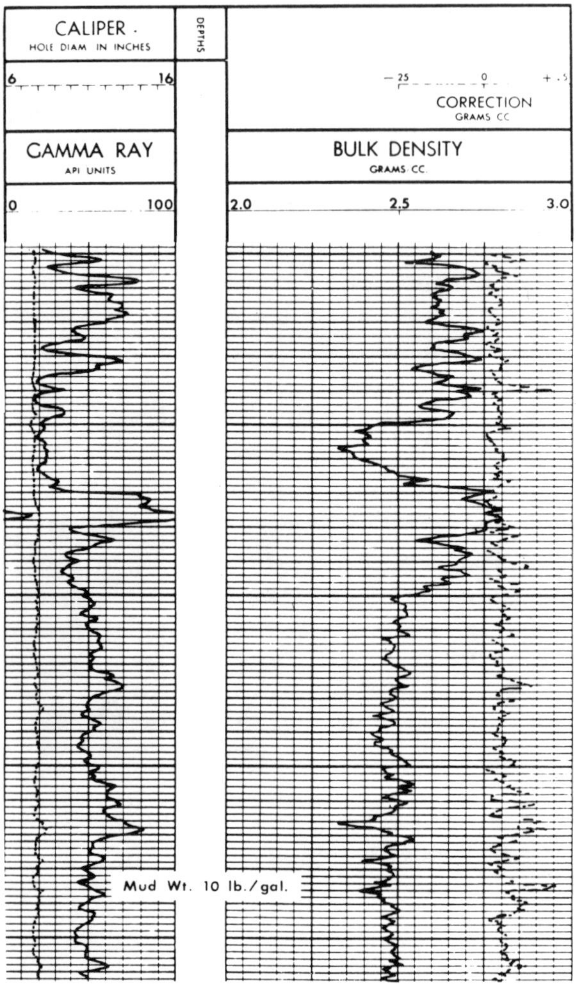

Fig. 6-5. A compensated formation density log presentation.

where ρ_{ma} = the density of the solid rock, usually 2.65 g cm^{-3} for sandstone, 2.71 for limestone, and above 2.85 for dolomite; and ρ_f = fluid density.

Fig. 6-5 is an example of a density log. The additional dotted curve on the right-hand side of the density log is the amount of correction ($\Delta\rho$) applied by the computer to the long-spaced reading to obtain the final bulk density.

Neutron logs

Two types of neutron logs are run in the oilfield. The two basic classifications of neutron logs used in the oil business are: (1) steady-state sources, and (2) pulsed sources. The former are used primarily as a porosity tool whereas the pulsed source tools are used to detect the existence of hydrocarbons, usually in cased-hole environments.

Neutron logs (steady-state)

Most steady-state neutron logs, hereafter called just neutron logs, used either Am-Be or Pu-Be sources. The detectors are predominately epithermal or thermal neutron detectors. Two particular types of neutron logs predominate today although a third is evident in many areas. The most popular and newest is the compensated neutron log which has a source and two detectors located on a mandrel. The near and far detector count rates are ratioed and give a linear response on semi-log plot similar to that shown in Fig. 6-6. This shows a considerable reduction in borehole effects over neutron logs using only one detector. The compensated neutron logs all measure thermal neutrons although some experimental tools being developed do measure epithermal neutrons.

The second neutron log of note is the side-wall neutron. This is a pad type device. The neutron source and detector are located on a pad which is

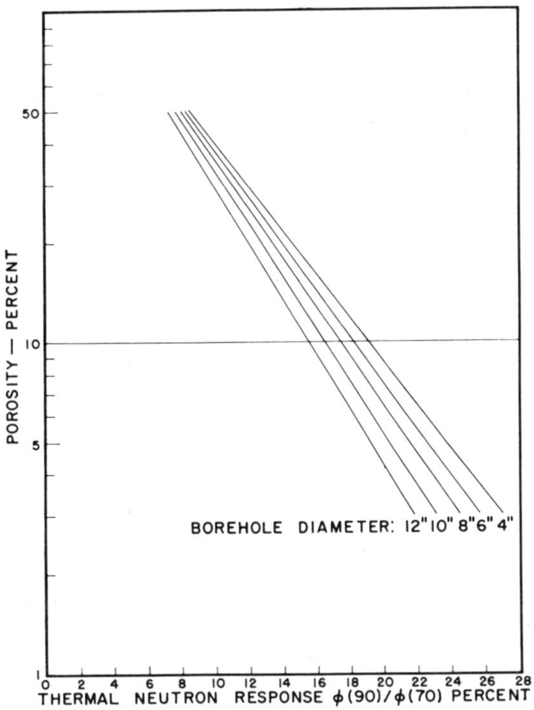

Fig. 6-6. Porosity response for a laboratory dual spaced neutron log.

pushed against the side of the hole in the same manner as with the density log. Side-wall neutrons devices all measure epithermal neutrons.

The side-wall neutrons are systematically being replaced with the compensated devices.

The third neutron log system, truly obsolete but not yet been totally replaced, is the single source and single detector on a mandrel. These are generally not calibrated in terms of porosities as are the other neutron logs due to the inability to define the exact location in the borehole at any given time.

Neutron logs are used primarily for porosity but also gain very wide use as a gas finding tool. Water and oil have very much the same hydrogen content and thus to the neutron log look the same, as the reactions are primarily between the hydrogen and neutrons. In gases the hydrogen content is low and accordingly gas shows up as low porosity. Fig. 6-7 shows an example of a neutron and density log run simultaneously and recorded on compatible porosity scales. The separation that occurs is due to the natural gas in the sandstone formation. Separation also occurs in shales but these are generally not considered gas reservoirs.

Pulsed neutron logging

Two different pulsed neutron logging techniques are in use in the petroleum business. The first measures the decay rate of thermal neutrons (thermal neutron life-time) in the formation and the other measures the inelastic collisions of carbon and oxygen in the formation. The former will

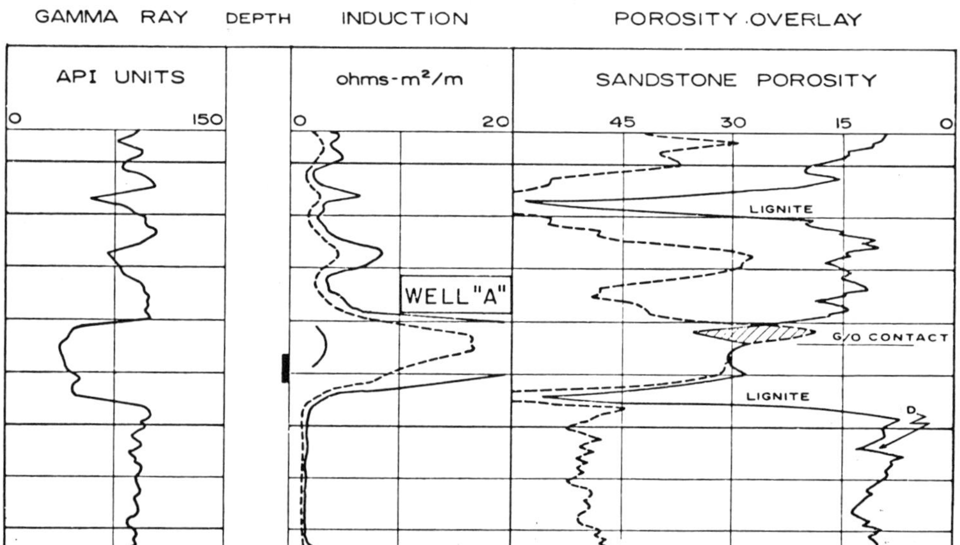

Fig. 6-7. An example of a simultaneous density (D) and neutron (N) log. Depth is in arbitrary units. G/O indicates the gas/oil contact.

be referred to as a pulsed neutron capture (PNC) log and the second will be called carbon-oxygen logging. Both of these techniques are used in case holes and are used for direct location of hydrocarbons. Unfortunately the data alone are not sufficient to make definitive decisions as lithology, and porosity logs must be run for an interpretation.

Both systems use high-voltage power supplies and accelerators. The neutrons are created by deuterium ions striking a tritium target. The power supplies range from classical Van der Graaf generators to solid-state power supplies. It should be remembered that all these logging tools are roughly $3\frac{1}{2}$ inches in diameter and a good portion of this diameter is a steel pressure housing, so that the electronic components occupy something less than 3 inches in diameter.

Pulsed neutron capture logs. Pulsed neutron capture logs which run under the trade names of neutron life-time and thermal decay-time logs measure the macroscopic thermal neutron absorption cross-section of the materials which controls the life-time of thermal neutrons in a material. The source is pulsed and the gamma rays of capture are measured during the off-period of the source. Immediately following the cessation of neutron production, there is a period of thermal neutron decay which is a borehole signal and after about 400 micro-seconds, the decay is controlled primarily by the formation characteristics (Fig. 6-8). The rate of thermal decay is made during this latter part of the die-away so as to effectively borehole compensate for the measurement.

The measurement of macroscopic thermal neutron absorption capture cross-section (cross-section) is related to porosity and water saturation as shown in equation (4):

$$\Sigma = \phi(\Sigma_w S_w + \Sigma_h S_h) + (1 - \phi)\Sigma_{ma} \qquad (4)$$

where Σ_{ma} = the cross-section of the solid rock; ϕ = the porosity; Σ_h, Σ_w = the cross-section of the hydrocarbons and water, respectively; S_h, S_w = the saturation of the hydrocarbons and water, respectively; and Σ = the formation cross-section.

The variation in Σ_w is a function of the salinity of the water with distilled water and oil having about the same capture cross-section ($22\ cm^{-1} \times 10^{-3}$). It can be seen that the effectiveness of the measurement relies on a significant difference between the water and hydrocarbon cross-sections. In oil zones, this is accomplished by having salty waters and in gas zones this is present in all cases as the natural gases have very low cross-sections relative to either fresh or salty water. The matrix cross-sections for various formations are very close to being the same with field clean sandstones being in the 9—12 sigma unit range, limestones 8—11 and dolomites 10—12. Shales have very high capture cross-sections and thus are easily distinguished from other rocks.

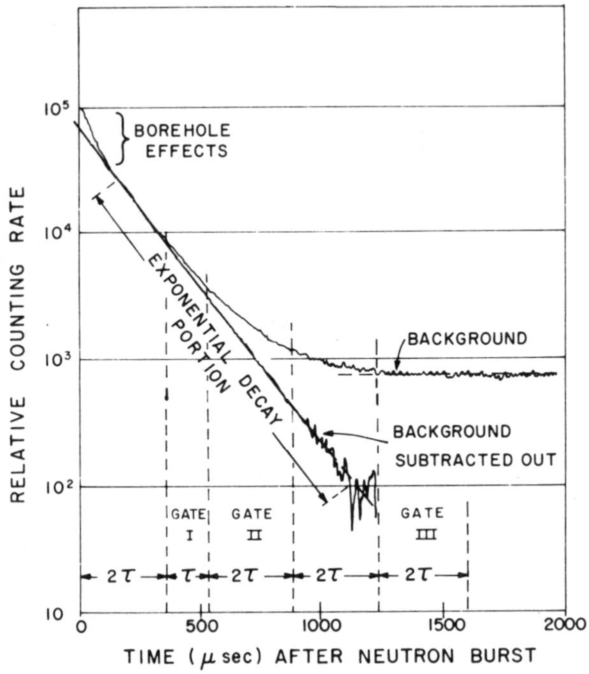

Fig. 6-8. Thermal neutron die-away curve obtained with PNC log.

The interpretations of the PNC logs is dependent on knowing the formation water salinity, porosity, and the expected hydrocarbon (gas or oil). These must be obtained from other sources.

The newer PNC logs have two detectors and simultaneously measure capture cross-section and the ratio of the far and near detector count rates. This ratio is relatable to porosity. In oil zones this significantly helps the interpretation. In gas zones where neutron logs see gas as low porosity it has been found that by presenting the count rate from the near and far detectors, as shown in Fig. 6-9, the presence of gas can be observed by the relative position of the two curves. This is a purely empirical technique.

PNC logs are used to find hydrocarbons in old wells that have never been logged and to monitor the producing zones to determine the efficiency of production and to help evaluate production problems where it is uncertain whether the well has become depleted or is mechanically being obstructed from producing hydrocarbons.

Carbon-oxygen logging. In formations where the water is fresh the PNC logs cannot distinguish between oil and water. To cover this special condition the carbon-oxygen log was developed. A pulsed source similar to that used for PNC logs is used. The gamma rays produced by inelastic scattering of the 14-

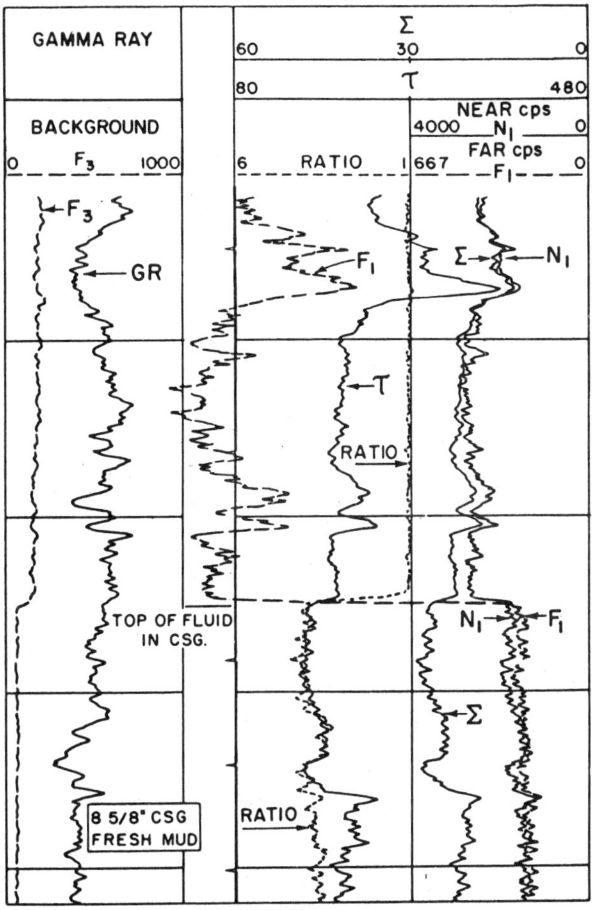

Fig. 6-9. An example of a PNC log with two detectors.

MeV neutrons are detected by a scintillation spectrometer. The role of the instrument is to detect the carbon/oxygen ratio in sandstones which is directly related to the relative abundances of hydrocarbons and water. The measurement of carbon/oxygen ratio is recorded on a station by station level. The stationary measurements are for periods up to 10 min. The station by station measurement is made necessary by the low counting rates and the need for very precise numbers. For example, in a water zone the measured carbon/oxygen ratio is typically around 1.65 and if the same zone is hydrocarbon-producing the number ranges from 1.7 to 1.85.

Fig. 6-10 shows a spectra of frequency versus gamma-ray energy for a tank filled with oil versus a tank filled with water. The carbon, oxygen and the escape peaks are obvious and the approximate window width used is shown.

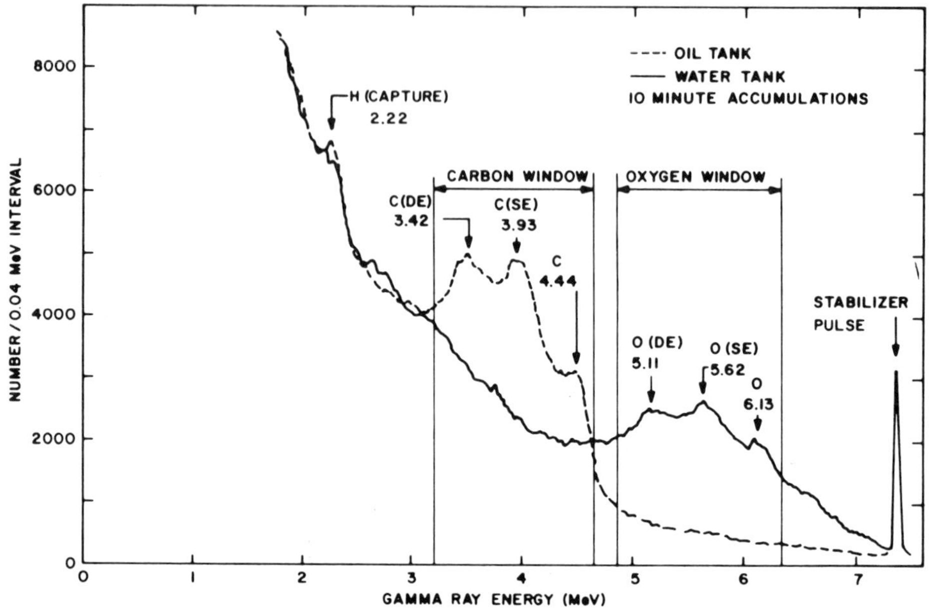

Fig. 6-10. Carbon-oxygen spectra in tanks of oil and water. DE = dual escape peak; SE = single escape peak.

This is an extreme case in that it would represent a formation with 100% porosity and 100% either oil or water saturation. Typically we are looking at 10—30% porosity and water saturations in the 20—100% range. Real formation spectra show considerably less separation between oil and water.

The only commercial carbon-oxygen log available today uses a thallium-activated sodium iodide crystal in a dewar flask. The photomultiplier is rigorously stabilized and the gating circuits are severely limited to be sure that high-quality pulse data is transmitted over the very long cable.

The carbon-oxygen log at the present time is not usable in carbonates, i.e., limestone or dolomite, because of the carbon in the rock material itself. It is necessary to be able to identify the type of lithology of the formation being investigated if it is not already known from geological drill cuttings. To accomplish this end a silicon/calcium ratio is run with the carbon-oxygen log. The silicon and calcium are measured either from gamma rays from thermal neutron capture interactions or from inelastic neutron interactions. Those that are run with thermal neutron capture interactions are salinity sensitive as the chlorine in calcium spectral shapes overlap. The putting together of the silicon/calcium ratio and the carbon/oxygen ratio allows for the identification of sands and sands with small amounts of calcite cement and the identification of the fluids inside. Fig. 6-11 shows a plot based on the use of both ratios.

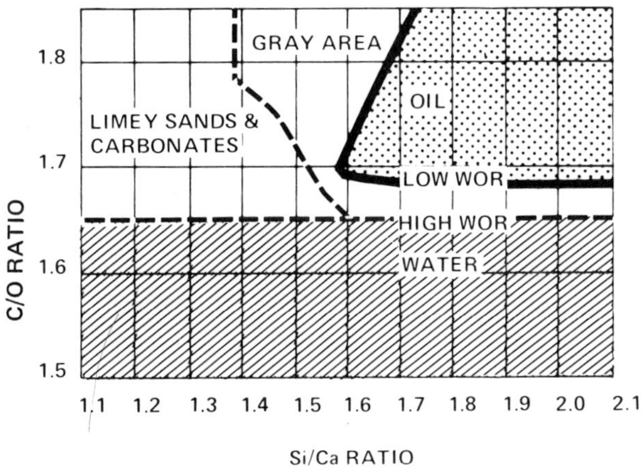

Fig. 6-11. Method for predicting fluid content of limy sandstones using carbon/oxygen and silicon/calcium ratios. WOR = water/oil ratio.

CONCLUSIONS

Nuclear well logging techniques play a major and very necessary role in well logging for petroleum.

REFERENCES

Culver, R.B., Hopkinson, E.C. and Youmans, A.H., 1973. Carbon-oxygen (C/O) logging instrumentation. *48th Annu. Soc. Pet. Eng., AIME Meet., September, Las Vegas, Nev.*, SPE 4640.

Hilchie, D.W., undated. *Neutron Lifetime Log Interpretation.* Dresser Atlas, Houston, Texas.

Hilchie, D.W., 1974. Neutron accelerator logging. *Conf. on Neutron Accelerators, October, North Texas State University, Denton, Texas.*

Hilchie, D.W., 1975. Nuclear well logging. *Am. Nucl. Soc. Meet., June, New Orleans, La.*

Schlumberger Log Interpretation, I. Principles. Schlumberger Ltd., New York, N.Y. (no date).

Schlumberger Log Interpretation, II. Applications. Schlumberger Ltd., New York, N.Y., (no date).

Schultz, W.E. and Smith, H.D., 1973. Laboratory and field evaluation of a carbon-oxygen well logging system. *48th Annu. Soc. Pet. Eng., AIME Meet., September, Las Vegas, Nev.*, SPE 4638.

Wichman, P.A., McWhirter, V.C. and Hopkinson, E.C., 1975. Field results of natural gamma-ray spectralog. *Soc. Prof. Well Log Analysts, 16th Annu. Symp., June, New Orleans, La.*

Chapter 7

THE POTENTIAL OF PLOWSHARE FOR RESOURCE DEVELOPMENT

JOHN TOMAN and HOWARD A. TEWES

INTRODUCTION

It was only natural that success in development of nuclear explosives should lead to investigations into practical use for these highly compact explosive sources. The 1956 war over the Suez Canal, fought by Britain, France, and Israel against Egypt, provided the motivation for initiation of studies into non-military uses of nuclear explosives. The possibility of excavating an alternate canal through Israel with nuclear explosives was discussed at the "First Plowshare Symposium" convened at the Livermore installation of the University of California Radiation Laboratory (now the Lawrence Livermore Laboratory, or LLL) in February 1957. In July 1957 Project Plowshare was formally established at the LLL to investigate industrial uses of nuclear explosives. From 1957 to 1970, the major thrust of the effort centered on excavation and the feasibility of excavating a sea-level canal across the Isthmus of Panama (Atlantic-Pacific Interoceanic Canal Study Commission, 1970). A summary of cratering experiments carried out as part of the Plowshare Program, as well as possible industrial applications of nuclear cratering are given by Toman (1970a, b).

American industrial interest in Plowshare was concentrated on underground or completely contained applications of nuclear explosives because of the lack of commercially attractive nuclear excavation projects within the Continental U.S. The potential applications which appeared to be economically attractive when compared with conventional technology included the use of nuclear cavities for underground storage of natural gas and hydrocarbon liquids, stimulation of natural gas reservoirs by rock breakage, the production of shale oil by the in-situ retorting concept, and chemical mining of deeply buried deposits of copper ore.

The most simple and straightforward application is that of storage, but no U.S. experiments have been conducted for this specific purpose. While gas well stimulation is inherently a more complex application of peaceful nuclear explosions (PNEs), three experiments have been carried out in this branch of the technology. Analysis of the results from these experiments has indicated that with careful planning, this application could be commercially attractive.

A substantial calculational and laboratory effort has been expended to assess the feasibility of in-situ retorting of shale oil within a nuclear chimney; a program of similar scope has been carried out in conjunction with industry to assess the economic and technological characteristics of nuclear chemical copper mining. These two potential resource recovery applications of PNEs are technically complex, and no full scale field tests have been conducted.

At the present time, the Plowshare Program is only very modestly funded by the Federal government, probably due mainly to the current socio-political climate. Consequently, no field experiments are currently planned. It appears that, for Plowshare technology to be applied on any scale to resource extraction in the United States, it must display very significant economic superiority over any competing conventional technology. In the event that social attitudes change in the next decade or so, or if the energy crisis becomes so severe that all potential energy resource recovery methods are pursued, the use of PNE technology may be reevaluated. The following overview of the use of nuclear explosives for resource development is an attempt to place the potential of this technology in perspective. The reference list presented at the end of this chapter is not intended to be exhaustive but is adequate for anyone desiring to perform additional background research.

PHENOMENOLOGY

The large yield of energy produced by an explosive with a relatively small volume is the major factor favoring the use of nuclear explosives for industrial rock crushing or earth moving applications. As an example, a 100-kt nuclear explosive designed for contained (non-cratering) applications has a volume of less than 1 m^3, while the chemical explosive such as TNT required to produce the same energy release has a volume on the order of 50,000 m^3.

In this brief discussion of nuclear explosives, we will confine ourselves to their utilization for contained applications. While the Soviets have proposed using nuclear explosives to remove the overburden from deeply buried ore bodies (see section "U.S.S.R. program"), we will not deal with this proposed method of explosive utilization in this section.

Cavity and chimney formation (Rawson, 1966)

In order to illustrate the effects of a contained nuclear detonation, we shall consider a particular example in which a 100-kt explosive is detonated at a depth of 900 m. The schematic representation shown in Fig. 7-1 indicates the development of the explosion effects resulting from such a detonation.

In less than a microsecond following the detonation, the total energy of

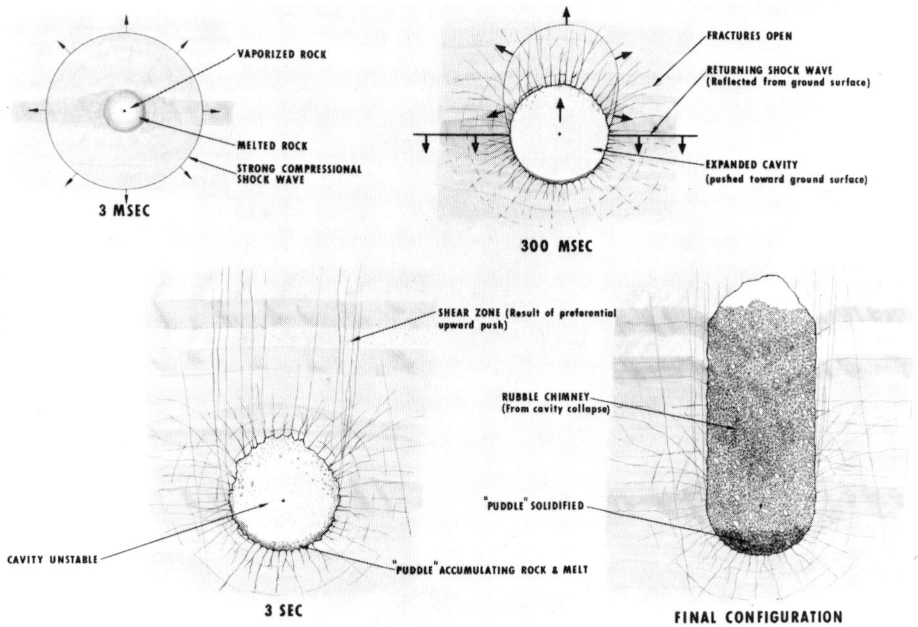

Fig. 7-1. Cavity growth and collapse history of contained nuclear explosions.

4.2×10^{21} ergs is released. Approximately 6000—9000 tonnes of rock, as well as the explosive itself, are vaporized. The temperature and pressure of this sphere of "rock gas" are initially on the order of 13,000,000°K and 10 TPa, respectively (Johnson and Violet, 1958; Johnson et al., 1959). A strong spherical shock front moves radially outward from this gas bubble deep underground — melting, crushing, displacing and cracking the surrounding rock. Well over 90% of the energy of this shock wave is deposited locally in the form of heat and work on the surrounding rock — typically, a total of about 100,000 tonnes of rock eventually are melted and/or vaporized. The fraction of the total explosion energy which is eventually manifested as an elastic or seismic wave ranges from about 1% to as low as a few hundredths of a percent.

The original bubble of "rock gas" expands, forming a cavity. The expansion phase continues until the gas pressure is approximately equal to the overburden pressure (in this example, about 22.5 MPa). Cavity growth is completed in a few hundred milliseconds to a radius of about 45 m. At this time, the walls of the cavity are lined with the 100,000 tonnes of melted rock (as noted above), and the cavity gas temperature is a few thousand degrees Kelvin.

The cavity persists for a period of time which is somewhat dependent on the mechanical properties of the rock overburden; during this time interval, the "rock gas" condenses, the temperature and pressure drop, and the molten rock lining the cavity flows down to form a puddle at the bottom of the cavity. Finally, the fractured rock above the explosion-generated cavity collapses into the spherical void, resulting in a roughly cylindrical column of broken rock which is referred to as the "chimney".

If the rock in which the explosive is detonated is relatively tight and compact, collapse leads to random packing of the rock fragments with a consequent reduction of the bulk density of the rock "bed" in the chimney region, together with a corresponding increase in the effective porosity and permeability of the rubble column. On the other hand, if the detonation environment can be characterized as a porous and compactable rock (such as the tuffaceous alluvium of the Nevada Test Site), little or no decrease in bulk density occurs with collapse. When the roof of the cavity collapses, the overlying rock continues to fail progressively upward until the void volume is translated to the ground surface and a subsidence crater is formed. The roofs of cavities in plastic unfractured rock (such as exists in many kinds of layered salt or in salt domes) are more stable and collapse very slowly, if at all.

The observed values for cavity radius and chimney height vary as a function of explosive yield, depth of burial, rock type and overburden density. Assuming that the cavity is formed by adiabatic expansion of high-pressure gas to overburden pressure, cavity radii can be calculated by using the approximate relation (Boardman et al., 1964):

$$R_c = CW^{1/3}/(\rho h)^{1/4} \, , \tag{1}$$

where R_c = cavity radius (m); C = constant (appropriate to the detonation medium); W = energy yield of explosion (kt); ρ = average overburden density; and h = depth of overburden (m).

On the basis of data from a large number of nuclear detonations in different media (including such dissimilar substances as tuff, granite, dolomite and salt), C displays an average value of about 66 (individual values of C range from 55 to 82). Cavity size is thus seen to be relatively insensitive to rock type.

The chimney height observed where the detonation medium is competent (i.e., where collapse does not continue upward to the ground surface) can be expressed by the simple relation:

$$H = kR_c \tag{2}$$

where H = chimney height (m), measured from the original location of the nuclear explosive to the apex of the chimney; k = constant (which ranges from about 3 to 6).

Chimney collapse and the resultant chimney characteristics are much more variable than is the cavity radius, R_c, since chimney size and shape are strongly influenced by inhomogeneities in the rock mass, such as bedding planes, fracture frequencies, and jointing characteristics.

Note also that equation (2) does not apply to detonations taking place in a plastic medium, such as salt, where chimneys appear to be minimal or non-existant.

In the rock outside the cavity-chimney region, the detonation-induced shock wave characteristically produces a zone of increased permeability in the rock environment. It has been calculated that this zone of enhanced permeability should extend to a distance as great as $4R_c$, as measured from the detonation point. Because this region displaying greater permeability than the surrounding rock can only be characterized with great difficulty (requiring the drilling of several exploratory wells surrounding the nuclear cavity-chimney system), it has not as yet been possible to physically measure the increase in permeability as a function of distance from the shot point. It should be emphasized that the extent of this zone of increased permeability is critical to the economic evaluation of the potential of peaceful nuclear explosions (especially for the recovery of natural gas). Based on modeling gas flow from nuclear chimneys in natural gas reservoirs, enhanced permeability appears to extend to a lateral distance of $3\ R_c$.

Seismic effects (Holzer, 1971)

From the inception of the Plowshare Program it has been recognized that understanding and predicting the ground motion generated by a nuclear explosion and any associated building or structural damage was necessary if nuclear explosions for peaceful purposes were to achieve their maximum usefulness and employment (Griggs, 1958). Over the past decade, methods have been developed that enable us to estimate, with reasonable accuracy, the ground motion from single, relatively high yield, detonations. Ground motion parameters which appear to best correlate with damage have been identified.

The approach used has involved the generation of a response spectrum (Nadolsky, 1969) which represents the structure response to a particular ground displacement. The simplifying assumption is made that the response of a structure to such a ground displacement can be approximated by the envelope formed by the maximum response of a series of simple harmonic oscillators which are tuned to different frequencies and damped to a fixed fraction of the critical value (Clough, 1967). The generation of this response spectrum can be carried out using a digital computer; a typical spectrum is shown in Fig. 7-2 which also shows the approximate period intervals in which buildings of various heights have been found to resonate.

For one- or two-story buildings, the value of the pseudo-absolute acceleration (PSAA) * in the 5—20 Hz range (periods of 0.2—0.05 second) seems to

* PSAA is a function of both the displacement of a building by a seismic pulse and the square of the natural response frequency of the building.

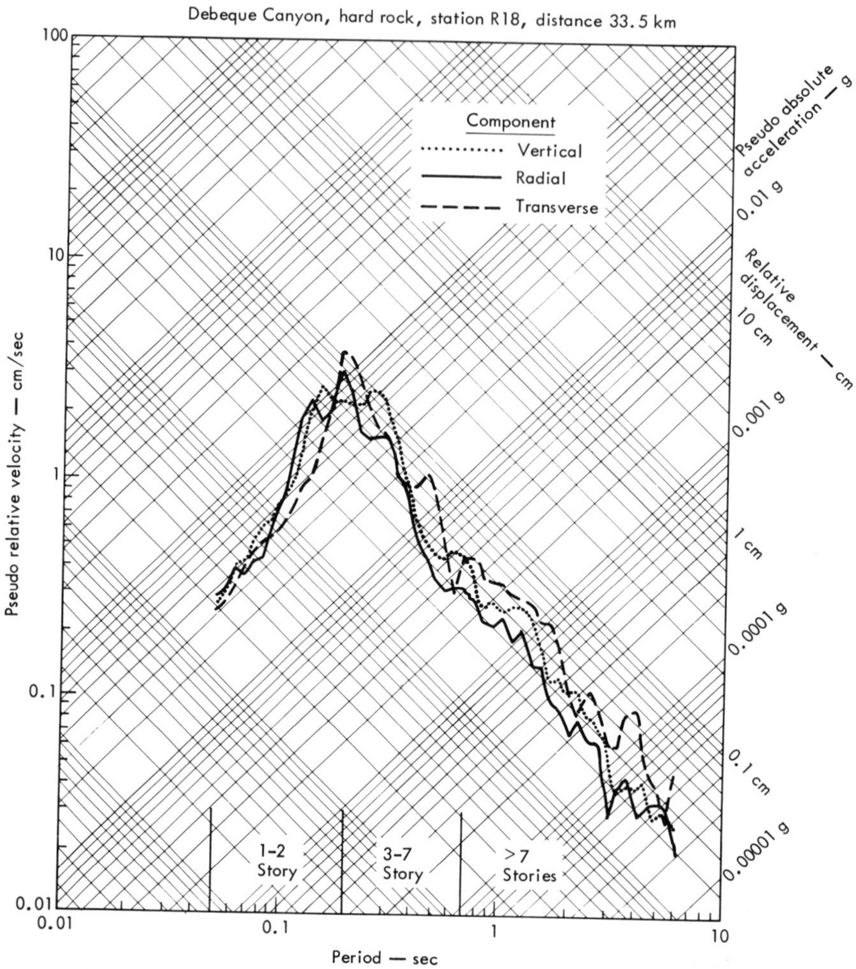

Fig. 7-2. Sample spectrum, with regions of building response.

correlate better with observed damage than any other motion parameter (Nadolsky, 1969). The ground motion measurements made on the Rulison gas stimulation experiment (U.S. Atomic Energy Commission, 1969b), together with thorough structural or architectural seismic damage documentation (U.S. Atomic Energy Commission, 1971) has empirically established this correlation.

This Rulison experience (Rizer, 1970; Farhoomand and Scholl, 1971) is summarized in Fig. 7-3, together with some results from the Salmon experiment and also from several detonations carried out at the Nevada Test Site. This figure shows that at PSAA of 1*g* in the frequency range of 5—20 Hz,

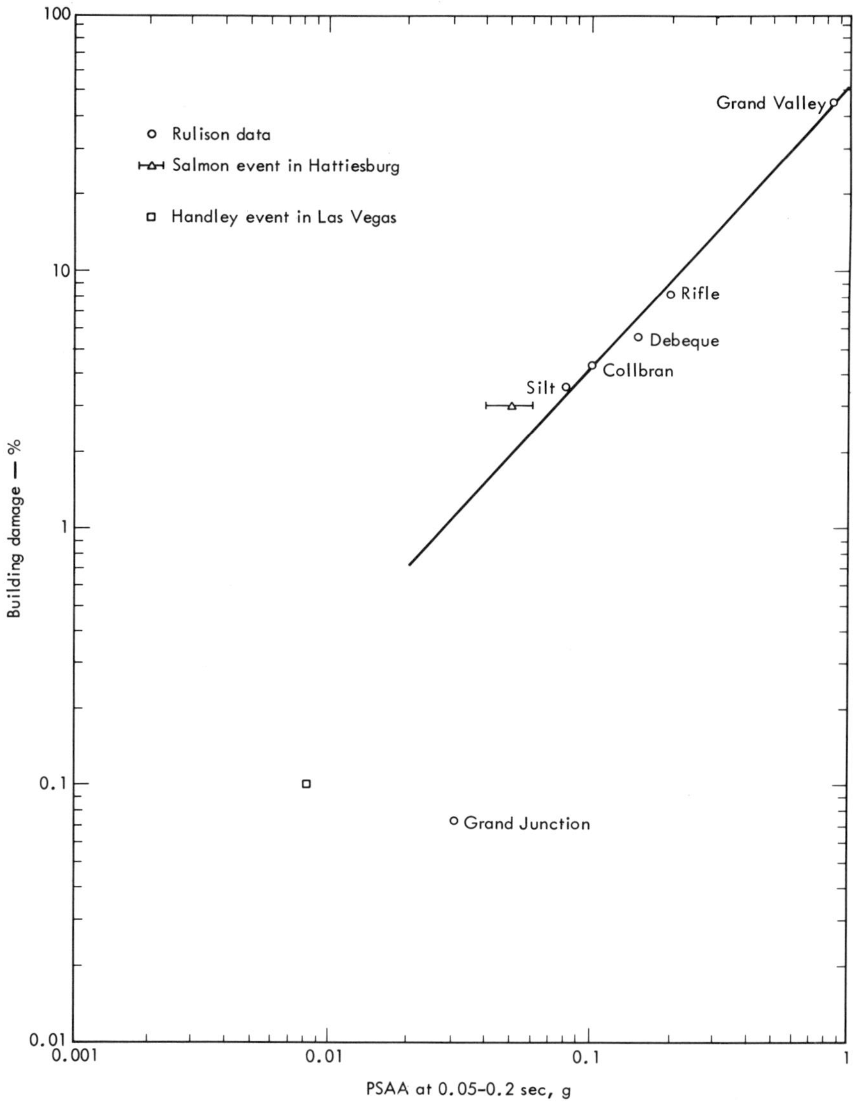

Fig. 7-3. Correlation between damage and pseudo-absolute acceleration.

over 50% of one- or two-story structures can be expected to sustain some damage. At about 0.01*g*, damage will be sporadic; at lower *g* loadings, little or no damage should be anticipated. Thus, Fig. 7-3 can be used as a quick and convenient aid in estimating building damage for one- or two-story structures in the vicinity of a planned nuclear detonation, provided that the response spectra of these structures can be predicted.

Radioactivity production, distribution, and transport (Tewes et al., 1974; Levy, 1972)

The specific radionuclides produced by a nuclear detonation used for underground resource development are much the same irrespective of the design of the nuclear explosive which is used; however, the relative amounts of the various radioactivities formed is critically dependent on the fission-to-fusion ratio of the explosive. Table 7-I shows the fission product radioactivity present 180 days after the detonation of a 100-kt all-fission explosive. Should a 100-kt thermonuclear explosive with a 5-kt fission trigger be employed, the amounts of all the radionuclides shown in Table 7-I should be reduced twentyfold. With regard to the radioactive products formed when

TABLE 7-I

Radionuclide source at D + 180 days after detonation of a 100-kt all-fission explosive (assumes ^{235}U fission)

Nuclide	Fission products		Predominant location of radioactivity
	half-life	activity (Ci)	
^{85}Kr	10.8 yr	2200	gas
^{89}Sr	52 days	220,000	chimney — on rock surfaces
^{90}Sr	28 yr	15,000	chimney — on rock surfaces
^{91}Y	59 days	360,000	chimney — on rock surfaces
^{95}Zr	65 days	440,000	melt — in vitreous matrix
^{95}Nb	35 days	800,000	melt — in vitreous matrix
^{103}Ru *	40 days	105,000	chimney — on rock surfaces
^{106}Ru *	1.0 yr	38,000	chimney — on rock surfaces
^{115m}Cd	43 days	120	chimney — on rock surfaces
^{123}Sn	125 days	600	chimney — on rock surfaces
^{124}Sb	60 days	50	chimney — on rock surfaces
^{125}Sb	2.7 yr	6000	chimney — on rock surfaces
^{125m}Te	58 days	1100	chimney — on rock surfaces
^{127m}Te *	109 days	7000	chimney — on rock surfaces
^{129m}Te *	34 days	1700	chimney — on rock surfaces
^{137}Cs *	30 yr	18,000	chimney — on rock surfaces
^{140}Ba *	12.8 days	900	chimney — on rock surfaces
^{141}Ce	33 days	130,000	melt — in vitreous matrix
^{143}Pr	13.6 days	1600	melt — in vitreous matrix
^{144}Ce *	284 days	360,000	melt — in vitreous matrix
^{147}Pm	2.6 yr	70,000	melt — in vitreous matrix
^{151}Sm	87 yr	500	melt — in vitreous matrix
^{155}Eu	1.8 yr	3000	melt — in vitreous matrix

* Relatively short-lived radioactive daughters are in transient equilibrium with these fission products; however, the total activity (in curies) listed for these fission products represents only that from the parent nuclide.

TABLE 7-II

Soil activation products (assumes Hardhat granite with 1 wt.% FeS_2, 1 wt.% $CuFeS_2$, and 5 wt.% H_2O)

Nuclide	Half-life	Activity (Ci)	Predominant location of radioactivity
^{3}H	12 yr	< 2000 **	water
^{14}C	5730 yr	7	gas
^{32}P	14 days	170	chimney — on rock surfaces
^{35}S	88 days	15,000	chimney — on rock surfaces
^{37}Ar	35 days	7200	gas
^{39}Ar	270 yr	8	gas
^{45}Ca	155 days	20,000	melt — in vitreous matrix
^{46}Sc	84 days	1400	melt — in vitreous matrix
^{54}Mn	303 days	180	melt — in vitreous matrix
^{55}Fe	2.6 yr	49,000	melt — in vitreous matrix
^{59}Fe	45 days	1700	melt — in vitreous matrix
^{60}Co	5.3 yr	120	melt — in vitreous matrix
^{65}Zn	245 days	1000	melt — in vitreous matrix
^{85}Sr	64 days	160	melt — in vitreous matrix
^{89}Sr *	52 days	550	melt — in vitreous matrix
^{95}Zr *	65 days	90	melt — in vitreous matrix
^{95}Nb *	35 days	150	melt — in vitreous matrix
^{134}Cs	2 yr	1500	chimney — on rock surfaces
^{141}Ce *	33 days	130	melt — in vitreous matrix
^{152}Eu	12 yr	600	melt — in vitreous matrix
^{160}Tb	72 days	830	melt — in vitreous matrix
^{170}Tm	130 days	2900	melt — in vitreous matrix
^{192}Ir	74 days	900	melt — in vitreous matrix

* Also formed as a fission product.

** Published values for gas-stimulation fission explosive detonated in Wagon Wheel environment.

neutrons emitted from the explosive are captured in the surrounding rock, the amounts of these products are influenced both by the composition of the rock and by the number of neutrons escaping from the nuclear explosive. Since both fission and thermonuclear explosives emit * approximately 2×10^{23} neutrons per kiloton of yield (Lessler et al., 1971), the same amounts of the various activation products are produced for a given total yield, irrespective of the type of explosive. Thus, the list of radioactivities given in Table 7-II represents rock activation produced by either a fission or a thermonuclear explosion. The one radioactive species shown in Table 7-II

* Assuming that the explosives do not employ neutron "shields" to diminish the number of neutrons released into the surroundings.

which is sensitive to the type of explosive used is ^{3}H (tritium). Since approximately 2 g of this nuclide is formed per kiloton of thermonuclear yield (Tewes, 1968), use of a fusion explosive of the type previously noted (5-kt fission trigger) would produce some 2,000,000 Ci of tritium, as against the less than 2000 Ci produced in a typical gas-well environment by the specially designed gas-stimulation explosive (Rubin et al., 1972). (It is necessary to minimize the tritium production associated with a nuclear explosion used to enhance the recovery of a petroleum product, inasmuch as some fraction of the tritium will be chemically incorporated into the petroleum product.)

It should be noted that the radioactive species formed by neutron absorption which are listed in Table 7-II represent only those resulting from activation of the rock surrounding the explosive. Additional radioactivity can be produced by neutron activation of the materials used in the construction of the explosive itself. The large amounts of such nuclides as ^{181}W, ^{188}W, and ^{182}Ta which have been observed in debris from nuclear cratering experiments (Nordyke and Williamson, 1965; Tewes, 1970; Lessler, 1968) indicate that care must be used in the design of a thermonuclear explosive to prevent the formation of radionuclides which could adversely affect the feasibility of certain mineral resource recovery projects.

The distribution of the various radioactive nuclides formed by a nuclear detonation is dependent on the chemical characteristics of each of these species at the time during which chimney formation takes place (Borg, 1973). That is, those radioactive substances which act in a refractory manner will largely be found in the so-called "puddle glass" or rock melt at the bottom of the chimney, while those which display some degree of volatility will be found to a greater or lesser extent to be deposited on the surfaces of the chimney rubble or in the void volume between the rubble fragments. Table 7-III gives a simple categorization of various typical radionuclides which could be found in debris from a nuclear detonation.

Those radioactive isotopes that are described as volatile appear to behave as such because either they themselves or one of their precursors in a radio-

TABLE 7-III

Chemical behavior of selected radioactive species

	Volatile	Non-volatile
Fission products	^{89}Sr, ^{90}Sr, ^{91}Y ^{103}Ru, ^{106}Ru ^{137}Cs	^{95}Zr, ^{95}Nb, ^{97}Zr ^{144}Ce ^{147}Nd, ^{151}Sm
Activation products	^{32}P	^{45}Ca, ^{54}Mn ^{56}Mn, ^{55}Fe, ^{59}Fe
Fusion products	^{3}H	

active decay chain exists in a volatile chemical form for some time during the distribution process. The former situation may be exemplified by such species as ^{103}Ru or ^{3}H while the latter is typified by the mass-90 decay chain: $^{90}Br \rightarrow {}^{90}Kr \rightarrow {}^{90}Rb \rightarrow {}^{90}Sr$, where, of course, the ^{90}Kr is a noble gas and the ^{90}Br and ^{90}Rb may be volatile under the conditions prevailing at the time of cavity collapse.

The refractory radioactive substances, in an analogous way, must themselves be refractory in nature or must be the end product of a radioactive decay chain all the members of which are also refractory.

On the basis of this chemical differentiation, the predominant location of those nuclides listed in Tables 7-I and 7-II — within the glass matrix or on the rubble surface — can be established as shown in the fourth column of these tables.

Regarding the question of groundwater transport of radionuclides present in the cavity-chimney system resulting from a contained underground nuclear detonation, the radioactive species remaining in the melt region will be either chemically part of the melt or occluded in it. The melt will tend to protect the radioactivity from leaching out appreciably, at least until temperatures and pressures are reached that would break down the melt matrix. On the other hand, the radioactivity in the chimney rubble region will have condensed on the rock surfaces and be more readily accessible to leaching by groundwater. This has been borne out experimentally.

In a flooded nuclear chimney the concentration in solution of a particular radioactive species will depend not only on its physical availability for leaching but also on its behavior in the prevailing chemical environment. Such factors as the temperature and pressure, the composition of the groundwater (including its pH), the solubility of the chemical forms in which the given species is present, and the ion-exchange behavior between the given species and the chimney rubble may all be important in determining the ultimate concentration of the species in solution in the chimney.

The radioactive species tritium is of special interest with respect to groundwater contamination. Tritium will exchange with the normal hydrogen in water and, therefore, tritium will be in the groundwater as part of the water molecules.

Because of the slowness of most groundwater movement, one rarely has to consider radioactive species with half-lives less than a few months to a year in evaluating the problem of radioactive contamination. In addition, when a contained nuclear explosion occurs below the water table, rock is displaced outward from the region of the explosion, creating more void space in the explosion region and resulting in a depression being formed in the potentiometric surface. * Until equilibrium is restored in the potentiometric surface,

* The potentiometric surface is the maximum elevation at which ground water can be found, and is defined by the height to which water will rise in a well under hydrostatic pressure.

the water flow will be inward toward the explosion zone, minimizing chances of radioactive contamination of the nearby groundwater. When the water-table level is restored, normal groundwater flow will resume. The restoration of the water-table level may take many months or even years. In a Plowshare application in which the chimney might be deliberately flooded, this time scale can be considerably shortened. However, it is unlikely that one could not afford to wait until the short-lived activities decayed (about six months) before flooding the chimney.

The parameters that affect the transport of radioactive contamination in the groundwater system are numerous and can extend over a very large range of values under field conditions. Thus, in order to make a realistic evaluation of the groundwater radioactive problem for a given Plowshare site, it is necessary to make a thorough study of the groundwater chemistry and of the hydrology of the water-bearing formations in the vicinity of the proposed project. Whether or not sorption plays an important role, ultimately the fate of any radioactive species that is present in the groundwater depends on the flow of the groundwater, where it goes, what paths it takes, how fast it moves, etc.

Formation types in which it is especially difficult to predict radionuclide dispersion include: (1) aquifers that contain narrow sections of very high permeability, (2) formations with strongly anisotropic permeability where the principal direction of flow may tend away from the regional hydraulic gradient and instead tend in the direction of maximum permeability (i.e., along local induced gradients), and (3) aquifers with more or less continuous variation of flow parameters and/or heterogeneous sections, resulting in large dispersive effects or simply anomalous flow patterns. Unless such deviations from ideal behavior are measured and properly taken into account, radioactive contamination either (1) may arrive where it is expected but much sooner than expected, (2) may arrive where it is not expected at all, or (3) may spread and travel in a much broader band than expected. In the latter case, although the concentration front may arrive sooner than expected, the radioactivity concentration will be lower, due to spreading over a larger volume of water.

For a given groundwater flow rate, the transport of a radioactive contaminant can be affected quite differently, depending on whether the flow is through a homogeneous porous medium or principally through many very small fractures. Flow through small fractures is more likely to have a greater velocity variance, resulting in more of a dispersive effect. Also, in flow through fractures, sorption and desorption takes place along the surfaces of the fractures, which will often have sorptive characteristics different from those of the bulk rock. Unless experimental conditions are carefully chosen, the fraction of a radioactive species sorbed on rock relative to that remaining in a solution in contact with the rock, as measured on the bulk rock in laboratory experiments, may be quite different from that prevailing under field conditions.

Such peaceful applications of nuclear explosions as waste disposal and in-situ ore leaching, because of the chemical reactions occurring in the chimney, may pose problems that are not encountered in chimneys with ordinary groundwater. The chemistry of the chimney environment is likely to have a bearing on whether or not radioactive contamination of the groundwater is a serious problem.

Although some work has been done on the groundwater contamination problem, much remains to be done to improve the base from which we may treat the problem in experiments where conditions are outside our experience.

SUMMARY OF PNE CONTAINED UNDERGROUND EXPERIMENTS CONDUCTED THROUGH 1975 *

The current capabilities for predicting the effects of deeply buried nuclear explosions have been basically derived from data on several hundred tests conducted at the Energy Research and Development Agency's (formerly AEC) Nevada Test Site as part of the nuclear weapons test program. A number of nuclear explosive experiments conducted specifically to address industrial applications have added to this wealth of data.

Project Gnome (Rawson, 1963) the first Plowshare experiment, was conducted December 10, 1961, near Carlsbad, New Mexico. A 3.1-kt explosive detonated at a depth of 361 m in a bedded salt formation produced a partially standing cavity with a radius of about 17.4 m. Project Handcar (Werth, 1970) was an experiment at the Nevada Test Site to study the effects of a nuclear detonation in a carbonate rock. A 12-kt explosive detonated at a depth of 402 m on November 5, 1964, produced a chimney with a radius of 21 m and a height of 71 m.

Three nuclear tests (Projects Gasbuggy, Rulison, and Rio Blanco) have been executed to evaluate the potential of PNEs for stimulating the flow of natural gas (methane) from submarginal reservoirs. Project Gasbuggy (Holzer, 1970) used an explosive with a yield of 29 kt buried at a depth of 1292 m in the San Juan Basin of New Mexico. It was detonated on December 10, 1967 and produced a chimney with a radius of approximately 25 m and a height of 102 m. Production data indicate that rock was fractured laterally out to a distance of about 90 m from the explosive location. Since this was the first test in a gas reservoir, its primary purpose was to determine the effect of the chimney and fractured rock beyond the chimney on the amount and quality of gas produced.

The Project Rulison (Reynolds, 1972) explosive with a yield of 43 kt was

* Scientific experiments, such as those used to produce very heavy isotopes, will not be treated in this discussion.

detonated on September 10, 1969, at a depth of 2568 m in the Piceance Basin of Colorado. A primary objective was to extend the knowledge of stimulation effects to the deeper reservoirs which are typical of the Rocky Mountain basins. Rulison was the most deeply buried nuclear explosion to have ever taken place in the United States. Data from this event verified the validity of extrapolating predictive techniques for estimating cavity and chimney dimensions to detonation depths far beyond existing experience. It is estimated that a chimney was formed with a radius of 23 m and a height of 82 m. Computer modeling of the chimney and reservoir to match gas production test data required that increased permeability to gas flow extend to a range of about 70 m. This is assumed to be the range of rock fracturing. A basic question relative to nuclear fracturing is whether the fractures produced by the detonation permanently increase the medium permeability or whether the great lithostatic pressures in deep reservoirs sooner or later seal whatever fractures are formed. Modeling of the chimney/reservoir system has not indicated a reduction in permeability or the sealing of fractures.

A recognized need for using multiple explosives to fracture the 600—1200 m thick vertical gas-bearing intervals found in the Rocky Mountain reservoirs was first investigated with Project Rio Blanco (Luetkehans et al., 1975). Three explosives with yields of 30 kt each were detonated simultaneously in a single well at depths of 1780 m, 1899 m, and 2039 m. The site was located in Rio Blanco County, Colorado, and the detonation date was May 17, 1973. Dimensions of the chimneys and the extent of fracturing were as expected for three individual explosives of the yield employed, but there is no evidence of enhanced fracturing from the dynamic interaction between the explosion-generated shock waves. The reservoir modeling technique, identical to that used for the Rulison evaluation, gave results consistent with a zone of enhanced permeability to a range of 3 cavity radii laterally from the explosive location on the basis of gas production test data. This ratio is similar to that determined for both Gasbuggy and Rulison. However, there is no vertical flow of gas between the three chimneys as was predicted. The reason for this behavior is yet to be established, but is not critical to the technical and economic feasibility of gas stimulation with nuclear explosives.

Developments were incorporated in the Rio Blanco test to prepare for a commercial technology. The experiment used a nuclear explosive system which, for the first time, was designed specifically for industrial applications. The outside diameter of the explosive canister was slightly less than 20 cm, and incorporated design features to allow its use in a hostile temperature and pressure environment. Also, the monitoring and firing of the explosives was accomplished from a remote location using a microwave relay system rather than electrical cable. A single coaxial cable from the wellhead to depth was used to monitor the status of components in each of the three explosive canisters and to detonate the explosives. The low production of tritium was a major advance over previously used nuclear explosives. In contrast to

Gasbuggy and Rulison, which produced about 40,000 Ci and 10,000 Ci respectively, less than 1000 Ci were produced by each of the three Rio Blanco explosives.

DESCRIPTION OF POTENTIAL APPLICATIONS RELATED TO ENERGY RESOURCES *

Storage

The construction and use of nuclear cavities for storage of natural gas or liquid hydrocarbons is the simplest and most straightforward application of nuclear explosions. Unlike resource extraction applications in which the location of the resource dictates where explosives must be detonated, storage locations are more flexible, but will tend to be as close to market areas as seismic damage considerations will allow. The primary and least expensive storage for natural gas is in depleted oil and gas fields, but by the mid-1960's, most of this storage was in use. In comparison to other conventional storage possibilities such as above-ground tanks, nuclear cavities appear to be very attractive. In fact, the only other conventional storage which appears to compare favorably with nuclear cavities from an economic standpoint is storage in washed-out salt cavities. The obvious disadvantage to the preferred conventional storage techniques of depleted reservoirs and washed-out salt cavities is that there is not much flexibility in location. In 1967, a feasibility study, Project Ketch (U.S. Atomic Energy Commission, 1967) conducted by Columbia Gas Systems, Inc., the Lawrence Livermore Laboratory, and the AEC gave favorable results, and execution of a field experiment was planned. Project sponsors were not prepared for the public opposition which was generated when people with technical credentials raised objections relative to public safety. After extensive public hearings, the initial objectors stated publicly that they believed the experiment could be conducted safely. However, by this time it was too late — public opposition continued and the project was abandoned at the request of Columbia Gas Corporation.

For several years following the abandonment of Project Ketch, very little industrial interest was shown in storage. In 1970, the El Paso Natural Gas Company conducted a feasibility study of nuclear gas storage for a site in Arizona in which the economics were again shown to be favorable (Randolph, 1971). The project was not pursued because of the imminent perceived shortage of natural gas.

---

* While it has been proposed to use PNE for the deep underground disposal of high-level nuclear wastes from the operation of commercial nuclear reactors (Cohen et al., 1971; Cohen et al., 1974), this application is not directly related to energy resource development, and will not be discussed in this presentation.
Only applications which have been studied by the U.S. are discussed in this section.

TABLE 7-IV

Comparison of different refinery feed-stock storage methods (Van der Harst and Knutson, 1969)

Method	Cost * (U.S. $/bbl)
Concrete sea-bottom oil storage	11.00—14.00
Mined caverns	5.00—11.00
Leached salt cavities	1.00— 2.50
Steel tanks and offshore platforms	2.00— 5.00
Steel tanks, onshore	2.00— 4.00
Nuclear cavities	0.77— 1.70

* 1970 dollars.

Currently, the critical status of energy availability and the threat of future OPEC embargoes may argue for the large-scale storage of crude oil for national security reasons. Refineries maintain only a few weeks supply of feed-stocks or finished products on hand. If the U.S. were to maintain a three-month stored reserve of refinery feed-stock, storage volume for about 1.5 billion barrels would be required, based on current consumption. Development of large off-shore storage facilities to accommodate "super-tankers" may also become a requirement in the next decade. Nuclear storage may be the most economic solution as shown in Table 7-IV above, developed by the CER Geonuclear Corporation (Van der Harst and Knutson, 1969). It is not believed that inflationary factors will alter these relative costs which are expressed in 1970 dollars. There are several unique advantages to nuclear storage. One is that the cavities would be deeply buried and virtually immune to enemy attack. For liquified natural gas (LNG), storage in a nuclear cavity offshore appears to be desirable from a safety standpoint. LNG is dangerous to handle and an explosion aboard a ship in port or in onshore storage facilities could be disastrous in a highly populated area. There are no major technical questions which have to be resolved prior to undertaking a full-scale test of nuclear storage of liquid hydrocarbons. The degree to which stored products become contaminated with specific radionuclides can be investigated in such a test, but contamination is likely to be very small. * The integrity of the chimney through a number of loading-unloading cycles, the nature of wetting of the chimney rubble by water and fuel, the transport of fine particles, and the growth of bacteria are other areas which need to be investigated.

* This is borne out by Soviet statements (Kedrovskiy et al., 1975).

Oil shale

The fact that a petroluem-like product could be obtained by heating a certain type of rock has been known for hundreds of years. Significant deposits of this rock, now called oil shale, exist in the U.S. and some 23 foreign countries. Within the U.S., the Green River Formation in Colorado, Utah, and Wyoming is estimated to contain 1.8 trillion barrels of shale oil. For comparison, the discovered U.S. oil in place is 433 billion barrels of which 155 billion barrels has already been produced. Considering the oil which is yet to be discovered, the total amount of future recoverable oil is estimated to be 250 billion barrels. In other words, one-third of U.S. oil has been produced and about two-thirds is left. With these reserves and an aggressive drilling program, domestic production could reach about 17 million barrels per day in the 1990's but would then decline precipitously. The difference between supply and demand would make the development of these oil shales a critical issue if the U.S. is to maintain a high degree of independence from foreign oil. Conservation, development of nuclear power reactors, solar energy, and geothermal energy, as well as increased use of coal will help somewhat, but the transportation sector consumes the majority of oil. It is unlikely that there will be a drastic reduction in the use of automobiles, since most vehicle miles are dedicated toward earning a living or family business. Only about 33% of vehicle miles are used for social and recreational purposes. It appears the only real alternative to massive exploitation of the oil shale is the development of vehicles which either are grossly more economical or use alternate fuels such as methanol (produced from coal) or hydrogen.

Oil shale is a fine-grained non-porous sedimentary rock which is more similar to limestone than shale. The organic material in the oil shale (which is called kerogen) yields petroleum hydrocarbons upon destructive distillation at temperatures of approximately 500°C. Oil shale is graded according to the amount of shale oil which it contains. Commercial surface retorting processes generally utilize oil shales in the 30—35 gal/t category. The basic methods of producing shale oil include: (1) surface or underground mining followed by retorting in above-ground facilities, (2) breaking the oil shale in place (in situ) and retorting in this below-ground configuration. In general, mining and surface retorting will be more expensive than in-situ processes and will have the additional significant environmental problem of disposing of large volumes of spent shale. In one in-situ process called RISE ("rubble in-situ extraction", Rothman, 1975), a large underground column of rubble is produced conventionally by mining and block caving techniques, and retorted in place using combustion in an air stream. About 20% of the rock will have to be mined and brought to the surface to provide the necessary volume for the rubble when it bulks. This mined material could be retorted in surface facilities if desired. With a nuclear chimney, the void volume is

TABLE 7-V

Comparison of estimated costs for various oil shale recovery methods (Rothman, 1975)

Method	Cost (U.S.$/bbl)	Capital investment for 100,000 bbl/day * (U.S.$ 1,000,000)
Mining and surface retorting	12	1000—1500
Rubblization by mining and in-situ retorting (RISE)	8	600
Nuclear fracturing and in situ retorting	5	300

* Cost (in 1975 dollars) includes upgrading to refinery feed-stock standards. Current costs are probably significantly higher, but relative costs would be unchanged.

formed by compressing the surrounding rock and slightly displacing the ground surface. The resulting chimney of rubble which forms could then be retorted in situ without the need for any underground mining. Estimated costs (Rothman, 1975) for various oil shale recovery methods are shown in Table 7-V.

In comparing the environmental consequences of recovering oil from oil shale for a 1 million barrels per day industry, a 20-mile × 30-mile region in the northern part of the Piceance Basin of Colorado was evaluated. This is one of the thickest and richest sections of the Green River Formation and contains 720 billion barrels in oil shale thickness greater than 120 m of which 600 billion barrels is in oil shale greater than 250 m in thickness. This area of 600 square miles is a small fraction of the 43,000 km^2 underlain by the Green River Formation. Table 7-VI compares the resource in this area amenable to the various methods of recovery, as well as some key environmental factors.

A major difficulty facing any recovery method used in this area involves groundwater. The large volumes of highly saline groundwater in the region of the shale deposits obviously cannot be disposed of by pumping into streams. Possible solutions include reinjecting this saline water back underground, or possibly even building a pipeline to the Great Salt Lake. The in-situ methods are not yet developed for commercial use, but studies and small-scale tests are proceeding on variants of this process. Although there are many technical and economic issues which must be resolved for the in-situ processes in general, development of a commercial process would resolve many issues directly applicable to the nuclear method. There would still remain some significant technical uncertainties in the nuclear method which can be evaluated only by a full scale field test. These uncertainties include the size and shape of the chimney, the distribution of rubble size, the minimum spacing allowable between chimneys to maintain the integrity of each

TABLE 7-VI

Comparison of the major methods that have been considered for recovering oil from oil shale for production of 1 million barrels per day

Recovery method and shale preparation	Oil shale resource			Environmental impact		
	grade (gal/t)	oil (10^9 bbl)		annual disposal required (10^9 t)	population influx (10^3)	water use (10^3 acre-ft)
		in place	recoverable			
Surface						
Underground mining, room and pillar	30	100	54	0.5	130	176
Open-pit mining	20	700	660	1.2	140	173
In situ						
Mining rubblization (RISE process)	20	700	350	0.23	100	96
Nuclear fracturing	20	650	200	none	55	65

chimney during the retorting phase, and the pressure of residual gases in the chimney(s) which were produced by thermal decomposition of carbonate-containing rocks. Preliminary studies of ground motion and the fate of the radionuclides associated with a nuclear chimney indicate that occupational and public safety can be maintained by conservative engineering design of the project. However, no experiments have been conducted to date and none are contemplated in the near future because of prevailing attitudes regarding nuclear technology. The primary deterrents to mining and surface retorting are the high cost of the operation, the large capital investment, and the environmental consequences of disposing of very large volumes of spent oil shale. Although the cost per barrel is comparable to foreign oil, the industry could be seriously affected by a drop in foreign oil prices if it were not protected by government price guarantees.

Gas stimulation

Measured reserves of natural gas in the U.S. amount to about 266 trillion cubic feet (TCF) with an additional 380 TCF indicated and inferred from geologic evidence. The current consumption is about 23 TCF per year. The above reserves are those which are economically recoverable at current prices. Another source of natural gas which is indicated and inferred from geologic evidence, but for which recovery is subeconomic at current prices, is that within tight gas sands located in Colorado, Utah and Wyoming. The esti-

Fig. 7-4. Nuclear gas-reservoir stimulation. The reservoir is modeled after the Green River Basin in Wyoming.

mated amount of gas in place is 600 TCF of which 300 TCF may be produced with advanced technology. These resources as identified by the FPC Natural Gas Technology Task Force (Federal Power Commission, 1973) were considered to be amenable to stimulation by either nuclear explosive or hydraulic fracturing techniques. In general, these reservoirs are very deep (between 1500 and 5000 m), contain alternating layers or lenses of sandstones and shale in thick sequences (600—1200 m), the gas-bearing sandstones have very low permeabilities (less than 15 microdarcies compared to several millidarcies in economically-produced reservoirs), the pay sands are not interbedded with high-permeability aquifers, and the reservoirs are in remote areas.

The concept for nuclear stimulation is shown in Fig. 7-4. Multiple explosives, each 100 kt in yield, would be placed in a single wellbore and spaced so that all of the gas-bearing sandstones are fractured within a reservoir as thick as 900—1200 m. The explosives would be detonated sequentially from the bottom-up with a sufficient time delay to insure that the maximum ground motion generated is that expected from a single 100-kt explosive. The large nuclear chimney or wellbore accounts for 60% of the stimulation effect and fracturing of the reservoir rock outside the chimney accounts for

40%. If the spacing of the sequentially detonated explosives results in good vertical permeability, reentry to the top of the top chimney would be sufficient to produce the natural gas from all of the chimneys. Otherwise, a separate production well must be constructed alongside the chimneys in the fractured region. To develop a reservoir adequately, nuclear wells would be spaced at 320-acre intervals or two per square mile.

The alternate method for exploiting the submarginal natural gas reservoirs is termed "massive hydraulic fracturing" (MHF). Hydraulic fracturing has been used by industry for over 20 years to overcome wellbore damage caused by drilling fluids plugging the pores of the rock and preventing gas flow into the wellbore. Fractures a few meters long are adequate to complete a commercial well in a producing reservoir. The creation of vertical fractures which extend about 150 m radially from the wellbore and span the 900—1200 m thick reservoirs were deemed to be feasible by the FPC Natural Gas Technology Task Force. In fact, fractures greater than 1500 m in length have been produced by Amoco in the Wattenberg field of Colorado. In this case a single gas-bearing sand about 12 m thick was successfully stimulated in this reservoir. The major technical uncertainty in this method is the minimum spacing between individual fractures to span the desired vertical interval of 600—1200 m. If successive fractures are too close, the second may break into the first and result in no new stimulation; if too far apart, many sandstone lenses would be by-passed and a sizeable fraction of the resource may be left in place. Another uncertainty which requires additional field testing is the selection of a suitable fracturing fluid. Such a fluid must have adequate viscosity to carry the proppant out to the extremities of the fracture; must not chemically or physically damage the gas-bearing sands; must have the lowest practical viscosity so as to minimize the energy wasted in forcing the crack to open wider than necessary; and can be produced out of the fracture (i.e., removed so as not to interfere with gas flow through the crack).

Wells produced by massive hydraulic fracturing will probably not be spaced closer than 80-acre intervals to produce the optimum percentage of recoverable gas per section.

A comparison of wellhead gas costs for either the nuclear stimulation method or massive hydraulic fracturing is shown in Table 7-VII for the Piceance, Uinta, and Green River Basins. The costs are for 1980 using 1972 dollars escalated at 3.5% per year. The inflationary factors are now known to be inadequate, but overall these costs are believed to be the best available. The greatest uncertainty may not be the use of undeveloped technologies, but rather may be in the quality of reservoirs which are assumed to exist in the three Rocky Mountain basins. A major conclusion from Project Rio Blanco is that these reservoirs are highly variable and that conventional techniques of reservoir evaluation may not be adequate to describe the quality of reservoirs with very low permeabilities. The reservoir capacity (permeability times thickness of gas-bearing sands) established by production of gas from

TABLE 7-VII

Comparison of wellhead gas costs (cents/MCF * at wellhead) (Federal Power Commission, 1973)

		Discounted cash flow rate of return **					
		10%		15%		20%	
		N ***	H ***	N	H	N	H
High production case							
Uinta Basin	1980	19	20	25	23	31	26
	1990	26	27	34	31	42	35
	2000	36	37	46	42	56	47
Piceance Basin	1980	36	33	45	38	54	42
	1990	49	45	61	51	73	57
	2000	65	61	81	69	98	77
Green River Basin	1980	52	54	66	63	80	72
	1990	71	73	90	85	108	97
	2000	95	98	121	114	145	130
Low production case							
Uinta Basin	1980	45	37	55	43	65	49
	1990	60	50	74	58	87	66
	2000	81	67	99	78	117	89
Piceance Basin	1980	79	62	93	72	107	82
	1990	107	84	127	97	144	110
	2000	144	113	170	131	194	148
Green River Basin	1980	52	54	66	63	80	72
	1990	71	73	90	85	108	97
	2000	95	98	121	114	145	130

* MCF = million cubic feet.
** Current-dollar basis, escalation included.
*** N = nuclear, H = hydraulic.
Principal economic factors incorporated into these projections are:
cost escalation factors: 4.0%/yr through 1975, 3.5%/yr 1976—1980, 3.0%/yr after 1980,
ad valorem tax rate 5.5%, severance tax 1%,
federal income tax 48%, depletion allowance 22%, royalty rate 12.5%.
An approximate 10c/MCF field processing charge for treating gas from nuclear-stimulated wells is not in these wellhead costs.

the nuclear well at Rio Blanco is 6 to 10 times lower than was estimated pre-shot in the vicinity of the top explosive. In the vicinity of the bottom Rio Blanco explosive, the reservoir capacity is judged to be about $\frac{1}{2}$ of the pre-shot estimate. However, the lateral extent of the gas bearing sandstone lenses appears to be very limited. The reservoir capacity and the size of the gas-bearing lenses are key factors in the volume of gas which may be produced from a well over a reasonable period of time. The use of a nuclear chimney as a technique for determining reservoir characteristics may be its most important use in the foreseeable future.

MHF well completion tests by the El Paso Natural Gas Company in the Pinedale unit in the northern part of the Green River Basin in Wyoming have produced similar disappointing results (Randolph, 1976). When data from these experiments are extrapolated to the prototype MHF wells envisioned in the Federal Power Commission study, the total 25-year gas production is a factor of 4 to 5 below that calculated by the natural gas survey for this basin, due mainly to the fact that the reservoir capacity is less than had been previously estimated.

U.S.S.R. PROGRAM (Nordyke, 1975)

Over the past several years, through a series of international meetings sponsored by the International Atomic Energy Agency (IAEA), bilateral meetings between U.S.S.R. and U.S. scientists, and in a number of publications (U.S. Atomic Energy Commission, 1969a; Kedrovskiy, 1970a, b), the Soviet Union has disclosed an active program for developing uses for peaceful nuclear explosions in their national economy. The results of 15 experimental and industrial explosions have been described (see Table 7-VIII and 7-IX); these projects were designed to develop seven different applications in the petroleum, gas and minerals industries, as well as for water resources development.

That the Soviet PNE program is even more active than that indicated in Tables 7-VIII and 7-IX can be concluded on the basis of the large number of unidentified seismic events that have been reported since 1965 outside of the normal nuclear-weapon test areas (see Table 7-X).

A brief discussion of past experiments and proposed applications of PNE reported by the Soviets will be given in the following subsections.

Excavation experiments and applications (Table 7-VIII; Myasnikov et al., 1970; Kirlev et al., 1975)

The Soviet PNE excavation program includes both water resource development and development of underground orebodies by means of overburden removal. However, experimental detonations have been reported only in the first of these two areas of interest.

Two prototype crater lip dams have been constructed, with the first (designated "1003") not located on a watercourse and presumably designed primarily to collect data on the configuration of a nuclear crater. The second of these experiments, however, was planned as a practical demonstration of

TABLE 7-VIII

Soviet excavation PNE applications

Application	Explosive yield	Comments
Water resource development		
1003	1.1-kt	cratering shot in siltstone
1004	~125-kt	crater in river produced two lakes, 1.7×10^7 m^3 (13,000 acre-ft); "proven technology"
Proposed reservoir	two 150-kt	to form 3×10^7 m^3 (24,000 acre-ft) reservoir
T-1	0.2-kt	cratering shot in sandstone, calibration for T-2
T-2	three 0.2-kt	row-charge cratering shot "model of Pechora-Kama"
Proposed Pechora-Kama Canal	250 explosives	divert Pechora River into Kama River and thence to Caspian Sea
Pechora-Kama row crater	three 15-kt	experiment at southern end of Pechora-Kama Canal alignment to gain data on cratering characteristics and stability in saturated, alluvial medium
Overburden removal		
Proposed mining project	~1-Mt row charge	will remove 9×10^8 m^3 of overburden at 5 kopecks per cubic meter

the use of a nuclear crater as a dam (see Fig. 7-5). A second Soviet use of PNE for water resource development has been in the area of canal construction. As in the case of the crater lip dam application, construction of a working prototype was preceded by experimental verification of the concept. In this case, the verification was provided by two experiments, both of which used nuclear explosives with individual yields of 0.2 kt. The first of these involved the detonation of a single charge, while the second used three charges to create a short channel in a sandstone medium.

A working prototype canal construction project has been recently reported by the Soviets, and is intended to serve as a small portion of the large (112 km long) Pechora-Kama Canal. Fig. 7-6 shows a picture of a portion of this canal cut, and Fig. 7-7 indicates the location of the proposed canal.

While the Soviets have described a very large overburden removal project for the mining of an extensive nonferrous metal deposit, there is no indication to date that any prototype detonations have been carried out. The scale

Fig. 7-5. Aerial view of "1004" crater and reservoir.

Fig. 7-6. View of prototype Pechora-Kama Canal nuclear cut.

Fig. 7-7. Location of Soviet Pechora-Kama Canal.

of the project is extremely large, with a total overburden of 2.3×10^9 m^3 which must be removed (9×10^8 m^3 are to be removed using nuclear explosives).

TABLE 7-IX

Soviet contained PNE applications

Application	Explosives	Comments
Control of runaway wells		
Urtabulak	30-kt	\$75 million worth of gas lost over 3 years prior to nuclear explosion
Nearby gas field	40-kt	"proven technology"
Oil stimulation		
Field A	two 2.3-kt and one 8-kt	U.S. economic analysis indicates 26% internal rate of return
Field B	two 8-kt	"proven technology"
Proposed Field C	three 20—30-kt	designed to break barrier so underlying water will push oil out
Gas stimulation		
Undescribed	—	statement that such an application was carried out
Proposed gas condensate field	three 40-kt	expect gas production increase of about 100 MCF */day
Underground storage of oil or gas		
Salt dome A	1.1-kt	salt dome — leaked water and radioactivity
Salt dome B	25-kt	10^6-bbl storage at one-sixth the cost of surface gas storage and one-third the cost of washed cavities
Unidentified cavity	—	tested with oil and gas at 6 MPa (60 atm)
Gas condensate storage facility	15-kt	300,000-bbl storage facility in industrial use at a gas condensate deposit — working pressure 8 MPa (80 atm)
Proposed — layered salt	two 35-kt	2×10^6-bbl storage for gas condensate
Proposed — tuff under permafrost	three 40-kt	2.5-BCF ** storage for gas at 7 MPa (70 atm)
Mineral development		
"Granddaddy shot"	—	granite shot similar to Hardhat
Proposed ore breaking	1.8-kt	will break 10^6 m^3 of ore in situ

* MCF = million cubic feet.
** BCF = billion cubic feet.

Contained experiments and applications (Table 7-IX; Kirlev et al., 1975; Kedrovskiy et al., 1975)

Soviet contained PNE applications can be divided into five categories, each of which will be briefly discussed below.

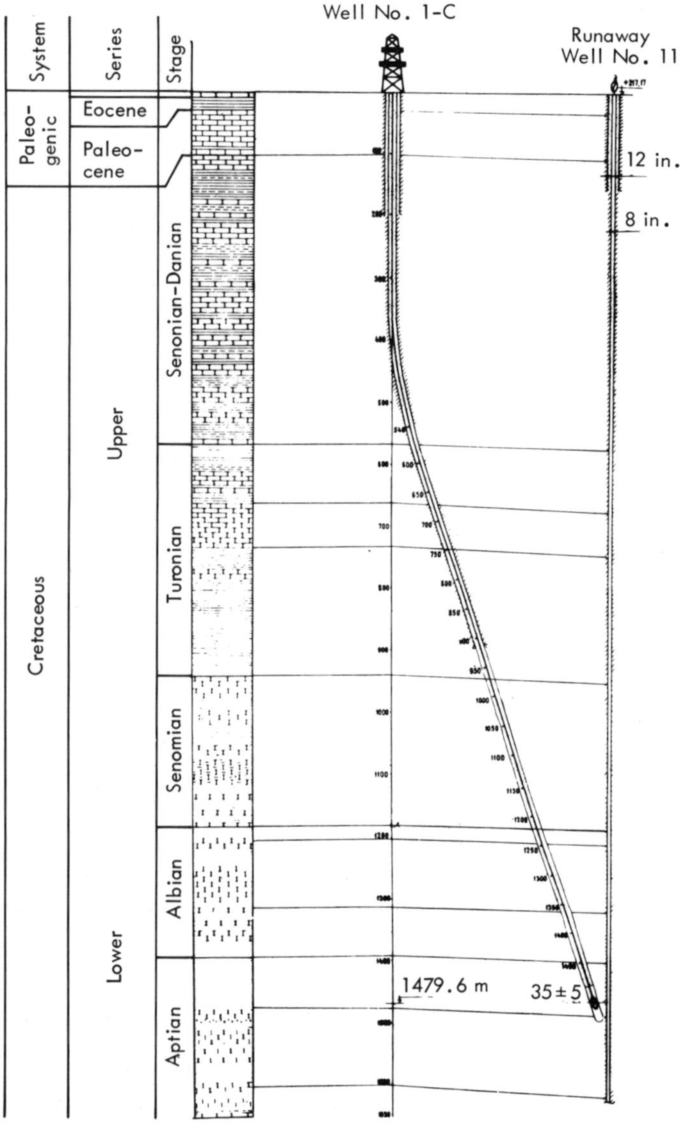

Fig. 7-8. Section of Urtabulak runaway gas well control project.

Control of runaway gas wells

The Soviets have reported the successful control of two runaway gas wells; in both cases, a nuclear detonation was used to seal the wellbore deep underground. For this technique to be successful, it is necessary to detonate the nuclear explosive in a relatively plastic medium (such as salt or clay) which will deform rather than crack or rupture, and will thus transfer the pressure generated by the nuclear explosion to the wellbore requiring closure. In both instances where this technique was used, the nuclear explosive was placed less than 40 m from the casing of the runaway well. The configuration used in the first well control project ("Urtabulak") is shown in Fig. 7-8. Note that this application has never been considered under the U.S. Plowshare Program, possibly because of the existence of a relatively sophisticated conventional runaway well control technology in this country.

Oil stimulation

A goal of the Soviet petroleum industry is the recovery of at least 50% of the in-place oil in producing fields. Generally, this goal has not been achieved, even when secondary recovery techniques, such as water flooding, have been applied. In an attempt to achieve their recovery goals, on two occasions the Soviets have turned to nuclear detonations to alter the characteristics of the petroleum reservoir.

In the first instance, the so-called "Field A", a total of three nuclear explosions were carried out to increase production from a gas drive reservoir. Following the detonations, not only did the flow of oil increase by almost 35%, but also the recovered oil contained significantly less dissolved gas than had been observed prior to the explosions. An economic analysis of the "Field A" project (using Soviet data, but employing U.S. oil pricing and explosive service costs) indicates that the internal rate of return should be on the order of 30% for such a project when routinely conducted. It is of interest to note that the Soviets have reported fracturing of the reservoir medium at distances as much as 10 to 20 times the cavity radii away from the detonation sites. This is significantly larger than the 3 to 4 cavity radii reported from U.S. experiments, and no reason for this apparent discrepancy has been suggested. The second field to be stimulated by nuclear detonations (designated "Field B") was a carbonate reef deposit which was being produced by water injection. Two nuclear explosives, detonated about a week apart, apparently succeeded in enhancing the production from this field by some 30—60%. In both oil stimulation products, the Soviets reported that the amount of radioactivity present in the oil produced after the respective detonations was "insignificant", and the recovered oil is in general use.

In addition, a third oil stimulation project has been proposed which will use three 20—30-kt explosives to fracture an acquitard, allowing the underlying water to rise under an existing oil deposit and drive the oil from the formation.

TABLE 7-X

Presumed nuclear events in the Soviet Union occurring away from normal test sites (through 1973 only)

NEIS or SIPRI data *							Event identified as presumed explosion by:		Location
Identification No.	date	origin time (GMT)	Lat. °N	Long. °E	Magnitude M_b	Depth (km)	SIPRI	USAEC	
1	65-01-15	05:59:59	49.89	78.97	6.0	0	no	yes	Semipalatinsk area **
2	66-04-22	02:58:04	47.86	47.72	4.9	0	yes	no	north of Caspian
3	66-09-30	05:59:53	38.8	64.5	5.1	33	yes	no	Bukhara
4	67-10-06	07:00:03	57.69	65.27	4.7	N	yes	no	east of Urals
5	68-05-21	03:59:12	38.916	65.159	5.4	13	yes	no	Bukhara
6	68-07-01	04:02:02	47.922	47.950	5.5	N	yes	yes	north of Caspian
7	69-09-02	04:59:57	57.415	54.860	4.9	0	yes	no	Urals region
8	69-09-08	04:59:56	57.365	55.108	4.9	0	yes	yes	Urals region
9	69-09-26	06:59:56	45.890	42.472	5.6	0	yes	yes	west of Caspian
10	69-12-06	07:02:57	43.832	54.783	5.8	0	yes	yes	east of Caspian
11	70-06-25	04:59:52	52.201	55.692	4.9	0	yes	no	north of Caspian
12	70-12-12	07:00:57	43.851	54.774	6.1	0	yes	yes	Caspian region
13	70-12-23	07:00:57	43.827	54.846	6.1	0	yes	yes	Caspian region
14	71-03-23	06:59:56	61.287	56.466	5.6	0	yes	yes	northern Urals

15	71-07-10	16:59:59	64.168	55.183	5.3	0	yes	yes	western slope of Urals
16	71-09-19	11:00:07	57.777	41.098	4.5	N	yes	no	Urals region
17	71-10-04	10:00:03	61.613	47.116	5.1	13 ± 29	yes	no	western Russia
18	71-10-22	05:00:00	51.575	54.536	5.3	6 ± 18	yes	yes	southern Urals
19	71-12-22	06:59:56	47.872	48.222	6.0	0	yes	yes	north of Caspian
20	72-07-09	—	49.9	35.2	5.0	—	yes	no	north of Black Sea
21	72-07-14	—	55.8	47.4	3.5	—	yes	no	north of Caspian
22	72-08-20	02:59:58	49.462	48.179	5.7	0	yes	yes	north of Caspian
23	72-09-04	07:00:04	67.689	33.445	4.6	7 ± 30	yes	no	western Russia
24	72-09-21	09:00:01	52.127	51.994	5.1	28 ± 19	yes	yes	southern Urals
25	72-10-03	08:59:58	46.848	45.010	5.8	0	yes	yes	northwest of Caspian
26	72-11-24	09:00:08	52.779	51.067	4.7	N	yes	no	western Russia
27	72-11-24	09:59:58	51.843	64.152	5.2	0	no	yes	southeastern Urals
28	73-08-15	01:59:58	42.711	67.410	5.3	0	***	yes	northwest of Tashkent
29	73-08-28	02:59:58	50.550	68.395	5.3	0	***	yes	northern Kazakh Desert
30	73-09-19	02:59:57	45.635	67.850	5.2	0	***	yes	central Kazakh Desert
31	73-09-30	04:59:57	51.608	54.582	5.2	0	***	yes	southern Urals
32	73-10-26	05:59:58	53.656	55.375	4.8	0	***	yes	southern Urals

* NEIS = National Earthquake Information Service (formerly NOAA and USC&GS); SIPRI = Stockholm International Peace Research Institute.

** This event was near the Semipalatinsk testing area but fell well outside the area of all previous Soviet events at the time it occurred. Significant amounts of atmospheric radioactivity were reported in association with this event leading to speculation that it was a PNE nuclear cratering event.

*** Date of event is after publication of SIPRI report.

Gas stimulation

On several occasions, Soviet PNE experts have stated that they have carried out a nuclear gas well stimulation experiment, but no details have been made available. In addition, the Soviet Union has proposed to use three 40-kt explosives to stimulate production in a gas/gas condensate field.

Underground storage of petroleum products

The Soviets have expressed a great deal of interest in this PNE application, and have expended a large fraction of their effort in the overall experimental program to bring this application to fruition as a proven industrial usage of PNE. Their economic analyses have indicated that gas condensate can be stored in nuclear cavities at one sixth the cost required for surface storage, and at one third the cost of storage in washed cavities (constructed in bedded salt or in salt domes).

In preliminary experiments, two cavities in salt have been constructed, one using a 1.1-kt explosive, and the other using a 25-kt explosive. The experimental results are schematically shown in Figs. 7-9 and 7-10, respectively. In both cases, the detonations were extensively instrumented and a number of postevent exploratory drill holes were used to confirm cavity sizes and the extent of fracturing.

In addition to these two experiments, two prototype industrial gas condensate storage cavities have been constructed. The yield of the nuclear explosive used to produce the first of these was not revealed; however, a 15-kt explosion was used in the second project. In both cases, the cavities are reported to be in routine industrial use, and the radioactive contamination of the stored condensate is "below the level of concern".

The Soviets have also proposed two other storage projects, one for gas condensate and the second for natural gas. The former would use two 35-kt explosives detonated in layered salt, while the latter application calls for the detonation of three 40-kt explosives in a tuffaceous environment overlain by permafrost. There is no indication to date that either of these two projects has been carried out.

Mineral development

In order to determine the characteristics of a nuclear cavity-chimney system, the Soviets detonated a low-yield nuclear explosive a number of years ago in granite. The resulting structure was extensively explored, with the fracture zone being thoroughly investigated. The experiment apparently resembled in many ways the U.S. Project Hardhat (a 5-kt explosive detonated in granite).

The Soviets have also proposed an ore-breaking project which would utilize a 1.8-kt nuclear explosive to fracture a 60—80 m thick orebody which dips into a mountainside at a 25° to 35° angle. A vertical and horizontal latticework grid of mined drifts to reflect the shock wave and enhance ore breakage has been planned for this prototype industrial project.

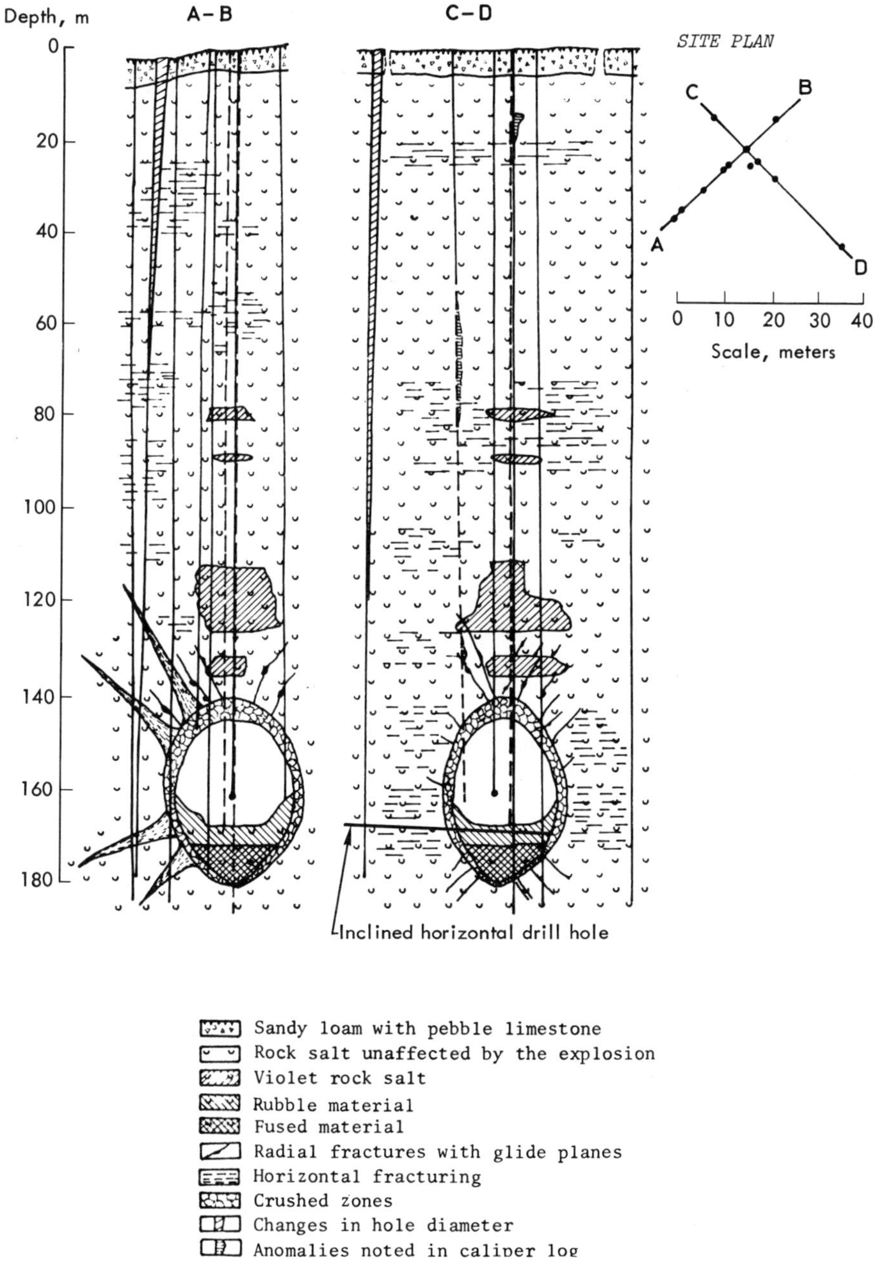

Fig. 7-9. Cross-section of 1.1-kt storage experiment in salt.

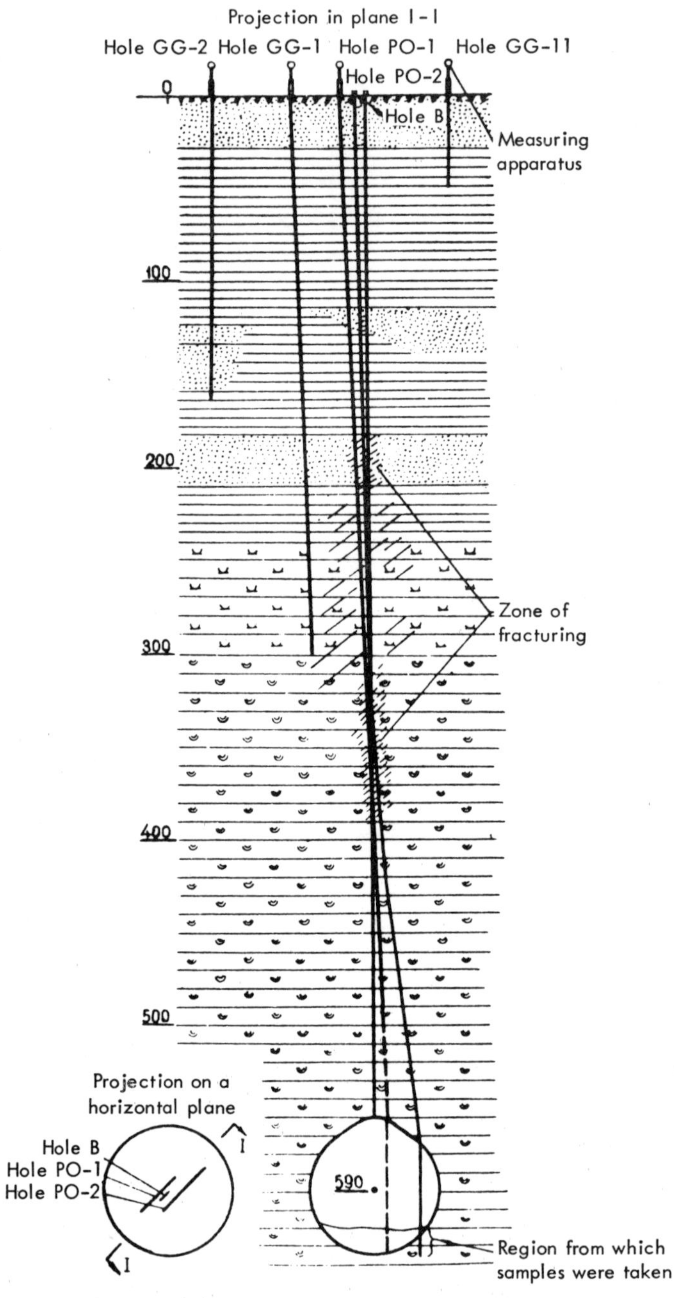

Fig. 7-10. Profile of 25-kt explosion in a salt dome.

OBSTACLES TO THE DEVELOPMENT OF A COMMERCIAL PLOWSHARE PROGRAM IN THE U.S. (Crump, 1975; Gahin, 1973)

A number of problems exist with respect to development of a viable commercial industry in the U.S. which is based on Plowshare technology. However, these can be generally placed into three categories: economic, both from the standpoint of funding of experimental programs and of the evaluation of a mature technology; political, in the sense of regulatory constraints on the rate of technology development; and social, from the standpoint of safety and environmental problems as perceived by at least a segment of the public. These areas will be briefly discussed in the following subsections; however, it should be emphasized that the presentation is not intended to be comprehensive.

Economic obstacles

Prior to its dissolution early in 1975, it had been the policy of the Atomic Energy Commission (AEC) to require the participation of an industrial partner in Plowshare resource development experiments conducted away from the Nevada Test Site. As an example of the type of arrangement which was required by the AEC, for the Rio Blanco project the AEC furnished the nuclear explosives, together with everything needed to detonate them and to ensure their security prior to firing. The AEC was also responsible for radiological safety, while the industrial partner was responsible for everything else, including drilling of emplacement and exploratory wells, installation of required utilities, etc. Should the division of financial responsibility for the construction of the prototype liquid metal fast breeder reactor (LMFBR) be made in an analogous way, ERDA's contribution might include only the fabrication of the fuel elements and the fielding and operation of a radiological monitoring network — with the electric utilities being required to pay all other engineering and construction costs. Under these circumstances, it is likely that the development of the LMFBR would proceed at even a slower rate than we are experiencing at the present time. Thus, AEC fiscal policy has not been especially encouraging to industry with regard to Plowshare experimental projects.

A second economic problem grows out of the first. With relatively meager prototype experimental project experience on which to base extrapolation to a large scale industrial application, the projections of economic viability become somewhat shaky, despite encouraging results of paper studies. Clearly, more extensive field experience is required in order that valid economic analyses of Plowshare resource recovery technologies can be made.

A final problem is one which also can be categorized as a political obstacle: due to the need to safeguard nuclear explosive design information, it is not possible to publish a firm estimate of explosive costs — either for single explosives or for the large numbers which would be needed for commercial applications. Thus, a part of the basic data needed for a realistic economic evaluation of commercial Plowshare applications is not generally available.

Political obstacles

These include the lack of required federal legislation, the presence of regulatory roadblocks, and the lack of firm decisions on the as yet unresolved questions relating to federal, state and local spheres of interest. While it is not possible to explore all these areas in depth, examples of each can be given to indicate the nature of the existing obstacles. In the case of needed federal legislation, under existing law, ERDA does not have the authority to provide nuclear explosives and firing services for commercial ventures. Legislation correcting this deficiency has been introduced in several sessions of Congress, but has never been passed. Concurrent with this is the fact that, as noted above, the costs of the nuclear explosive and firing services will have to be established by statute (or the responsibility for this will have to be appropriately delegated to ERDA).

From the standpoint of potential regulatory roadblocks, one such problem involves the fact that there now exists no provision for the sale of industrial products containing small amounts of residual radioactivity from a commercial Plowshare application. In order for such products to be legally salable without each purchaser having to obtain a license to receive radioactive material, the Nuclear Regulatory Commission (NRC) would have to conduct rulemaking hearings as provided by law, and publish the product standards for radioactivity in the Federal Register. The complete procedure from application to published rule can take as much as several years.

Finally, the role and scope of individual states and local governments in regulating the application of Plowshare technologies must be resolved. As an example of this area of concern, in November 1974 the voters of Colorado passed a constitutional amendment which requires a state-wide referendum to approve the detonation of a nuclear explosive within the borders of the state. The constitutionality of this action is still open to question, but would have to be resolved before a commercial Plowshare project could take place in Colorado.

Social obstacles

The problems in this sector can be exemplified by the passage of the amendment to the Colorado constitution, discussed above. The public perception of danger from peaceful nuclear explosives is almost visceral in nature, since the vision of the mushroom cloud elicited by a vocal opposition is a symbol of death and disaster.

However, Project Rio Blanco was conducted (May 17, 1973) with the support of a majority of the local populace as a result of repeated public informational meetings and close contact with the project sponsors. Polls conducted by the Grand Junction Sentinel, both pre- and post-detonation, showed that most of the people responding favored continuing investigations to develop the technology. The importance of public dissemination of factual information can be seen in the passage in November 1974 of the Colorado amendment to require referendum approval of future nuclear explosions. Environmental groups in Denver framed the amendment and maintained a well organized drive to gain public support. There was, however, no organized group to provide factual information regarding the environmental and safety issues being raised about Plowshare projects. Government employees (of ERDA and of the University of California, Lawrence Livermore Laboratory) were prohibited by law from engaging in "political" activities in the State of Colorado. Major oil companies were reluctant to become involved publicly because their public image at this time was already tarnished. The outcome of the statewide vote was obvious long before the election.

The experience from Project Rio Blanco shows clearly that public acceptance of Plowshare technologies is possible if the public is well informed and financially protected from damage.

The lack of government support (both legislative and executive) and industrial reluctance to become involved in a controversial technology makes it improbable that advances will be made in development of Peaceful Nuclear Explosions in the near future. It would appear that only an extreme shortfall of mineral and energy resources can provide the necessary impetus for increasing public acceptance of Plowshare technologies.

REFERENCES

Atlantic-Pacific Interoceanic Canal Study Commission, 1970. *Interoceanic Canal Studies.*

Boardman, C.R., Rabb, D.D. and McArthur, R.D., 1964. Responses of four rock mediums to contained nuclear explosions. *J. Geophys. Res.*, 69: 3457—3469.

Borg, I., 1973. *Long-Lived Radionuclides Contained in Glasses Produced by a Nuclear Explosion in Granodiorite (Piledriver Event).* Lawrence Livermore Laboratory, University of California, UCRL-74588.

Clough, R.W., 1967. Earthquake response of structures. In: R.L. Wiegel (Editor), *Earthquake Engineering.* Prentice-Hall, New York, N.Y.

Cohen, J.J., Lewis, A.E. and Braun, R.L., 1971. *Use of a Deep Nuclear Chimney for the In Situ Incorporation of Nuclear Fuel-Reprocessing Waste in Molten Silicate Rock.* Lawrence Livermore Laboratory, University of California, UCRL-51044.

Cohen, J.J., Braun, R.L., Schwartz, L.L. and Tewes, H.A., 1974. *Economic and Environmental Evaluation of Nuclear Waste Disposal by Underground In Situ Melting.* Lawrence Livermore Laboratory, University of California, UCRL-51713.

Crump, J.R., 1975. *PNE (Peaceful Nuclear Explosion) Activity Predictions for Arms Control Planning.* Gulf Universities Research Consortium, GURC-143.

Farhoomand, I. and Scholl, R.E., 1971. *Statistical Correlation of Observed Ground Motion and Low-Rise Building Damage — Project Rulison.* U.S. Atomic Energy Commission, JAB 99-59.

Federal Power Commission, 1973. *National Gas Survey, Volumes I and II.*

Gahin, F.S., 1973. *Plowshare Technology Assessment. Public Participation (with Reference to Nuclear Stimulation of Natural Gas).* Western Interstate Nuclear Board, PB-231 017.

Griggs, D.T., 1958. Surface motion from deep nuclear shots. In: H. Zodtner (Editor), *Industrial Uses of Nuclear Explosives.* Lawrence Livermore Laboratory, University of California, UCRL-5253.

Holzer, A., 1970. *Gasbuggy in Perspective.* Lawrence Livermore Laboratory, University of California, UCRL-72175.

Holzer, A., 1971. *Ground Motion Effects from Nuclear Explosions; A Review of Damage Experience and Prediction Methods.* Lawrence Livermore Laboratory, University of California, UCRL-51062.

Johnson, G.W. and Violet, C.E., 1958. *Phenomenology of Contained Nuclear Explosions.* Lawrence Livermore Laboratory, University of California, UCRL-5124 Rev. 1.

Johnson, G.W., Higgins, G.H. and Violet, C.E., 1959. *Underground Nuclear Detonations.* Lawrence Livermore Laboratory, University of California, UCRL-5636.

Kedrovskiy, O.L., 1970a. Industrial uses of underground nuclear explosions. In: I.D. Morokhov (Editor), *Nuclear Explosions for Peaceful Purposes.* Atomizdat, Moscow.

Kedrovskiy, O.L., 1970b. Prospective applications of underground nuclear explosions in the National Economy of the U.S.S.R. In: *Peaceful Nuclear Explosions: Proceedings of a Panel, Vienna, 2—6 March, 1970.* Translated for Lawrence Livermore Laboratory, University of California, UCRL-Trans-10477.

Kedrovskiy, O.L., Myasnikov, K.V., Leonov, A.E., Romadin, N.M., Borodnev, V.F. and Nikiforov, G.A., 1975. The application of contained nuclear explosions for creating underground reservoirs and testing their operation for the storage of gas condensate. In: *French and Russian Presentations to the Meeting of the International Atomic Energy Agency's Technical Committee on the Peaceful Uses of Nuclear Explosions, Vienna, 20—24 January, 1975*, Paper TC-1-4/13. Translated for the Atomic Weapons Research Establishment, AWRE-Trans-69.

Kirlev, V.V., Kedrovskiy, O.L., Valentinov, Yu.A., Myasnikov, K.V., Nikiforov, G.A., Prozorov, L.B. and Potapov, V.K., 1975. Multiple nuclear explosions for excavation in alluvial rocks. In: *French and Russian Presentations to the Meeting of the International Atomic Energy Agency's Technical Committee on the Peaceful Uses of Nuclear Explosions, Vienna, 20—24 January, 1975*, Paper TC-1-4/14. Translated for the Atomic Weapons Research Establishment, AWRE-Trans-69.

Levy, H.B., 1972. *On Evaluating the Hazards of Groundwater Contamination by Radioactivity from an Underground Nuclear Explosion.* Lawrence Livermore Laboratory, University of California, UCRL-51278.

Lessler, R.M., 1968. *Schooner Preshot Analysis.* Lawrence Livermore Laboratory, University of California, UCRL-50530 Rev. 1.

Lessler, R.M., Green, J.B., Jr. and Holzer, F., 1971. *Tritium Production in Plowshare Applications.* Lawrence Livermore Laboratory, University of California, UCRL-73170 Rev. 1.

Luetkehans, G.R., Toman, J. and DiBona, B.G., 1975. The Rio Blanco experiment — objectives, design, and execution. *Nucl. Technol.*, 27: 539—558.

Myasnikov, K.V., Prozorov, L.B. and Sitnikov, I.E., 1970. Mechanical effects of single and multiple underground nuclear cratering explosions and the properties of the excavations dug by them. In: I.D. Morokhov (Editor), *Nuclear Explosions for Peaceful Purposes.* Atomizdat, Moscow. Translated for Lawrence Livermore Laboratory, University of California, UCRL-Trans-10517.

Nadolsky, M.E., 1969. Architectural damage to residential structures from seismic disturbances. *Bull. Seismol. Soc. Am.*, 59: 487—502.

Nordyke, M.D., 1975. A review of Soviet data on the peaceful uses of nuclear explosions. *Ann. Nucl. Energy*, 2: 657—673.

Nordyke, M.D. and Williamson, M.M., 1965. *The Sedan Event*. U.S. Atomic Energy Commission, PNE-242F.

Randolph, P.L., 1971. Natural gas storage. *Trans. Am. Nucl. Soc.*, 14: 688—689.

Randolph, P.L., 1976. MHF research in the Green River Basin. In: *Proceedings of the Symposium on Stimulation of Low Permeability Reservoirs*. Colorado School of Mines, Golden, Colo.

Rawson, D.E., 1963. Review and summary of some Project Gnome results. *Trans. Am. Geophys. Union*, 44: 129—135.

Rawson, D.E., 1966. *Industrial Applications of Contained Nuclear Explosions*. Lawrence Livermore Laboratory, University of California, UCRL-14756.

Reynolds, M., Jr., 1972. Project Rulison: summary of results and analyses. *Nucl. Technol.*, 14: 187—193.

Rizer, G.C., 1970. *A Method for Predicting Seismic Damage to Residential-Type Structures from Underground Nuclear Explosions*. Lawrence Livermore Laboratory, University of California, UCRL-50959.

Rothman, A.J., 1975. *Research and Development on Rubble In Situ Extraction of Oil Shale (RISE) at Lawrence Livermore Laboratory*. Lawrence Livermore Laboratory, University of California, UCRL-76448.

Rubin, B., Schwartz, L. and Montan, D., 1972. *An Analysis of Gas Stimulation Using Nuclear Explosives*. Lawrence Livermore Laboratory, University of California, UCRL-51226.

Tewes, H.A., 1968. *Radioactivity from Nuclear Excavation Explosives*. Lawrence Livermore Laboratory, University of California, UCRL-71323.

Tewes, H.A., 1970. Results of the Schooner Excavation Experiment. In: *Proceedings of the Symposium on Engineering with Nuclear Explosives*. American Nuclear Society/U.S. Atomic Energy Commission, CONF-700101, Vol. I, pp. 306—333.

Tewes, H.A., Levy, H.B. and Schwartz, L.L., 1974. *Nuclear Chemical Copper Mining and Refining*. Lawrence Livermore Laboratory, University of California, UCRL-51345 Rev. 1.

Toman, J., 1970a. Summary of nuclear excavation applications. In: *Proceedings of the Symposium on Engineering with Nuclear Explosives*. American Nuclear Society/U.S. Atomic Energy Commission, CONF-700101, Vol. I, pp. 245—279.

Toman, J., 1970b. Results of cratering experiments. In: *Peaceful Nuclear Explosions: Proceedings of a Panel, Vienna, 2—6 March, 1970*. IAEA, Vienna, Paper IAEA-PL-388/16, pp. 345—375.

U.S. Atomic Energy Commission, 1967. *Project Ketch: A Feasibility Study on Creating Natural Gas Storage with Nuclear Explosions*. U.S. Atomic Energy Commission, PNE-1200.

U.S. Atomic Energy Commission, 1969a. *Review of Possible Peaceful Applications of Nuclear Explosions in the National Economy of the Soviet Union*. U.S. Atomic Energy Commission, AEC-TR-7121.

U.S. Atomic Energy Commission, 1969b. *Observed Seismic Data, Rulison Event*. U.S. Atomic Energy Commission, NVO-1163-197.

U.S. Atomic Energy Commission, 1971. *Structural Response Studies for Project Rulison*. U.S. Atomic Energy Commission, JAB 88-78.

Van der Harst, L. and Knutson, C.F., 1969. Why nuclear geostorage systems for petroleum? In: *Proceedings of the Symposium on Engineering with Nuclear Explosives*. American Nuclear Society/U.S. Atomic Energy Commission, CONF-700101, Vol. II, pp. 1322—1342.

Werth, G.C., 1970. *Handcar Explosion in Dolomite*. Lawrence Livermore Laboratory, University of California, UCRL-50951.

Appendix

FUNDAMENTALS OF ATOMIC AND NUCLEAR PHYSICS

JEROME G. MORSE

Although elementary in nature for nuclear-oriented geologists and engineers, this brief review is intended to augment the background of those distant from this field and aid in the understanding of the preceding chapters. Extended treatments of this subject are available elsewhere (Glasstone, 1950; Evans, 1965; Lapp and Andrews, 1972; Bowler, 1973).

ATOMIC STRUCTURE

An atom is composed of a positively charged nucleus surrounded by negatively charged particles, called electrons, which are sufficient in number to balance the positive nuclear charge. In terms of the classical concept, a particle is a relatively concentrated structure which has inertia, or resistance to any change in its velocity. Mass is a measure of this inertia. Atomic diameters are of the order of a few angstroms (1 Å = 10^{-8} cm). Virtually all of an atom's mass is found in its nucleus, whose diameters approximate 10^{-4} Å. With its mass concentrated in so small a volume, nuclear density is an extremely large number. For example, if nuclei were closely packed side by side, the density of matter would be about 10^{12} g cm^{-3} (500 million tons per cubic centimeter).

The nucleus

Atomic nuclei are composed of two types of primary particles, called protons and neutrons. The proton, which has a mass of 1.00728 atomic mass units (amu), carries a single unit positive charge which is equal in magnitude to the negative electronic charge of 4.8025×10^{-10} electrostatic units (esu). The simplest nucleus is that of the hydrogen atom, which contains only a single proton. Neutrons have no electric charge; the mass of this particle is 1.00867 amu. There are also secondary nuclear particles of lesser mass, such as mesons, etc., but these play little or no role in the subjects treated in this book.

All nuclei, other than those of hydrogen, contain a mixture of neutrons and protons, with the proportion of neutrons generally increasing with the atomic weight of the element. The total number of nucleons (protons plus neutrons) in an atomic nucleus is the mass number, often designated A. The atomic number of an element is equal to the number of protons, or net positive charges, contained in its nucleus, and this is called Z. If X is the normal chemical symbol of a given element, in nuclear parlance, the nucleus of X is described as ${}^{A}_{Z}X$. For example, the largest naturally occurring element is uranium, written as ${}^{238}_{92}U$. Z is equal to 92, indicating its number of protons. Since A equals 238, then $238-92=146$, with the latter representing the number of neutrons in its nucleus.

Recent but not yet fully confirmed data suggest the existence of plutonium which may have resulted from a naturally occurring nuclear reactor in Africa.

Atomic numbers of elements range from 1 to 106, although lawrencium (103) is the last one named formally.

Using the above methodology, one can predict that the mass of the helium nucleus (${}^{4}_{2}He$) is twice that of the proton plus twice the mass of the neutron. However, the measured mass of the helium nucleus was found to be 4.0015 amu, the atomic mass less that of two electrons, with each of the latter contributing 0.00055 amu. Performing a simple calculation, one finds (American Institute of Physics, 1972):

$$
\begin{aligned}
\text{mass of protons} &= 2(1.00728) = 2.01456 \text{ amu} \\
\text{mass of neutrons} &= 2(1.00867) = 2.01734 \text{ amu} \\
&\qquad\qquad\qquad\;\; \overline{4.03190 \text{ amu}} \\
\text{mass of He nucleus} &= 4.00151 \text{ amu} \\
\text{mass defect} &= 0.03039
\end{aligned}
$$

This discrepancy, or defect, exceeds possible instrument errors in measurement and is explained by Einstein's law which defines the equivalence of mass and energy. In the process of fusing two protons and two neutrons to form this nucleus, radiant or heat energy has been liberated with the amount equivalent to the mass defect. Conversely, the energy input required to remove constituent particles from the nucleus is called the binding energy, and it will be treated shortly.

The energy equivalent of the mass defect is calculated from Einstein's equation:

$$E = mc^2 \tag{1}$$

where c = the velocity of light (2.998×10^{10} cm s^{-1}), m = mass defect in amu, and 1 amu = 1.660×10^{-24} g. The energy equivalent is: E = (0.3039 amu) $\times$ (1.660×10^{-24} g/amu)(2.998×10^{10} cm s^{-1})^2 = 4.534×10^{-5} erg.

The electron volt (eV) and million electron volts (MeV) are useful energy units in nuclear science with the former defined as the amount of energy acquired by an electron when accelerated through a potential difference of 1 volt.

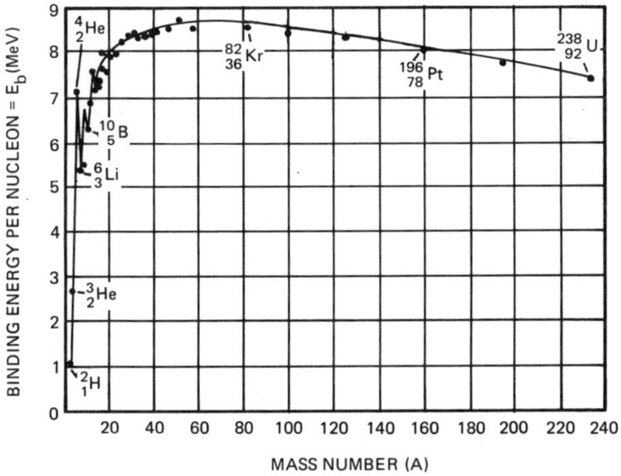

Fig. A-1. Variation of binding energy per nucleon with mass number (Lapp and Andrews, 1972).

1 eV = 1.602×10^{-12} erg, and therefore the energy defect for the helium nucleus is 28.3 MeV.

Since there are four nucleons in the He nucleus, the binding energy per nucleon, which offers a comparative measure of nuclear stability is 28.3/4 or 7.1 MeV. Binding energies for other nuclei can be calculated in a similar manner. The average binding energy per nucleon, as seen in Fig. A-1, varies from 6 to 8 MeV throughout the periodic table, except for a few of the lightest nuclei. The higher the energy per nucleon, the more stable the nucleus. In the process of nuclear fission, large amounts of energy are liberated when the heavy nuclei are broken down into those of medium mass. Even greater amounts of energy are released in nuclear fusion when light nuclei, such as hydrogen, are combined into heavier particles as occurs in regions of the sun, as seen in the above example.

Electrons

This subject is treated more extensively in the chapter on X-ray techniques (Chapter 3) and is covered here only briefly. An electron is a negatively charged particle whose mass is 1/1820 of that of the proton, or 0.00055 amu. Since the atom is electrically neutral, the number of electrons must equal the number of protons. Electrostatic forces maintain electrons in orbital configurations at varying distances from the nucleus. Chemical properties of the elements are determined largely by the number of electrons and their arrangement.

ISOTOPES AND RADIOACTIVITY

The atomic number (Z) of an element defines the number of protons in its nucleus, the number of orbital electrons and hence its chemical identity. When atoms of a given element (same Z) have different mass numbers (A), they are called isotopes of this element. Varying A at constant Z, indicates a change in the number of neutrons in the nucleus. Although the chemical behavior of isotopes of a given element is the same for each, nuclear behavior is often quite different. The element uranium occurs in nature in at least three isotopic forms, having mass numbers of 234, 235, and 238. Their abundances are given in Table A-I.

Radioactivity

The majority of elements in their naturally occurring forms are stable; that is, their nuclei are unaffected in normal chemical processes. Nuclear instability was first reported by Becquerel in 1896 after he discovered that emanations from uranium salts caused blackening of photographic plates, even when the latter were protected from visible and ultraviolet light. Later, it was shown that the nucleus was the source of these radiations. It was also shown that elements whose Z is greater than 83 (Po, Ra, Th, etc.) consist almost entirely of unstable or radioactive isotopes.

Radioactivity is a spontaneous process in which the nucleus undergoes decay at a definite rate. This decay is accompanied generally by the emission of an electrically charged particle or electromagnetic radiation or both. Often the products of decay are also radioactive and therefore, a series of disintegrations must take place before an atomic species is formed having a stable nucleus. The latter is called the ground state and is the end result of energy emission.

In addition to naturally occurring radioactive substances, radioactivity may be induced artificially by bombardment of stable elements with charged particles, neutrons, etc. This area is covered in the chapter "Radioactivation Methods" (Chapter 5) and is also treated here in section "Compound nucleus" and onwards.

TABLE A-I

Isotopic composition of natural uranium

Isotope	Abundance (%)
$^{234}_{92}U$	0.006
$^{235}_{92}U$	0.712
$^{238}_{92}U$	98.28

Half-life

The rate of radioactive decay follows a simple law. For a given radionuclide, each nucleus has a definite probability of decaying in a unit of time. This probability has a constant value which is characteristic of the particular radioisotope. At any point in time, it may be said that a constant fraction of the radioactive nuclei is disintegrating. This constant fraction of decay is peculiar to each individual species and is independent of the chemical or physical state of the element at all readily accessible temperatures and pressures. Control or alteration of the decay rate does not seem to be possible.

At a given instant of time, the rate of decay is directly proportional to the number of radioactive atoms of the isotope present (N), and is given by:

$$-\mathrm{d}N/\mathrm{d}t = \lambda N \qquad (2)$$

where t is the time and λ is the proportionality or decay constant of the species. Rearranging the equation:

$$\mathrm{d}N/N = -\lambda t \qquad (3)$$

and integrating between $N = N_0$ at $t = 0$, and $N = N$ at time t:

$$\log_e(N_0/N) = \lambda t$$

or:

$$N = N_0 \ e^{-\lambda t} \qquad (4)$$

From the above it is seen that radioactive decay is an exponential process with the actual rate determined by the decay constant (λ) and the number of radionuclei present (see Fig. A-2). (Note: although the $\mathrm{d}N/\mathrm{d}t$ notation is used here, it actually refers to very large values of N. The Δ notation, $\Delta N/\Delta T$, describes the situation more precisely.)

Illustrative example

A platinum sample, in which radioactivity was induced, has a measured activity of 10^3 disintegrations per second (dps). Calculate the activity of the sample 1 day later.

Solution — Assume sample measured is ^{197}Pt:

$t_{1/2} = 18$ h
from equation (3), $\lambda = 0.693/18$ h $= 0.0385$ h^{-1}
from equation (4), $N/N_0 = e^{-\lambda t}$
$\lambda t = (0.0385$ h$^{-1})(24$ h$) = 0.924$

(Note: units used for t must be consistent)

$e^{-\lambda t} = 0.397$
$N_0 = 10^3$ dps

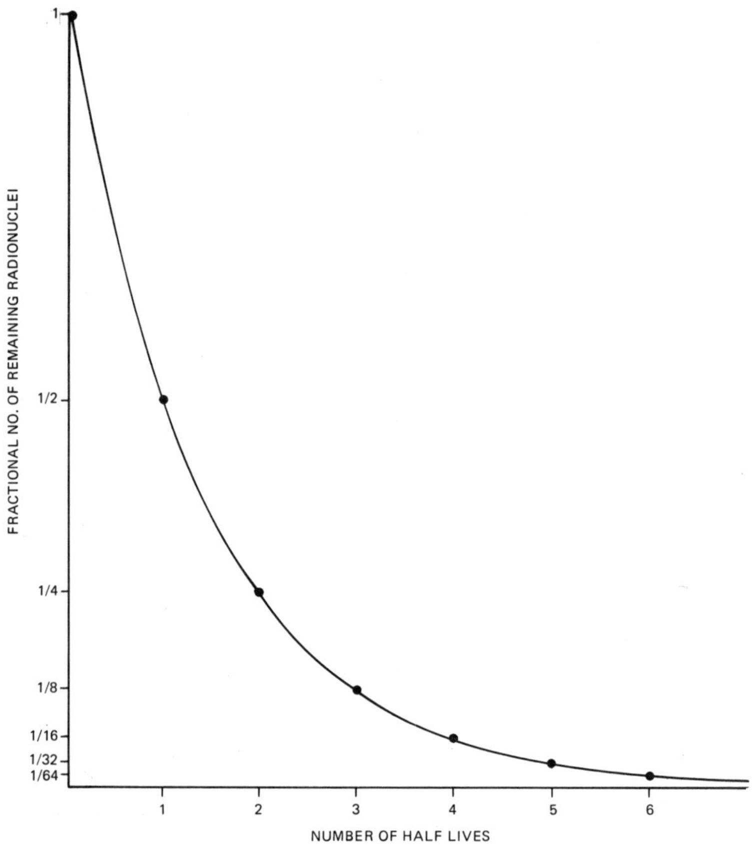

Fig. A-2. Half-life expressed as exponential function of radioactive decay.

Answer: $N = 4 \times 10^2$ dps

(Note: if one half-life (18 h) had elapsed, the activity would have been 5×10^2 dps.)

PROPERTIES OF RAYS AND PARTICLES EMITTED BY RADIONUCLIDES

The radioactive decay of primary interest here is emission from the nucleus of alpha particles (α), beta particles (β), and/or gamma rays (γ). The first of these are helium nuclei, the second are electrons and the third consists of quanta of electromagnetic radiations whose energies are usually greater than those of X-rays. A description of their properties follows.

Alpha radiation (α)

Many of the heavy radioelements spontaneously emit these ions (^4_2He) becoming other elemental species. These processes are generally accompanied by the release of electromagnetic energy, and some examples are:

$$^{238}_{92}U \rightarrow {}^{234}_{90}Th + {}^4_2He$$

$$^{226}_{88}Ra \rightarrow {}^{222}_{86}Rn + {}^4_2He$$

Note that in the emission of an α particle, Z reduces by 2 and A by 4 in the product nuclide. The alpha particle is ejected from the nucleus at high velocities, often as great as 10^9 cm s^{-1}, or 90% of the velocity of light. Although they are heavy and energetic, they penetrate only short distances in matter removing electrons from this matter to become helium atoms. Two rules of thumb are: (1) most alphas are stopped by a sheet of paper; (2) their range in air is about 1 cm for each MeV of energy. There is a rough inverse relationship between the half-life of the radionuclide and its energy (and therefore the range of alpha particles). The shorter the half-life, the further is the alpha particle range.

Beta radiation (β)

Electrons ejected by radioactive nuclei are called beta particles and these are emitted at velocities approaching that of light. Their origin within the nucleus results from the transformation of a neutron to a proton and an electron, and this emission increases Z of the product by 1 with no change in A:

$$^1_0n \rightarrow {}^1_1H + {}^{\ 0}_{-1}e$$

Examples are:

$$^{234}_{90}Th \rightarrow {}^{234}_{91}Pa + {}^{\ 0}_{-1}e$$

$$^{214}_{82}Pb \rightarrow {}^{214}_{83}Bi + {}^{\ 0}_{-1}e$$

To account for the conservation of energy in this process, Pauli postulated that a neutrino accompanies the emission of a beta particle, with this new particle containing almost negligible mass and no electrical charge. The neutrino, as seen in Fig. A-3 carries about two-thirds of the total energy released in beta decay. The penetrating power of betas is generally much greater than that of the alpha particle, but they are usually absorbed in a few millimeters of metal, such as aluminum. Table A-II lists their ranges in air, which as expected are energy-dependent.

Positive beta particles (β^+ or e^+) called positrons also exist and they behave in a manner similar to negative betas. Within a short time of its emission, however, a positron will probably combine with an electron of which a pre-

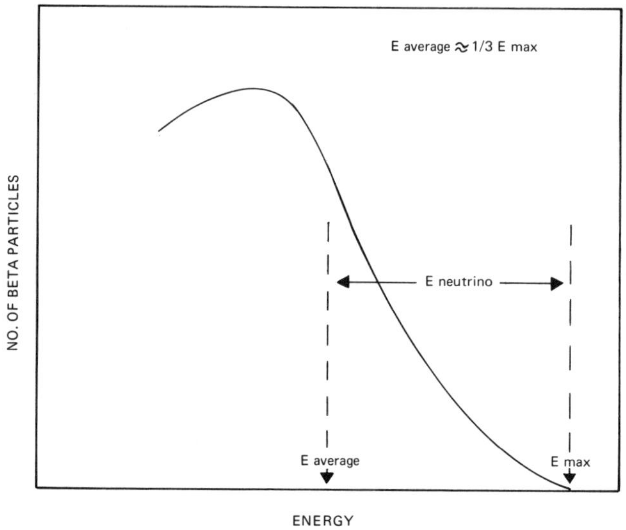

Fig. A-3. Energy spectrum of beta particles.

ponderance exists largely in the orbit of the nucleus. The attraction of these oppositely charged particles for each other leads to electrical neutralization and annihilation. The energy liberated is called annihilation radiation, and its value is calculated using the Einstein relationship given earlier. With the total mass of the electron and the positron taken as 0.00110 amu, the energy released is 1.02 MeV. To conserve momentum, two photons of electromagnetic radiation are formed moving in opposite directions, each with energy of 0.51 MeV, and with the properties of gamma radiation.

When electrons of energy greater than 1 MeV pass through matter, continuous X-rays are produced. These are called bremsstrahlung (German for "slowing down" radiation) and the efficiency of conversion of electron kinetic energy into these X-rays increases with both electron energy and the

TABLE A-II

Approximate ranges of beta particles in air

Energy (MeV)	Ranges (m)
0.1	0.11
0.5	1.5
1.0	3.7
2.0	8.5
3.0	13

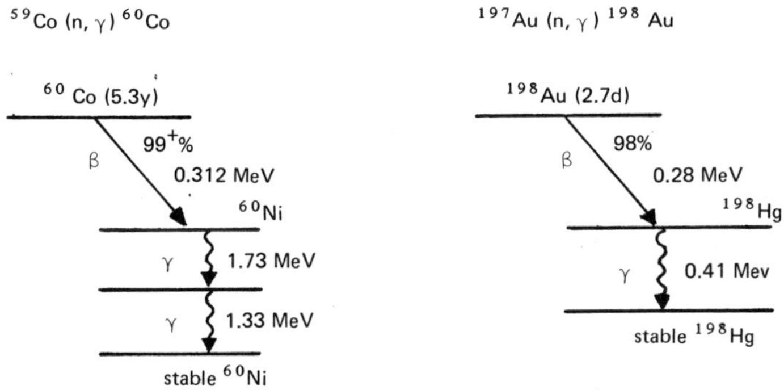

Fig. A-4. Decay schemes (simplified) of beta-gamma emitters.

atomic number of the matter traversed. The bremsstrahlung energy spectrum derived from a monoenergetic source is continuous and it extends from the energy of the initial electron down to zero.

Gamma rays (γ)

These radiations are the same physically as X-rays in that they are quanta (of discrete energy values) or photons of electromagnetic radiation; gamma rays, however, arise from energy changes originating within the nucleus. For example, nuclei resulting from alpha or beta emission are often left in an excited state, and the excess energy is then liberated as gamma radiation, as shown in Fig. A-4. Further, X-ray energy is of the order of thousands of electron volts (keV) but most gammas are of the order of millions of electron volts (MeV) in magnitude.

Gamma photons have considerably greater penetrating power than do the particulates mentioned earlier. A second important difference is that these photons have no true range. Reduction in intensity of a stream of gamma photons incident on a slab of material such as lead, is described by an exponential relationship:

$$I/I_0 = e^{-\mu x} \tag{5}$$

where I_0 = the initial intensity of the gamma beam; I = the intensity of the beam traversing a medium of thickness x (cm); and μ = the linear absorption coefficient (a function of the γ energy and the medium).

The thickness of matter needed to halve the intensity is:

$$I/I_0 = \tfrac{1}{2} = e^{-\mu x_{1/2}} \tag{6}$$

$$x_{1/2} = \log_e 2/\mu = 0.693/\mu \tag{7}$$

TABLE A-III

Linear absorption coefficients, μ (cm^{-1}), and half-thicknesses, $x_{1/2}$ (cm), of common materials

Energy (MeV)	Water		Aluminum		Iron		Lead	
	μ	$x_{1/2}$	μ	$x_{1/2}$	μ	$x_{1/2}$	μ	$x_{1/2}$
0.5	0.090	7.7	0.23	3.0	0.63	1.1	1.7	0.40
1.0	0.067	10.3	0.16	4.3	0.44	1.6	0.77	0.90
2.0	0.048	14.4	0.12	5.8	0.33	2.1	0.51	1.36
3.0	0.038	18.2	0.09	7.7	0.30	2.3	0.47	1.47

Illustrative example

A situation arises in a field exploration program wherein the measurement of a sample's activity must be made in an area of relatively high background gamma radiation. Thin lead sheet is available to shield both the sample and the detector so that the measurement can be made under conditions of reduced interference. If the background gamma source is known to have an energy level of 0.5 MeV, how thick a lead shield is needed to reduce the background flux by a factor of 10?

Solution — Using equation (5) in logarithmic form, we have:

$\log_e(I/I_0) = \mu x$

when:

$x = x_{1/2}\,,\ I/I_0 = 0.5$

$x = x_{1/10}\,,\ I/I_0 = 0.1$

writing one equation for each value of x, we have:

$\log_e 0.5 = -\mu x_{1/2}$

$\log_e 0.1 = -\mu x_{1/10}$

eliminating μ between the equations, we find:

$x_{1/10} = x_{1/2}(\log_e 0.1/\log_e 0.5)$

Answer: From Table A-III, $x_{1/2} = 0.40$ cm for $E = 0.5$ MeV and $x_{1/10} = 1.33$ cm.

(Note: 0.4-cm-thick sheet of lead will reduce the flux to 50%; a second 0.4-cm sheet reduces the flux to 25%; a third 0.4-cm sheet reduces the flux to 12.5%; etc.)

Three of the more important ways in which gamma photons interact with an absorbing material are: the photoelectric effect, the Compton effect and

The photoelectric process. The γ-ray is completely absorbed. An electron is ejected with the γ-rays energy minus the electron's binding energy.

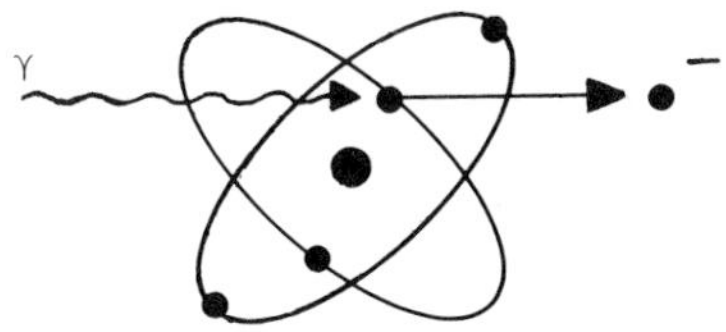

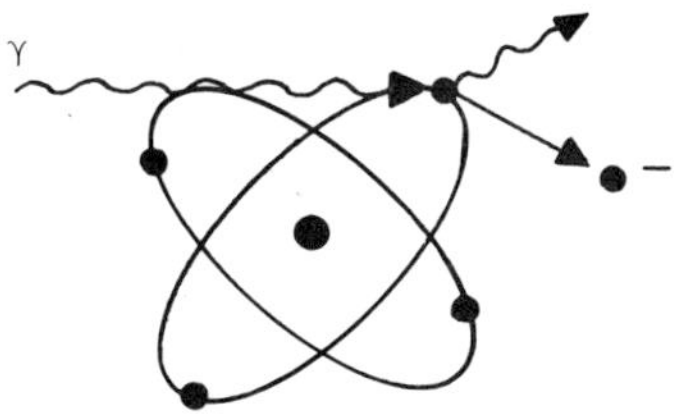

The Compton recoil process. A γ-ray of lower energy proceeds in a new direction. An electron is ejected with the energy difference.

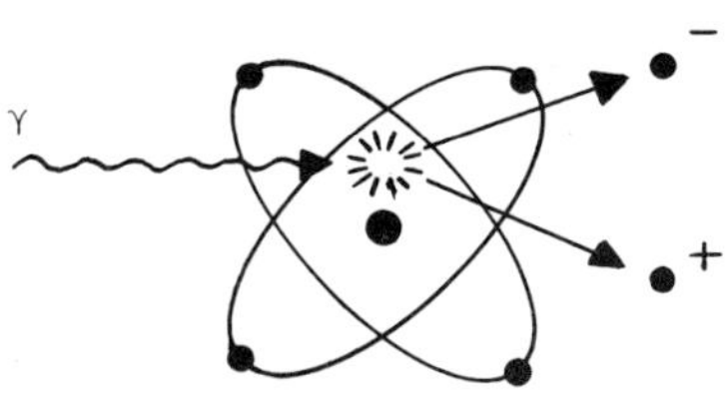

The pair production process. A γ-ray is annihilated. An electron and a positron are created and share the γ-ray's energy minus 1.02 Mev.

Fig. A-5. Gamma photons interacting with matter.

pair production. These will be described briefly below and are illustrated in Fig. A-5.

Photoelectric effect. The photoelectric effect involves the interaction of a gamma-ray photon with an orbital electron. If the photon energy exceeds the electron binding energy and this energy is transferred to the electron, the latter will be ejected from the atom. The kinetic energy of the ejected electron is approximately the difference between the initial photon energy and the binding energy. The photoelectron produced behaves as a beta particle and it may lead to the ionization of other atoms. Photoelectric absorption of gamma radiation is important at energies less than 1 MeV, and for absorbers of high Z.

Compton effect. The second type of interaction is the Compton effect, which also involves a photon hitting an orbital electron of the absorbing material, but here the collision is elastic, analogous to the impact of one billiard ball on another. On colliding, both momentum and energy are conserved. Part of the energy of the incident photon is transferred to the electron, and the photon is scattered or deflected from its original path. The magnitude of the effect is dependent on the number of orbital electrons in the absorber, therefore, it is directly proportional to Z, and like the photoelectric effect, it is more significant for materials of high atomic number. The important difference between the two is that in the photoelectric effect, the photon is completely absorbed, whereas in the Compton process, the photon is scattered and reduced in energy. After suffering several Compton collisions, the photon ultimately may be absorbed through photoelectric interaction. The Compton effect becomes dominant at gamma energies of 0.6—4 MeV.

Pair production. Pair production obtains when a gamma photon with energy greater than 1.02 MeV passes near the atomic nucleus. The photon may be annihilated in this strong electrical field with the subsequent formation of an electron-positron pair. This process becomes dominant at higher photon energies, particularly at those in excess of 5 MeV, whereas the photoelectric and Compton effects decrease with increasing photon energies. Pair production, like the photoelectric effect, results in the absorption of the gamma photon, but it also gives rise to the formation of two 0.51-MeV photons from the annihilation reaction.

UNITS OF RADIOACTIVITY

The curie

The most frequently used unit is the curie, which is defined as 3.7×10^{10} disintegrating atoms per second. Early work based this number on the decay rate of 1 gram of radium, later shown to be in error, but the original number is accepted as the basic dimension of the curie.

It is of interest to compute the actual decay rate of $^{226}_{88}Ra$ from its half-life:

$$t_{1/2} = 1620 \text{ yr}$$

$$\lambda = 0.693/(1620 \text{ yr})(365 \text{ days/yr})(24 \text{ h/day})(60 \text{ min/h})(60 \text{ s/min}) = 1.36 \times 10^{-11} \text{ s}^{-1}$$

$A \quad = 226$

$$\frac{\text{number of atoms}}{\text{gram atomic weight}} = \frac{\text{Avogadro's number}}{A} = \frac{6.02 \times 10^{23}}{226}$$

$$\begin{aligned}\text{disintegration rate} = -\mathrm{d}N/\mathrm{d}t &= \lambda t \\ &= 1.36 \times 10^{-11} \times 6.02 \times 10^{23}/226 \\ &= 3.61 \times 10^{10}\ \text{dps}\end{aligned}$$

Specific activity is often expressed as the ratio of the number of radioactive atoms to the total number of atoms of the element in the sample (curies per gram of material), and it is also a useful term.

Mention is made of the roentgen, defined as the quantity of X- or gamma radiation producing in 0.001293 g of air (1 cm^3 at standard temperatures and pressures, STP), ions carrying 1 esu of charge of either sign. Recall that the charge on an electron is 4.8×10^{-10} esu. The roentgen relates primarily to the effect produced by radiation.

Decay series

When the products of radioactive decay are also radioactive, a series or family of radioelements may exist simultaneously. If the parent nuclide in a chain has a relatively long half-life compared with that of its daughter, the amount of daughter (or daughters) present will be constant after a period of time. This phenomenon is called a steady state, as the product elements will decay as fast as they are formed.

$A \rightarrow B \rightarrow C \rightarrow \ldots \rightarrow X$, is an example of a decay chain in which a steady state, or secular equilibrium may be established.

Since $-\mathrm{d}N/\mathrm{d}t = \lambda N$ for all radioactive decay, and since the rate of decay of A is equal to the rate of formation of B, etc., in the above example, or $-\mathrm{d}N_A/\mathrm{d}t = \mathrm{d}N_B/\mathrm{d}t$, one can write: $\lambda_A N_A = \lambda_B N_B = \lambda_C N_C = \lambda_D N_D$, etc. It must be noted that the equation that $\mathrm{d}N_B/\mathrm{d}t = \lambda_A N_A - \lambda_B N_B$ describes the net rate of change of B with time. The first term of the right represents the rate of formation of B and the second term, the rate of decay of B. A similar relationship applies to C through X.

In natural radioactive series, such as those of uranium, thorium, etc., the very long parent half-life permits the assumption that N_A is virtually constant. A condition eventually obtains when $\lambda_A N_A = \lambda_B N_B$, so that the net rate of change of B is zero. Then A decays to form B at the same rate B decays to form C, and that the quantity of B in the system is constant. This is called a state of secular equilibrium. If the amount of parent nuclide remains essentially unchanged, all members of such a series except the last, will be in equilibrium with each other with the lapse of sufficient time.

THE NEUTRON

The existence of the neutron was first verified experimentally by Chadwick when he bombarded beryllium with high-energy alpha rays, giving the reaction:

$${}^{9}_{4}Be + {}^{4}_{2}He \rightarrow {}^{12}_{6}C + {}^{1}_{0}n$$

This may be written:

$${}^{9}_{4}Be(\alpha, n){}^{12}_{6}C$$

Hydrogen atoms in a paraffin target, on being struck by neutrons, were converted to protons, dislodged and detected.

Neutrons are particularly effective in engaging in reactions with nuclei by virtue of their lack of electrical charge, which enables passage through the electrostatic fields of the atom. They can move through matter for long distances without being stopped. Electrical or magnetic fields externally applied have no effect on their motion. This motion or energy is described in terms of electron volts.

Consider the case of high-energy neutrons (>1 MeV) passing through a volume of heavy water (D_2O). They lose energy due to multiple collisions with the deuterium nuclei to the extent that their energies will become that of the medium, and they will be in thermal equilibrium with the heavy water. Called thermal neutrons, their kinetic energy can be calculated. In a medium such as heavy water, their velocity at room temperatures is about 2200 m s^{-1}. Taking the neutron mass as 1.6×10^{-24} g, its kinetic energy is:

$$E = mv^2/2 = (1.66 \times 10^{-24})(2.2 \times 10^5)/2$$

$$4.0 \times 10^{-14} \text{ erg} = 0.025 \text{ eV}$$

Reactions can take place with thermal (0.025 eV) or higher-energy neutrons in several ways, of which the four most common are: neutron absorption or capture, transmutation, scattering, and fission. All of these are treated in greater depth in Chapter 5.

Since neutrons are produced only by nuclear reactions, sources of neutrons arise from the following:

(1) Using selected target nuclei, interactions with alpha or gamma radiation from radionuclides, or charged particles or gamma radiation from accelerating devices.

(2) Nuclear fission or fusion under controlled or uncontrolled conditions.

COMPOUND NUCLEUS

The concept of the compound nucleus was advanced by Bohr in 1936 to explain nuclear reactions in the MeV region, a detailed discussion of this is beyond the scope of this section. Here we shall consider briefly reactions which result when nuclei are bombarded with radiation, such as photons, neutrons and ions, and these reactions take place in a series of steps.

First, the incident particle, having finite kinetic energy, penetrates the potential barrier, strikes a nucleus and transfers some or all of its energy to it. Second, the kinetic and binding energies of the particle are added to the structure of the nucleus, resulting in a redistribution of its internal energy and it achieves an excited state. This is the compound nucleus. Third, this excited nucleus now emits either electromagnetic or particulate radiation, or both, or it may fission yielding these radiations along with fission fragments. The lifetime of a compound nucleus is approximately 10^{-12} to 10^{-14} second, which is considerably longer than that required for a particle to move one nuclear diameter, estimated as 10^{-21} second.

CROSS-SECTION

When incident particles of sufficient energy collide with a nucleus, the probability that a nuclear reaction will occur is a function of: the flux or number of particles per second; type and energies; and the number of target nuclei within the sample. This probability term is called the reaction cross-section and it is seen as the ability of a substance to remove particles from a beam of radiation. Cross-section data are readily available in the literature and are tabulated by reaction type, such as absorption, capture, scattering or fission, nature of the particle and its energy. The unit of cross-section, called a barn, is equivalent to 10^{-24} cm^2. It derives from the expression "hitting the side of a barn", with 1 barn the average geometric nuclear cross-section.

For neutron-induced nuclear reactions of a given energy level, the amount of radioactive species formed may be calculated from:

$$R = \sigma \phi n (1 - e^{-\lambda t}) \qquad (8)$$

where R = activity of the product (dps); σ = cross-section (cm^2); ϕ = flux of bombarding particles (number cm^{-2} s^{-1}); n = number of target nuclei per sample; λ = decay constant of the product radionuclide; and t = exposure time.

Illustrative examples

(1) An ore contains an estimated copper value of 1%. If a 10-g sample is exposed to a thermal neutron flux of 10^{11} n cm^{-2} s^{-1} (100 mg ^{252}Cf source) for 30 minutes, what is the resulting activation induced in the copper?

Solution — Consider the reaction ^{63}Cu(n, γ)^{64}Cu. The number (n) of ^{63}Cu nuclei are: natural abundance of ^{63}Cu = 0.691; atomic weight = 63.5; number of atoms per gram atomic weight = 6.02×10^{23}. Using equation (8):

$n = 10\text{ g }(0.01)(0.691)6.02 \times 10^{23}/63.5$
$\quad = 6.55 \times 10^{20}$ atoms
$\sigma = 4.5$ barns $= 4.5 \times 10^{-24}$ cm^2 thermal neutron cross-section
$\phi = 10^{11}$ n cm^{-2} s^{-1}
$T = 12.9$ h (half-life of ^{64}Cu)
$t = 0.5$ h
$\lambda = \log_e 2/T = 0.693/12.8 = 0.054$ h^{-1}
$\lambda t = 0.027$
$(1 - e^{-\lambda t}) = 0.03$
$R = \sigma\phi n(1 - e^{-\lambda t})$
$\quad = 4.5 \times 10^{-24} \times 10^{11} \times 6.55 \times 10^{20} \times 0.03$

Answer: $R = 8.8 \times 10^6$ dps

(2) Conversely, a 10-g crushed rock sample, taken from an orebody with known manganese horizons, is exposed to a thermal neutron flux of 10^{10} n cm^{-2} s^{-1} for a 10-minute period. For purposes of illustration, only the manganese is activated. Nuclear measurements attribute a decay rate to ^{56}Mn of 6.4×10^7 dps. Calculate the percentage of manganese by weight in the sample.

Solution — Consider the reaction ^{55}Mn(n, γ)^{56}Mn:

$R = 6.4 \times 10^6$ dps
^{55}Mn = 100% naturally abundant
$\sigma = 13.3$ barns $= 13.3 \times 10^{-24}$ cm^2
$\phi = 10^{10}$ n cm^{-2} s^{-1}
$T = 2.6$ h (half-life of ^{56}Mn)
$t = 10$ min $= 0.167$ h
$\lambda t = 0.045$
$(1 - e^{-\lambda t}) = 0.044$

Using equation (8), and f = fraction of Mn by weight in the sample:

$$n = \frac{\text{sample weight} \times (f) \times 6.02 \times 10^{23}}{\text{atomic weight of Mn}}$$
$\quad = R/\sigma\phi(1 - e^{-\lambda t})$
$\quad = 6.4 \times 10^6/(13.3 \times 10^{-24})(10^{10})(0.044)$
$\quad = 1.09 \times 10^{21}$
$f = (1.09 \times 10^{21})(55)/(10)(6.02 \times 10^{23}) = 0.00996$

Answer: $f = 1\%$ ^{55}Mn

ARTIFICIAL RADIOACTIVITY

Transmutation, the dream of alchemy to convert base metals into gold, or one element into another, was effected by Rutherford in 1919 when he bombarded nitrogen with alpha particles and produced an isotope of oxygen. In the shorthand notation indicated earlier, the reaction was $^{14}N(\alpha,p)^{17}O$. Rutherford's discovery is basic to some of the techniques described in this book, in that the process of transmutation, now no longer a dream, a stable element can be converted into a radioactive species, with the latter detectable down to very low concentrations. A few examples involving transmutation are:

$^{10}_{5}B(\alpha,n)^{13}_{7}N$ (^{13}N: β emitter, $t_{1/2}$ = 10 min)

$^{27}_{13}Al(\alpha,n)^{30}_{15}P$ (^{30}P: positron emitter; $t_{1/2}$ = 2.5 min)

$^{106}_{46}Pd(n,p)^{106}_{45}Rh$ (^{106}Rh: β emitter, $t_{1/2}$ = 30 s)

Alternatively, the product nuclei may result as radioactive without transmutation, such as:

$^{23}_{11}Na(n,\gamma)^{24}_{11}Na$ (^{24}Na: β emitter; $t_{1/2}$ = 15 h)

$^{59}_{27}Co(n,\gamma)^{60}_{27}Co$ (^{60}Co: β,γ emitter; $t_{1/2}$ = 5.3 yr)

A few examples of activation reactions are included below to show their general utility in field work.

(1) Use of thermal neutrons:

$^{197}_{79}Au(n,\gamma)^{198}_{79}Au$ (^{198}Au: β,γ emitter; $t_{1/2}$ = 64.8 h)

$^{23}_{11}Na(n,\gamma)^{24}_{11}Na$ (^{24}Na: β emitter; $t_{1/2}$ = 15 h)

(2) Fast neutrons:

$^{27}_{13}Al(n,p)^{27}_{12}Mg$ (^{27}Mg: β emitter; $t_{1/2}$ = 10 min)

$^{11}_{5}B(n,\alpha)^{8}_{3}Li$ (^{8}Li: β,γ emitter; $t_{1/2}$ = 0.83 s)

(3) Photoactivation:

$^{26}_{12}Mg(\gamma,p)^{25}_{11}Na$ (^{25}Na: β,γ emitter; $t_{1/2}$ = 1 min)

STATISTICAL CONSIDERATIONS IN MEASUREMENT OF RADIOACTIVITY

Brief mention is made regarding the importance of counting statistics in radioactivity measurements in order to reduce errors in data interpretation. Radioactive nuclei decay in a random manner and therefore this decay is governed by the laws of probability which are used to determine error limits in measured data. For field use in particular, these laws will enable the worker to optimize the counting time so that errors will be minimized.

Numerous texts on radioactivity measurements and counting statistics are available for detail on this subject (Chase and Rabinowitz, 1967).

REFERENCES

American Institute of Physics, 1972. *American Institute of Physics Handbook*. McGraw-Hill, New York, N.Y., 3rd ed.

Bowler, M.G., 1973. *Nuclear Physics*. Pergamon Press, Oxford.

Chase, G.D. and Rabinowitz, J.L., 1967. *Principles of Radioisotope Methodology*. Burgess Publishing Co., Minneapolis, Minn., Chapters 3, 4 and 5.

Evans, R.D., 1965. *The Atomic Nucleus*. McGraw-Hill, New York, N.Y.

Glasstone, S., 1950. *Sourcebook on Atomic Energy*. Van Nostrand, New York, N.Y.

Lapp, R.E. and Andrews, H.L., 1972. *Nuclear Radiation Physics*. Prentice-Hall, Englewood Cliffs, N.J.

SUBJECT INDEX